Praise for *The Alignment Problem*

"A new field has emerged that respond

nological shifts represented by our

defined world. In *The Alignment Pro*

surveys the 'AI Fairness' community, i

characters; some of its historical roots in science, philosophy, and activism; and crucially, many of its philosophical quandaries and limitations."

—Cathy O'Neil, author of *Weapons of Math Destruction*

"Meticulously researched and superbly written. . . . [T]he responsible— ethical, legal and beneficial—development and use of AI is not about technology. It is about us: how we want our world to be; how we prioritize human rights and ethical principles; who comprises this 'we.' This discussion can wait no longer." —Virginia Dignum, *Nature*

"Brian Christian, an accomplished technology writer, offers a nuanced and captivating exploration of this white-hot topic, giving us along the way a survey of the state of machine learning and of the challenges it faces."

—David A. Shaywitz, *Wall Street Journal*

"This is the book on artificial intelligence we need right now. Brian Christian takes us on a technically fluent (yet widely accessible) journey through the most important questions facing AI and humanity. It is thought-provoking and vital reading for anyone interested in our future." —Mike Krieger, cofounder of Instagram

"A fascinating, provocative, and insightful tour of all the ways that AI goes wrong and all the ways people are trying to fix it. Essential reading if you want to understand where our world is heading."

—Stuart Russell, professor of computer science, University of California, Berkeley, coauthor of *Artificial Intelligence: A Modern Approach*, and author of *Human Compatible*

"An abundantly researched and captivating book that explores the road humanity has taken to create a successor for itself—a road that's rich with surprising discoveries, unexpected obstacles, ingenious solutions, and, increasingly, hard questions about the soul of our species."

—Jaan Tallinn, cofounder of Skype and the Future of Life Institute

"*The Alignment Problem* should be required reading for anyone influencing policy where algorithms are in play—which is everywhere. But unlike much required reading, the book is a delight to read, a playful romp through personalities and relatable snippets of science history that put the choices of our present moment into context."

—Jennifer Pahlka, founder of Code for America and former deputy CTO of the United States

"*The Alignment Problem* is a profound book, exploring how advances in AI can lead to powerful systems unaligned with human values—from social media stealing our attention to future systems that may steal our future. An essential book for our time."

—Toby Ord, senior research fellow in philosophy, the University of Oxford and author of *The Precipice*

"Essential reading. Christian brings much needed clarity to a subject that is often talked about but little understood."

—Tim O'Reilly, founder and CEO of O'Reilly Media

"Incredibly engaging, beautifully written, socially important, and also an exceptionally clear and accurate introduction to the ideas of modern machine learning and cognitive science."

—Alison Gopnik, professor of psychology, University of California, Berkeley, and author of *The Gardener and the Carpenter*

"Brian Christian is a fine writer and has produced a fascinating book. AI seems destined to become, for good or ill, increasingly prominent in our lives. We should be grateful for this balanced and hype-free perspective on its scope and limits."

—Martin Rees, emeritus professor of cosmology and astrophysics,
University of Cambridge

"A deeply enjoyable and meticulously researched account of how computer scientists and philosophers are defining the biggest question of our time: how will we create intelligent machines that will improve our lives rather than complicate or even destroy them? There's no better book than *The Alignment Problem* at spelling out the issues of governing AI safely."

—James W. Barrat, best-selling author of *Our Final Invention*

The
Alignment
Problem

The
Alignment
Problem

MACHINE LEARNING
AND HUMAN VALUES

BRIAN CHRISTIAN

W. W. NORTON & COMPANY

Independent Publishers Since 1923

For information about permission to reproduce selections from this book, write to
Permissions, W. W. Norton & Company, Inc., 500 Fifth Avenue, New York, NY 10110

For information about special discounts for bulk purchases, please contact
W. W. Norton Special Sales at specialsales@wwnorton.com or 800-233-4830

Manufacturing by LSC Communications, Harrisonburg
Book design by Chris Welch
Production manager: Lauren Abbate

Library of Congress Cataloging-in-Publication Data

Names: Christian, Brian, 1984– author.
Title: The alignment problem : machine learning and human values / Brian Christian.
Description: First edition. | New York, NY : W. W. Norton & Company, [2020] |
Includes bibliographical references and index.
Identifiers: LCCN 2020029036 | ISBN 9780393635829 (hardcover) |
ISBN 9780393635836 (epub)
Subjects: LCSH: Artificial intelligence—Moral and ethical aspects. |
Artificial intelligence—Social aspects. | Machine learning—Safety measures. |
Software failures. | Social values.
Classification: LCC Q334.7 .C47 2020 | DDC 174/.90063—dc23
LC record available at https://lccn.loc.gov/2020029036

ISBN 978-0-393-86833-3 pbk.

W. W. Norton & Company, Inc., 500 Fifth Avenue, New York, N.Y. 10110
www.wwnorton.com

W. W. Norton & Company Ltd., 15 Carlisle Street, London W1D 3BS

5 6 7 8 9 0

For Peter
who convinced me

And for everyone
doing the work

I remember in 2000 hearing James Martin, the leader of the Viking missions to Mars, saying that his job as a spacecraft engineer was not to land on Mars, but to land on the model of Mars provided by the geologists.

—PETER NORVIG[1]

The world is its own best model.

—RODNEY BROOKS[2]

All models are wrong.

—GEORGE BOX[3]

CONTENTS

CONTENTS

The
Alignment
Problem

PROLOGUE

1935, Detroit. Walter Pitts is running down the street, chased by bullies.

He ducks into the public library to take shelter, and he hides. He hides so well that the library staff don't even realize he's there, and they close for the night. Walter Pitts is locked inside.[1]

He finds a book on the shelves that looks interesting, and he starts reading it. For three days, he reads the book cover to cover.

The book is a two-thousand-page treatise on formal logic; famously, its proof that 1+1=2 does not appear until page 379.[2] Pitts decides to write a letter to one of the authors—British philosopher Bertrand Russell—because he believes he's found several mistakes.

Several weeks go by, and Pitts gets a letter in the mail postmarked from England. It's Bertrand Russell. Russell thanks him for writing, and invites Pitts to become one of his doctoral students at Cambridge.[3]

Unfortunately, Walter Pitts must decline the offer—because he's only twelve years old, and in the seventh grade.

Three years later, Pitts learns that Russell will be visiting Chicago to give a public lecture. He runs away from home to attend. He never goes back.

———

A<small>t</small> Russell's lecture, Pitts meets another teenager in the audience, named Jerry Lettvin. Pitts only cares about logic. Lettvin only cares about poetry and, a distant second, medicine.[4] They become inseparable best friends.

Pitts begins hanging out around the University of Chicago campus, dropping in on classes; he still lacks a high school diploma and never formally enrolls. One of these classes is by the famed German logician Rudolf Carnap. Pitts walks into his office hours, declaring he's found a few "flaws" in Carnap's latest book. Skeptically, Carnap consults the book; Pitts, of course, is right. They talk awhile, then Pitts walks out without giving his name. Carnap spends months asking around about the "newsboy who knew logic."[5] Eventually Carnap finds him again and, in what will become a motif throughout Pitts's academic life, becomes his advocate, persuading the university to give him a menial job so he will at least have some income.

It's now 1941. Lettvin—still a poet first, in his own mind—has, despite himself, gotten into medical school at the University of Illinois, and finds himself working under the brilliant neurologist Warren McCulloch, newly arrived from Yale. One day Lettvin invites Pitts over to meet him. At this point Lettvin is twenty-one and still living with his parents. Pitts is seventeen and homeless.[6] McCulloch and his wife take them both in.

Throughout the year that follows, McCulloch comes home in the evenings and he and Pitts—who is barely older than McCulloch's own children—regularly stay up past midnight talking. Intellectually, they are the perfect team: the esteemed midcareer neurologist and the prodigy logician. One lives in practice—the world of nervous systems and neuroses—and the other lives in theory—the world of symbols and proofs. They both want nothing more than to understand the nature of truth: what it is, and how we know it. The fulcrum of this quest—the thing that sits at the perfect intersection of their two disparate worlds—is, of course, the brain.

It was already known by the early 1940s that the brain is built of neurons wired together, and that each neuron has "inputs" (dendrites) as well as an "output" (axon). When the impulses coming into a neuron exceed a certain threshold, then that neuron, in turn, emits a pulse. Immediately this begins to feel, to McCulloch and Pitts, like logic: the pulse or its absence signifying *on* or *off, yes* or *no, true* or *false.*[7]

They realize that a neuron with a low-enough threshold, such that it would fire if *any* of its inputs did, functioned like a physical embodiment of the logical *or.* A neuron with a high-enough threshold, such that it would only fire if *all* of its inputs did, was a physical embodiment of the logical *and.* There was nothing, then, that could be done with logic— they start to realize—that such a "neural network," so long as it was wired appropriately, could not do.

Within months they have written a paper together—the middle-aged neurologist and teenage logician. They call it "A Logical Calculus of Ideas Immanent in Nervous Activity."

"Because of the 'all-or-none' character of nervous activity," they write, "neural events and the relations among them can be treated by means of propositional logic. It is found that the behavior of every net can be described in these terms . . . and that for any logical expression satisfying certain conditions, one can find a net behaving in the fashion it describes."

The paper is published in 1943 in the *Bulletin of Mathematical Biophysics.* To Lettvin's frustration, it makes little impact on the biology community.[8] To Pitts's disappointment, the neuroscience work of the 1950s, notably a landmark study of the optic nerve of the frog—done by none other than his best friend, Jerry Lettvin—will show that neurons appear to be much messier than the simple "true" or "false" circuits he envisioned. Perhaps propositional logic—its *ands, ors,* and *nots*—was not, ultimately, the language of the brain, or at least not in so straightforward a form. This kind of impurity saddened Pitts.

But the impact of the paper—of those long conversations into the night at McCulloch's house—would be enormous, if not entirely in the way that McCulloch and Pitts envisioned. It would be the foundation for a completely new field: the project to actually *build* mechanisms out of these simplified versions of neurons, and see just what such "mechanical brains" could do.[9]

INTRODUCTION

In the summer of 2013, an innocuous post appeared on Google's open-source blog titled "Learning the Meaning Behind Words."[1]

"Today computers aren't very good at understanding human language," it began. "While state-of-the-art technology is still a ways from this goal, we're making significant progress using the latest machine learning and natural language processing techniques."

Google had fed enormous datasets of human language, mined from newspapers and the internet—in fact, *thousands* of times more text than had ever been successfully used before—into a biologically inspired "neural network," and let the system pore over the sentences for correlations and connections between the terms.

The system, using so-called "unsupervised learning," began noticing patterns. It noticed, for instance, that the word "Beijing" (whatever that meant) had the same relationship to the word "China" (whatever that was) as the word "Moscow" did to "Russia."

Whether this amounted to "understanding" or not was a question for philosophers, but it was hard to argue that the system wasn't capturing *something* essential about the sense of what it was "reading."

Because the system transformed the words it encountered into numerical representations called vectors, Google dubbed the system "word2vec," and released it into the wild as open source.

To a mathematician, vectors have all sorts of wonderful properties that allow you to treat them like simple numbers: you can add, subtract, and multiply them. It wasn't long before researchers discovered something striking and unexpected. They called it "linguistic regularities in continuous space word representations,"[2] but it's much easier to explain than that. Because word2vec made words into vectors, it enabled you to do *math with words*.

For instance, if you typed `China + river`, you got `Yangtze`. If you typed `Paris – France + Italy`, you got `Rome`. And if you typed `king – man + woman`, you got `queen`.

The results were remarkable. The word2vec system began humming under the hood of Google's translation service and its search results, inspiring others like it across a wide range of applications including recruiting and hiring, and it became one of the major tools for a new generation of data-driven linguists working in universities around the world.

No one realized what the problem was for two years.

In November 2015, Boston University PhD student Tolga Bolukbasi went with his advisor to a Friday happy-hour meeting at Microsoft Research. Amid wine sipping and informal chat, he and Microsoft researcher Adam Kalai pulled out their laptops and started messing around with word2vec.

"We were playing around with these word embeddings, and we just started randomly putting words into it," Bolukbasi says. "I was playing on my PC; Adam started playing."[3] Then something happened.

They typed:

`doctor – man + woman`

The answer came back:

`nurse`

"We were shocked at that point, and we realized there was a problem," says Kalai. "And then we dug deeper and saw that it was even worse than that."[4]

The pair tried another.

`shopkeeper – man + woman`

The answer came back:

housewife

They tried another.

computer programmer – man + woman

Answer:

homemaker

Other conversations in the room by this point had stopped, and a group had formed around the screen. "We jointly realized," says Bolukbasi, *"Hey, there's something wrong here."*

———

In judiciaries across the country, more and more judges are coming to rely on algorithmic "risk-assessment" tools to make decisions about things like bail and whether a defendant will be held or released before trial. Parole boards are using them to grant or deny parole. One of the most popular of these tools was developed by the Michigan-based firm Northpointe and goes by the name Correctional Offender Management Profiling for Alternative Sanctions—COMPAS, for short.[5] COMPAS has been used by states including California, Florida, New York, Michigan, Wisconsin, New Mexico, and Wyoming, assigning algorithmic risk scores—risk of general recidivism, risk of violent recidivism, and risk of pretrial misconduct—on a scale from 1 to 10.

Amazingly, these scores are often deployed statewide without formal audits.[6] COMPAS is a proprietary, closed-source tool, so neither attorneys, defendants, nor judges know exactly how its model works.

In 2016, a group of data journalists at ProPublica, led by Julia Angwin, decided to take a closer look at COMPAS. With the help of a public records request to Florida's Broward County, they were able to get the records, and the risk scores, of some seven thousand defendants arrested in 2013 and 2014.

Because they were doing their research in 2016, the ProPublica team had the equivalent of a crystal ball. Looking at data from two years prior, they actually *knew* whether these defendants, predicted either to reoffend or not, actually did. And so they asked two simple questions. One: Did the model actually correctly predict which defendants were indeed the "riskiest"? And two: Were the model's predictions biased in favor of or against any group in particular?

An initial look at the data suggested something might be wrong. They found, for instance, two defendants arrested for similar counts of drug possession. The first, Dylan Fugett, had a prior offense of attempted burglary; the second, Bernard Packer, had a prior offense of nonviolently resisting arrest. Fugett, who is White, was assigned a risk score of 3/10. Packer, who is Black, was assigned a risk score of 10/10.

From the crystal ball of 2016, they also knew that Fugett, the 3/10 risk, went on to be convicted of three further drug offenses. Over the same time period, Packer, the 10/10 risk, had a clean record.

In another pairing, they juxtaposed two defendants charged with similar counts of petty theft. The first, Vernon Prater, had a prior record of two armed robberies and one attempted armed robbery. The other defendant, Brisha Borden, had a prior record of four juvenile misdemeanors. Prater, who is White, was assigned a risk score of 3/10. Borden, who is Black, was assigned a risk score of 8/10.

From the vantage of 2016, Angwin's team knew that Prater, the "low-risk" defendant, went on to be convicted of a later count of grand theft and given an eight-year prison sentence. Borden, the "high-risk" defendant, had no further offenses.

Even the defendants themselves seemed confused by the scores. James Rivelli, who is White, was arrested for shoplifting and rated a 3/10 risk, despite having prior offenses including aggravated assault, felony drug trafficking, and multiple counts of theft. "I spent five years in state prison in Massachusetts," he told a reporter. "I am surprised it is so low."

A statistical analysis appeared to affirm that there was a systemic disparity.[7] The article ran with the logline "There's software used across the country to predict future criminals. And it's biased against blacks."

Others weren't so sure—and ProPublica's report, published in the spring of 2016, touched off a firestorm of debate: not only about COMPAS, not only about algorithmic risk assessment more broadly, but about the very concept of fairness itself. How, exactly, are we to define—in statistical and computational terms—the principles, rights, and ideals articulated by the law?

When US Supreme Court Chief Justice John Roberts visits Rensselaer Polytechnic Institute later that year, he's asked by university president Shirley Ann Jackson, "Can you foresee a day when smart machines—driven with artificial intelligences—will assist with courtroom fact-finding or, more controversially, even judicial decision-making?"

"It's a day that's here," he says.[8]

———

That same fall, Dario Amodei is in Barcelona to attend the Neural Information Processing Systems conference ("NeurIPS," for short): the biggest annual event in the AI community, having ballooned from several hundred attendees in the 2000s to more than *thirteen thousand* today. (The organizers note that if the conference continues to grow at the pace of the last ten years, by the year 2035 the *entire human population* will be in attendance.)[9] But at this particular moment, Amodei's mind isn't on "scan order in Gibbs sampling," or "regularizing Rademacher observation losses," or "minimizing regret on reflexive Banach spaces," or, for that matter, on Tolga Bolukbasi's spotlight presentation, some rooms away, about gender bias in word2vec.[10]

He's staring at a boat, and the boat is on fire.

He watches as it does donuts in a small harbor, crashing its stern into a stone quay. The motor catches fire. It continues to spin wildly, the spray dousing the flames. Then it slams into the side of a tugboat and catches fire again. Then it spins back into the quay.

It is doing this because Amodei ostensibly told it to. In fact it is doing exactly what he told it to. But it is not what he meant.

Amodei is a researcher on a project called Universe, where he is part of a team working to develop a single, general-purpose AI that can play hundreds of different computer games with human-level skill—a challenge that has been something of a holy grail among the AI community.

"And so I just, I ran a few of these environments," Amodei tells me, "and I was VPNing in and looking to see how each one was doing. And then just the normal car race was going fine, and there was like a truck race or something, and then there was this *boat* race." Amodei watches for a minute. "And I was looking at it, and I was like, 'This boat is, like, going around in circles. Like, what in the world is going on?!'"[11] The boat wasn't simply acting randomly; it wasn't wild or out of control. In fact, it was the opposite. It had *settled* on this. From the computer's perspective, it has found a nearly perfect strategy, and was executing it to a T. Nothing made sense.

"Then I eventually looked at the reward," he says.

Amodei had made the oldest mistake in the book: "rewarding A, while hoping for B."[12] What he *wanted* was for the machine to learn how to win the boat race. But it was complicated to express this rigorously—he would need to find a way to formalize complex concepts like track position, laps, placement among the other boats, and so on. Instead, he used what seemed like a sensible proxy: points. The machine found a loophole, a tiny harbor with replenishing power-ups where it could ignore the race entirely, do donuts, and rack up points . . . *forever.*

"And, of course, it's partially my fault," he says. "I just run these various games; I haven't looked *super* closely at the objective function. . . . In the other ones, score was sensibly correlated to finishing the race. You got points for getting power-ups that were always along the road. . . . The proxy of score that came with the game was good for the other ten environments. But for this eleventh environment, it wasn't good."[13]

"People have criticized it by saying, 'Of course, you get what you asked for,'" Amodei says. "It's like, 'You weren't optimizing for finishing the race.' And my response to that is, Well—" He pauses. "That's true."

Amodei posts a clip to his group's Slack channel, where the episode is instantly deemed "hilarious" by all concerned. In its cartoonish, destructive slapstick, it certainly is. But for Amodei—who now leads the AI safety team at San Francisco research lab OpenAI—there is another, more sobering message. At some level, this is *exactly* what he's worried about.

The real game he and his fellow researchers are playing isn't to try to win boat races; it's to try to get increasingly general-purpose AI systems to do what we want, particularly when what we want—and what we *don't* want—is difficult to state directly or completely.

The boat scenario is admittedly just a warm-up, just practice. The property damage is entirely virtual. But it is practice for a game that is, in fact, no game at all. A growing chorus within the AI community—first a few voices on the fringe, and increasingly the mainstream of the field—believes, if we are not sufficiently careful, that this is *literally* how the world will end. And—for today at least—the humans have lost the game.

———

This is a book about machine learning and human values: about systems that learn from data without being explicitly programmed, and about how exactly—and *what* exactly—we are trying to teach them.

The field of machine learning comprises three major areas: In *unsupervised* learning, a machine is simply given a heap of data and—as with the word2vec system—told to make sense of it, to find patterns, regularities, useful ways of condensing or representing or visualizing it. In *supervised* learning, the system is given a series of categorized or labeled examples—like parolees who went on to be rearrested and others who did not—and told to make predictions about new examples it hasn't seen yet, or for which the ground truth is not yet known. And in *reinforcement* learning, the system is placed into an environment with rewards and punishments—like the boat-racing track with power-ups and hazards—and told to figure out the best way to minimize the punishments and maximize the rewards.

On all three fronts, there is a growing sense that more and more of the world is being turned over, in one way or another, to these mathematical and computational models. Though they range widely in complexity—from something that might fit on a spreadsheet on the one hand, to something that might credibly be called *artificial intelligence* on the other—they are steadily replacing both human judgment *and* explicitly programmed software of the more traditional variety.

This is happening not only in technology, not only in commerce, but in areas with ethical and moral weight. State and federal law increasingly mandates the use of "risk-assessment" software to determine bail and parole. The cars and trucks on our freeways and neighborhood streets are increasingly driving themselves. We no longer assume that our mortgage application, our résumé, or our medical tests will be seen by human eyes before a verdict is rendered. It is as if the better part of humanity were, in the early twenty-first century, consumed by the task of gradually putting the world—figuratively and literally—on autopilot.

In recent years, alarm bells have gone off in two distinct communities. The first are those focused on the present-day ethical risks of technology. If a facial-recognition system is wildly inaccurate for people of one race or gender but not another, or if someone is denied bail because of a statistical model that has never been audited and that no one in the courtroom—including the judge, attorneys, and defendant—understands, this is a problem. Issues like these cannot be addressed within traditional disciplinary camps, but rather only through dialogue: between computer scientists, social scientists, lawyers, policy experts, ethicists. That dialogue has begun in a hurry.

The second are those worried about the future dangers that await as our systems grow increasingly capable of flexible, real-time decision-making, both online and in the physical world. The past decade has seen what is inarguably the most exhilarating, abrupt, and worrying progress in the history of machine learning—and, indeed, in the history of artificial intelligence. There is a consensus that a kind of taboo has been broken: it is no longer forbidden for AI researchers to discuss concerns of safety. In fact, such concerns have over the past five years moved from the fringes to become one of the central problems of the field.

Though there is a rivalry of sorts over whether the immediate or the longer-term issues should take priority, these two communities are united in their larger aims.

As machine-learning systems grow not just increasingly pervasive but increasingly powerful, we will find ourselves more and more often in the position of the "sorcerer's apprentice": we conjure a force, autonomous but totally compliant, give it a set of instructions, then scramble like mad to stop it once we realize our instructions are imprecise or incomplete—lest we get, in some clever, horrible way, precisely what we asked for.

How to prevent such a catastrophic divergence—how to ensure that these models capture our norms and values, understand what we mean or intend, and, above all, do what we want—has emerged as one of the most central and most urgent scientific questions in the field of computer science. It has a name: *the alignment problem.*

In reaction to this alarm—both that the bleeding edge of research is getting ever closer to developing so-called "general" intelligence and that real-world machine-learning systems are touching more and more ethically fraught parts of personal and civic life—has been a sudden, energetic response. A diverse group is mustering across traditional disciplinary lines. Nonprofits, think tanks, and institutes are taking root. Leaders within both industry and academia are speaking up, some of them for the first time, to sound notes of caution—and redirecting their research funding accordingly. The first generation of graduate students is matriculating who are focused explicitly on the ethics and safety of machine learning. The alignment problem's first responders have arrived at the scene.

This book is the product of nearly a hundred formal interviews and many hundreds of informal conversations, over the course of four years and many tens of thousands of miles, with researchers and thinkers from this field's young history and its sprawling frontier. What I found was a field finding its legs, amid exhilarating and sometimes terrifying progress. A story I thought I knew showed itself to be, by turns, more riveting, harrowing, and hopeful than I had understood.

Machine learning is an ostensibly technical field crashing increasingly on human questions. Our human, social, and civic dilemmas are becoming technical. And our technical dilemmas are becoming human, social, and civic. Our successes and failures alike in getting these systems to do "what we want," it turns out, offer us an unflinching, revelatory mirror.

This is a story in three distinct parts. Part one explores the alignment problem's beachhead: the present-day systems already at odds with our best intentions, and the complexities of trying to make those intentions explicit in systems we feel capable of overseeing. Part two turns the focus to reinforcement learning, as we come to understand systems that not only predict, but act; there are lessons here for understanding evolution, human motivation, and the delicacy of incentives, with implications for business and parenting alike. Part three takes us to the forefront of technical AI safety research, as we tour some of the best ideas currently going for how to align complex autonomous systems with norms and values too subtle or elaborate to specify directly.

For better or worse, the human story in the coming century is likely to be one of building such systems and setting them, one by one, in motion. Like the sorcerer's apprentice, we will find ourselves just one set of agents among many, in a world crowded—as it were—with brooms.

How, exactly, do we intend to teach them?

And what?

PART I

Prophecy

1

REPRESENTATION

In the summer of 1958, a group of reporters are gathered by the Office of Naval Research in Washington, D.C., for a demonstration by a twenty-nine-year-old researcher at the Cornell Aeronautical Laboratory named Frank Rosenblatt. Rosenblatt has built something he calls the "perceptron," and in front of the assembled press corps he shows them what it can do.

Rosenblatt has a deck of flash cards, each of which has a colored square on it, either on the left side of the card or on the right. He pulls one card out of the deck and places it in front of the perceptron's camera. The perceptron takes it in as a black-and-white, 20-by-20-pixel image, and each of those four hundred pixels is turned into a binary number: 0 or 1, dark or light. The four hundred numbers, in turn, are fed into a rudimentary neural network, the kind that McCulloch and Pitts had imagined in the early 1940s. Each of these binary pixel values is multiplied by an individual negative or positive "weight," and then they are all added together. If the total is negative, it will output a –1 (meaning the square is on the left), and if it's positive, it will output a 1 (meaning the square is on the right).

The perceptron's four hundred weights are initially random, and its outputs, as a result, are nonsense. But every time the system guesses "wrong," Rosenblatt "trains" it, by dialing up the weights that were too low and turning down the weights that were too high.

Fifty of these trials later, the machine now *consistently* tells left-side cards and right-side cards apart, including ones he hasn't shown it before. The demonstration itself is strikingly modest, but it signifies something grander. The machine is, in effect, learning from experience—what Rosenblatt calls a "self-induced change in the wiring diagram."[1]

McCulloch and Pitts had imagined the neuron as a simple unit of input and output, of logic and arithmetic, and they had shown the enormous power of such rudimentary mechanisms, in great enough numbers and suitably connected. But they had said next to nothing about how exactly the "suitably connected" part was actually meant to be achieved.[2]

"Rosenblatt made a very strong claim, which at first I didn't believe," says MIT's Marvin Minsky, coincidentally a former classmate of Rosenblatt's at the Bronx High School of Science.[3] "He said that if a perceptron was physically capable of being wired up to recognize something, then there would be a procedure for changing its responses so that eventually it would learn to carry out the recognition. Rosenblatt's conjecture turned out to be mathematically correct, in fact. I have a tremendous admiration for Rosenblatt for guessing this theorem, since it is very hard to prove."

The perceptron, simple as it is, forms the blueprint for much of the machine-learning systems we will go on to discuss. It contains a model *architecture:* in this case, a single artificial "neuron" with four hundred inputs, each with its own "weight" multiplier, which are then summed together and turned into an all-or-nothing output. The architecture has a number of adjustable variables, or *parameters:* in this case, the positive or negative multipliers attached to each input. There is a set of *training data:* in this case, a deck of flash cards with one of two types of shapes on them. The model's parameters are tuned using an optimization algorithm, or *training algorithm.*

The basic training procedure for the perceptron, as well as its many contemporary progeny, has a technical-sounding name—"stochastic gradient descent"—but the principle is utterly straightforward. Pick one of the training data at random ("stochastic") and input it to the model. If the output is exactly what you want, do nothing. If there is a difference between what you wanted and what you got, then figure out in which direction ("gradient") to adjust each weight—whether by literal turning of physical knobs or simply the changing of numbers in software—to lower the error for this particular example. Move each of them a little bit in the appropriate direction ("descent"). Pick a new example at random, and start again. Repeat as many times as necessary.

This is the basic recipe for the field of machine learning—and the humble perceptron will be both an overestimation and an underestimation of what is to come.

"The Navy," reports the *New York Times*, "revealed the embryo of an electronic computer today that it expects will be able to walk, talk, see, write, reproduce itself and be conscious of its existence."[4]

The *New Yorker* writes that the perceptron, "as its name implies, is capable of original thought." "Indeed," they write, "it strikes us as the first serious rival to the human brain ever devised."

Says Rosenblatt to the *New Yorker* reporter, "Our success in developing the perceptron means that for the first time a non-biological object will achieve an organization of its external environment in a meaningful way. That's a safe definition of what the perceptron can do. My colleague disapproves of all the loose talk one hears nowadays about mechanical brains. He prefers to call our machine a self-organizing system, but, between you and me, that's precisely what any brain is."[5]

That same year, *New Scientist* publishes an equally hopeful, and slightly more sober, article called "Machines Which Learn."[6] "When machines are required to perform complicated tasks it would often be useful to incorporate devices whose precise mode of operation is not specified initially," they write, "but which learn from experience how to do what is required. It would then be possible to produce machines to do jobs which have not been fully analysed because of their complexity. It seems likely that learning machines will play a part in such projects as the mechanical translation of languages and the automatic recognition of speech and of visual patterns."

"The use of the term 'learning machine' invites comparison with the learning of people and animals," the article continues. "The drawing of analogies between brains and machines requires caution to say the least, but in a general way it is stimulating for workers in either field to know something of what is happening in the other, and it is possible that speculation about machines which learn may eventually produce a system which is a true analogue of some form of biological learning."

The history of artificial intelligence is famously one of cycles of alternating hope and gloom, and the Jetsonian future that the perceptron seemed to herald is slow to arrive.

Rosenblatt, with a few years of hindsight, will wish the press had used a bit more caution in their reactions to his invention. The popular press "fell to the task with all of the exuberance and sense of discretion of a pack of happy bloodhounds," he says—while admitting, on his own behalf, a certain "lack of mathematical rigor in preliminary reports."[7]

Minsky, despite his "tremendous admiration" for Rosenblatt and his machine, begins "to worry about what such a machine could not do." In 1969, he and his MIT colleague Seymour Papert publish a book called *Perceptrons* that effectively slams the door shut on the entire vein of research. Minsky and Papert show, with the stiff formality of mathematical proof, that there are seemingly basic patterns that Rosenblatt's model simply will never be able to recognize. For instance, it is impossible to train one of Rosenblatt's machines to recognize when a card has an *odd* versus an *even* number of squares on it. The only way to recognize more complex categories like this is to use a network with multiple layers, with earlier layers creating a *representation* of the raw data, and the later layers operating on the representation. But no one knows how to tune the parameters of the early layers to make representations useful for the later ones. The field hits the equivalent of a brick wall. "There had been several thousand papers published on perceptrons up to 1969," says Minsky.

"Our book put a stop to those."[8]

It is as if a dark cloud has settled over the field, and everything falls apart: the research, the money, the people. Pitts, McCulloch, and Lettvin, who have all three moved to MIT, are sharply exiled after a misunderstanding with MIT's Norbert Wiener, who had been like a second father figure to Pitts and now won't speak to him. Pitts, alcoholic and depressed, throws all of his notes and papers into a fire, including an unpublished dissertation about three-dimensional neural networks that MIT tries desperately to salvage. Pitts dies from cirrhosis in May 1969, at the age of 46.[9] A few months later Warren McCulloch, at the age of 70, succumbs to a heart seizure after a long series of cardiopulmonary problems. In 1971, while celebrating his 43rd birthday, Frank Rosenblatt drowns in a sailing accident on the Chesapeake Bay.

By 1973, both the US and British governments have pulled their funding support for neural network research, and when a young English psychology student named Geoffrey Hinton declares that he wants to do his doctoral work on neural networks, again and again he is met with the same reply: "Minsky and Papert," he is told, "have proved that these models were no good."[10]

THE STORY OF ALEXNET

It is 2012 in Toronto, and Alex Krizhevsky's bedroom is too hot to sleep. His computer, attached to twin Nvidia GTX 580 GPUs, has been running day and night at its maximum thermal load, its fans pushing out hot exhaust, for two weeks.

"It was very hot," he says. "And it was loud."[11]

He is teaching the machine how to see.

Geoffrey Hinton, Krizhevsky's mentor, is now 64 years old and has not given up. There is reason for hope.

By the 1980s it became understood that networks with multiple layers (so-called "deep" neural networks) *could*, in fact, be trained by examples just as a shallow one could.[12] "I now believe," admitted Minsky, "that the book was overkill."[13]

By the late '80s and early '90s, a former postdoc of Hinton's named Yann LeCun, working at Bell Labs, had trained neural networks to identify handwritten numerals from 0 to 9, and neural networks found their first major commercial use: reading zip codes in post offices, and deposit checks in ATMs.[14] By the 1990s, LeCun's networks were processing 10 to 20% of all checks in the United States.[15]

But the field hit another plateau, and by the 2000s, researchers were still largely stuck fiddling with databases of handwritten zip codes. It was understood that, in *principle*, a big-enough neural network, with enough training examples and time, can learn almost anything.[16] But no one had fast-enough computers, enough data to train on, or enough patience to make good on that theoretical potential. Many lost interest, and the field of computer vision, along with computational linguistics, largely moved on to other things. As Hinton would later summarize, "Our labeled data-sets were thousands of times too small. [And] our computers were millions of times too slow."[17] Both of these things, however, would change.

With the growth of the web, if you wanted not fifty but five hundred thousand "flash cards" for your network, suddenly you had a seemingly bottomless repository of images. There was only one problem, which was that they usually didn't come with their category label readily attached. You couldn't train a network unless you knew what the network's output was supposed to be.

In 2005, Amazon launched its "Mechanical Turk" service, allowing for the recruiting of human labor on a large scale, making it possible to hire thousands of people to perform simple actions for pennies a click. (The service was particularly well suited to the kinds of things that future AI is thought to be able to do—hence its tagline: *artificial* artificial intelligence.) In 2007, Princeton professor Fei-Fei Li used Amazon Mechanical Turk to recruit human labor, at a scale previously unimaginable, to build a dataset that was previously impossible. It took more than two years to build, and had three *million* images, each labeled, by human hands, into more than five thousand categories. Li called it ImageNet, and released it in 2009. The field of computer vision suddenly had a mountain of new data to learn from, and a new grand challenge. Beginning in 2010, teams from around the world began competing to build a system that can reliably look at an image—dust mite, container ship, motor scooter, leopard—and say what it is.

Meanwhile, the relatively steady progress of Moore's law throughout the 2000s meant that computers could do in minutes what the computers of the 1980s took days to do. One further development, however, turned out to be crucial. In the 1990s, the video-game industry began to produce dedicated graphics processors called GPUs, designed to render complex 3D scenes in real time; instead of executing instructions with perfect precision one after another, as a traditional CPU does, they are capable of doing a great many simple and sometimes approximate calculations at once.[18] Only later, in the mid-2000s, did it come to be appreciated that the GPU could do a lot more than light and texture and shadow.[19] It turned out that this hardware, designed for computer gaming, was in fact tailor-made for training neural networks.

At the University of Toronto, Alex Krizhevsky had taken a class on writing code for GPUs, and decided to try it on neural networks. He applied himself to a popular image-recognition benchmark called CIFAR-10, which contained thumbnail-sized images that each belonged to one of ten categories: airplane, automobile, bird, cat, deer, dog, frog, horse, ship, or truck. Krizhevsky built a network and began using a GPU to train it to categorize CIFAR-10 images. Shockingly, he was able to train his network from a random starting configuration all the way to state-of-the-art accuracy. In *eighty seconds*.[20]

It is at this point Krizhevsky's labmate, Ilya Sutskever, takes notice and offers him what will become a kind of siren song. "I bet," Sutskever says, "you can make it work on ImageNet."

They build an enormous neural network: 650,000 artificial neurons, arranged into 8 layers, connected by 60 million adjustable weights. In his bedroom at his parents' house, Krizhevsky starts showing it pictures.

Step by step, piece by piece, the system gets a few percent more accurate.

The dataset—as big as it is, a few million pictures—isn't enough. But Krizhevsky realizes he can fake it. He starts doing "data augmentation," feeding the network mirror images of the data. That seems to help. He feeds it images that are cropped slightly, or tinted slightly. (A cat, after all, still looks like a cat when you lean forward or to the side, or go from natural to artificial light.) This seems to help.

He plays with different architectures—this number of layers, that number of layers—groping more or less blindly for what configuration might just happen to work best.

Krizhevsky occasionally loses the faith. Sutskever never does. Time and again he spurs Krizhevsky on. *You can make it work.*

"Ilya was like a religious figure," he says. "It's always good to have a religious figure."

Trying out a new version of the model, and training it until the accuracy maxed out, takes about two weeks, running twenty-four hours a day—which means that the project, though at some level frantic, also has a lot of downtime. Krizhevsky thinks. And tinkers. And waits. Hinton has come up with an idea called "dropout," where during training certain portions of the network get randomly turned off. Krizhevsky tries this, and it seems, for various reasons, to help. He tries using neurons with a so-called "rectified linear" output function. This, too, seems to help.

He submits his best model on the ImageNet competition deadline, September 30, and then the final wait begins.

Two days later, Krizhevsky gets an email from Stanford's Jia Deng, who is organizing that year's competition, cc'd to all of the entrants. In plain, unemotional language, Deng says to click the link provided to see the results.

Krizhevsky clicks the link provided and sees the results.

Not only has his team won, but they have *obliterated* the rest of the entire field. The neural network trained in his bedroom—its official name is "SuperVision," but history will remember it simply as "AlexNet"—made *half* as many errors as the model that came in second.

By the Friday of the conference, when it is time for the ImageNet Large Scale Visual Recognition Challenge workshop, the word has spread. Krizhevsky has been given the final talk slot of the day, and at 5:05 p.m. he takes his place up at the presenter's lectern. He looks around the room. In the front row is Fei-Fei Li; to the side is Yann LeCun. There is a majority, it seems, of the leading computer vision researchers in the world. The room is over capacity, with people standing along the aisles and walls.

"I was nervous," he says. "I was not comfortable."

And then, in front of the standing-room audience, not comfortable, Alex Krizhevsky tells them everything.

When Frank Rosenblatt was interviewed about his perceptron in 1958, he was asked what practical or commercial uses a machine like the perceptron might have. "At the moment, none whatever," he replied cheerfully.[21]

"In these matters, you know, use follows invention."

THE PROBLEM

On Sunday evening, June 28, 2015, web developer Jacky Alciné was at home watching the BET Awards when he got a notification that a friend had shared a picture with him through Google Photos. When he opened Google Photos, he noticed the site had been redesigned. "I was like, 'Oh, the UI's changed!' I remembered I/O [Google's annual software developer conference] happened, but I was curious; I clicked through."[22] Google's image recognition software had automatically identified groups of photos, and gave each a thematic caption. "Graduation," said one—and Alciné was impressed that the system had managed to identify the mortarboard and tassel on his younger brother's head. Another caption stopped him cold. The album cover was a selfie of Alciné and a friend of his. Alciné is Haitian-American; both he and his friend are Black.

"Gorillas," it said.

"So I thought— To be honest, I thought that *I* did something." He opened the album, expecting he had somehow misclicked or mistagged something. The album was full of dozens of photos of Alciné and his friend. And nothing else. "I'm like— This was seventy-plus photos. There's no way. . . . That's actually where I really realized what happened."

Alciné took to Twitter. "Google Photos," he wrote, "y'all fucked up. My friend's not a gorilla."[23]

Within two hours, Google+ chief architect Yonatan Zunger reached out. "Holy fuck," he wrote. "This is 100% Not OK."

Zunger's team deployed a change to Google Photos within another few hours, and by the following morning, only two photos were still mislabeled. Then Google took a more drastic step: they removed the label entirely.

In fact, three years later, in 2018, *Wired* reported that the label "gorilla" was *still* manually deactivated on Google Photos. That means that, years later, nothing will be tagged as a gorilla, including gorillas.[24]

Curiously, the press in 2018, just as in 2015, appeared to repeatedly mischaracterize the nature of the mistake. Headlines proclaimed, "Two Years Later, Google Solves 'Racist Algorithm' Problem by Purging 'Gorilla' Label from Image Classifier"; "Google 'Fixed' Its Racist Algorithm by Removing Gorillas from Its Image-Labeling Tech"; and "Google Images 'Racist Algorithm' Has a Fix But It's Not a Great One."[25]

Being himself a programmer and familiar with machine-learning systems, Alciné knew the issue wasn't a biased *algorithm*. (The algorithm was stochastic gradient descent, just about the most generic, vanilla, all-purpose idea in computer science: go through your training data at random, tune your model's parameters to assign slightly higher probability to the correct category for that image, and repeat as needed.) No, what he immediately sensed was that something had gone terribly awry in the training data itself. "I couldn't even blame the algorithm," he says. "It's not even the algorithm at fault. It did exactly what it was designed to do."

The problem, of course, with a system that can, in theory, learn just about anything from a set of examples is that it finds itself, then, at the mercy of the examples from which it's taught.

CALIBRATION AND THE HEGEMONY OF DESIGN

> The extent to which we take everyday objects for granted is the precise extent to which they govern and inform our lives.
>
> —MARGARET VISSER[26]

The single most photographed American of the nineteenth century— more than Abraham Lincoln or Ulysses S. Grant—was Frederick Douglass, the abolitionist author and lecturer who had himself escaped from slavery at the age of twenty.[27] This was no accident; for Douglass, the photograph was just as important as the essay or the speech. The photograph was just coming into its own through the daguerreotype in the 1840s, and Douglass immediately understood its power.

Before the photograph, representations of Black Americans were limited to drawings, paintings, and engravings. "Negroes can never have impartial portraits at the hands of white artists," Douglass wrote. "It seems to us next to impossible for white men to take likenesses of black men, without most grossly exaggerating their distinctive features."[28] One exaggeration, in particular, prevailed during Douglass's time. "We colored men so often see ourselves described and painted as monkeys, that we think it a great piece of good fortune to find an exception to this general rule."[29]

The photograph not only countered such caricatures but, further, made possible a kind of transcending empathy and recognition. "Whatever may be the prejudices of those who may look upon it," said Douglass of a photograph of the first Black US senator, Hiram Revels, "they will be compelled to admit that the Mississippi Senator is a man."[30]

But all was not entirely well. As photography became more standardized and mass-produced in the twentieth century, some began to feel that the field of photography was itself worthy of critique. As W.E.B. Du Bois wrote in 1923, "Why do not more young colored men and women take up photography as a career? The average white photographer does not know how to deal with colored skins and having neither sense of the delicate beauty or tone nor will to learn, he makes a horrible botch of portraying them."

We often hear about the lack of diversity in film and television— among casts and directors alike—but we don't often consider that this problem exists not only in front of the camera, not only behind the camera, but in many cases *inside* the camera itself. As Concordia University communications professor Lorna Roth notes, "Though the available academic literature is wide-ranging, it is surprising that relatively few of these scholars have focused their research on the skin-tone biases within the actual apparatuses of visual reproduction."[31]

For decades, she writes, film manufacturers and film developers used a test picture as a color-balance benchmark. This test picture became known as the "Shirley card," named after Shirley Page, a Kodak employee and the first model to pose for it.[32] It perhaps goes without saying that Shirley and her successors were overwhelmingly White. The chemical processing of film was tuned accordingly, and as a result cameras simply didn't take good photos of Black people.

(In video just as in photography, colors have for decades been calibrated to White skin. In the 1990s, Roth interviewed one of the camera operators on *Saturday Night Live* about the process of tuning the cameras before broadcast. He explained, "A good VCR person will have a color girl stand in front of the cameras and stay there while the technicians focus on her flesh tones to do their fine adjustments to balance the cameras. This color girl is always white.")[33]

Amazingly, Kodak executives in the 1960s and '70s described the major impetus for making film that was sensitive to a wider range of darker tones as having come not from the civil rights movement but from the *furniture and chocolate industries,* which complained that film wasn't properly showing the grains of darker woods, or the difference between milk and dark chocolate.[34]

Former manager of Kodak Research Studios Earl Kage reflects on this period of research: "My little department became quite fat with chocolate, because what was in the front of the camera was consumed at the end of the shoot." Asked about the fact that this was all happening against the backdrop of the civil rights movement, he adds, "It is fascinating that this has never been said before, because it was never Black flesh that was addressed as a serious problem that I knew of at the time."[35]

In time Kodak began using models of more diverse skin tones. "I started incorporating black models pretty heavily in our testing, and it caught on very quickly," recalls Kodak's Jim Lyon. "I wasn't attempting to be politically correct. I was just trying to give us a chance of making a better film, one that reproduced everybody's skin tone in an appropriate way."

By the 1990s, the official Kodak Shirley card now had three different models on it, of different races. Their Gold Max film—initially marketed with the claim that it could photograph a "dark horse in low light"—now featured in television commercials with diverse families. One depicts a Black boy in a bright white karate gi, smiling as he performs a kata and presumably receives his next belt. It says, "Parents, would you trust this moment to anything other than Kodak Gold film?"

Their original target audience had given them a problematic calibration measure. Now a new calibration measure had given them a new audience.

FIXING THE TRAINING SET

All machine-learning systems, from the perceptron onward, have a kind of Shirley card at their heart: namely, the set of data on which they were trained. If a certain type of data is underrepresented or absent from the training data but present in the real world, then all bets are off.[36]

As UC Berkeley's Moritz Hardt argues, "The whole spiel about big data is that we can build better classifiers largely as a result of having more data. The contrapositive is that less data leads to worse predictions. Unfortunately, it's true by definition that there is always proportionately less data available about minorities. This means that our models about minorities generally tend to be worse than those about the general population."[37]

Alciné's frustrated tweets the night of the incident echo exactly this sentiment. He's a software engineer. He instantly diagnoses what has gone wrong. Google Photos, he infers, just didn't have nearly as many pictures of Black people in it as pictures of White people. And so the model, seeing anything unfamiliar, was much more prone to error.

"Again, I can completely understand how that happens," Alciné tells me.[38] "Like if you take a picture of an apple, but only red apples, when it sees a green apple it might think it's a pear. . . . Little things like that. That I understand. But then, you're the world's— Your mission is to index the entire world's social knowledge, so how did you, like, just skip over an entire *continent* of people?"

The problems of the twentieth century appear to be repeating them-selves uncannily in the twenty-first. Fortunately, it seems that some of the solutions are, too. All it would take was someone willing to question exactly who and what were represented in these twenty-first-century "Shirley cards," anyway—and what a better one might look like.

When Joy Buolamwini was a computer science undergrad at Georgia Tech in the early 2010s, she was given an assignment to program a robot to play peekaboo. The programming part was easy, but there was one issue: the robot wouldn't recognize Buolamwini's face. "I borrowed my roommate's face to get the project done, submitted the assignment, and figured, 'You know what, somebody else will solve this problem.'"[39]

Later in her undergraduate studies, she traveled to Hong Kong for an entrepreneurship competition. A local startup was giving a demo of one of its "social robots." The demo worked on everyone in the tour group . . . except for Buolamwini. As it happened, the startup was using the *very* same off-the-shelf, open-source face-recognition code that she herself had used back at Georgia Tech.

In one of the first articles explicitly addressing the notion of bias in computing systems, the University of Washington's Batya Friedman and Cornell's Helen Nissenbaum had warned that "computer systems, for instance, are comparatively inexpensive to disseminate, and thus, once developed, a biased system has the potential for widespread impact. If the system becomes a standard in the field, the bias becomes pervasive."[40]

Or, as Buolamwini herself puts it, "Halfway around the world, I learned that algorithmic bias can travel as quickly as it takes to download some files off of the internet."[41]

After a Rhodes Scholarship at Oxford, Buolamwini came to the MIT Media Lab, and there she began working on an augmented-reality proj-ect she dubbed the "Aspire Mirror." The idea was to project empowering or uplifting visuals onto the user's face—making the onlooker trans-form into a lion, for instance. Again, there was only one problem. The Aspire Mirror only worked on Buolamwini herself when she put on a white mask.

The culprit is not stochastic gradient descent; it is, clearly, the sets of images on which these systems are trained. Every face-detection or face-recognition system has, behind it and implicitly within it, a set of images—typically tens or hundreds of thousands—on which the system was originally trained and developed. This training data, the Shirley cards of the twenty-first century, is often invisible, or taken for granted, or absent entirely: a pretrained model disseminated online almost never comes with its training data included. But it is very much present, and will permanently shape the behavior of a deployed system.

A major movement in rooting out bias, then, is trying to better expose, and better understand, the training datasets behind major academic and commercial machine-learning systems.

One of the more popular public-domain databases of pictures of faces, for instance, is what's known as the Labeled Faces in the Wild (LFW) dataset, painstakingly assembled in 2007 from online news articles and image captions by a team from UMass Amherst, and used by innumerable researchers thereafter.[42] The composition of this database was not deeply studied, however, until many years later. In 2014, Michigan State's Hu Han and Anil Jain analyzed the dataset and determined it was more than 77% male, and more than 83% White.[43] The most common individual in the dataset is the person who appeared most often in online news photos in 2007: then-president George W. Bush, with 530 unique images. In fact, there are more than twice as many images of George W. Bush in the LFW dataset as there are of all Black women, combined.[44]

The original 2007 paper describing the database noted that a set of images gathered from online news articles "clearly has its own biases," but these "biases" are considered from a technical, rather than social, standpoint: "For example, there are not many images which occur under extreme lighting conditions, or very low lighting conditions." Such lighting issues aside, the authors write, "the range and diversity of pictures present is very large."

Twelve years later, however, in the fall of 2019, a disclaimer suddenly appeared on the webpage of the Labeled Faces in the Wild dataset that takes a different view. It notes, "Many groups are not well represented in LFW. For example, there are very few children, no babies, very few people over the age of 80, and a relatively small proportion of women. In addition, many ethnicities have very minor representation or none at all."[45]

In recent years, greater attention has been paid to the makeup of these training sets, though much remains to be done. In 2015, the United States Office of the Director of National Intelligence and the Intelligence Advanced Research Projects Activity released a face image dataset called IJB-A, boasting, they claimed, "wider geographic variation of subjects."[46] With Microsoft's Timnit Gebru, Buolamwini did an analysis of the IJB-A and found that it was more than 75% male, and almost 80% light-skinned. Just 4.4% of the dataset were dark-skinned females.[47]

Eventually it became clear to Buolamwini that the "somebody else [who] will solve this problem" was—of course—her. She started a broad investigation into the current state of face-detection systems, which became her MIT thesis. She and Gebru set out first to build a dataset with a more balanced representation of both gender and skin tone. But where would they get their images from? Previous datasets, drawing from online news, for instance, were totally imbalanced. They decided on *parliaments*, compiling a database of the representatives of six nations: Rwanda, Senegal, South Africa, Iceland, Finland, and Sweden. This dataset was notably *un*diverse in things like age, lighting, and pose, with almost all subjects middle-aged or older, centered in the frame, and looking straight into the camera with a neutral or slightly smiling expression. But, measured by skin and by gender, it was arguably the most diverse machine-learning dataset assembled to date.[48]

With this parliamentarian dataset in hand, Buolamwini and Gebru looked at three commercially available face-classification systems— from IBM, Microsoft, and the Chinese firm Megvii, maker of the widely used Face++ software—and ran each through the paces.

Across the dataset as a whole, all three systems did reasonably well at correctly classifying the gender of the subject—approximately 90% for all three companies. In all three cases, the software was roughly 10 to 20% more accurate on male faces than female faces, and all were also roughly 10 to 20% more accurate on lighter faces than darker ones. But when Buolamwini did an intersectional analysis of the two, the starkest result by far appeared. All three systems were *dramatically* worse at classifying faces that were both dark-skinned *and* female. IBM's system, for instance, had an error rate of only 0.3% for light-skinned males, but 34.7% for dark-skinned females: more than a *hundredfold* greater.

The abolitionist and women's rights activist Sojourner Truth is arguably best known for her 1851 speech "Ain't I a Woman?" Buolamwini poignantly echoes this question into the twenty-first century, pointing to photos of Truth that are miscategorized, again and again and again, by contemporary commercial face-classification software, as male.[49]

On December 22, 2017, Buolamwini reached out to the three firms with her results, explaining that she would be presenting them at an upcoming conference, and giving each an opportunity to respond. Megvii did not respond. Microsoft responded with a generic statement: "We believe the fairness of AI technologies is a critical issue for the industry and one that Microsoft takes very seriously. We've already taken steps to improve the accuracy of our facial recognition technology and we're continuing to invest in research to recognize, understand and remove bias."[50] IBM, however, was another story entirely. They responded the same day, thanked Buolamwini for reaching out, replicated and confirmed her results, invited her to both their New York and Cambridge campuses, and within a matter of weeks announced a new version of their API with a *tenfold* improvement in the error rate for dark-skinned females.[51]

"Change is possible," she says. There was no fundamental obstacle, technological or otherwise, to equalizing this performance gap; it just took someone asking the right questions.

Buolamwini and Gebru's work highlights the skepticism we ought to feel when a company announces that their system is, say, "99% accurate": Accurate on what? Accurate for whom? It's a reminder, too, that every machine-learning system *is* a kind of parliament, in which the training data represent some larger electorate—and, as in any democracy, it's crucial to ensure that everyone gets a vote.[52]

Bias in machine-learning systems is often a direct result of the data on which the systems are trained—making it incredibly important to understand who is represented in those datasets, and to what degree, before using them to train systems that will affect real people.

But what do you do if your dataset is as inclusive as possible—say, something approximating the entirety of written English, some hundred billion words—and it's the world *itself* that's biased?

THE DISTRIBUTIONAL HYPOTHESIS: WORD EMBEDDINGS

You shall know a word by the company it keeps.

—J. R. FIRTH[53]

Let's say you find a message in a bottle, washed up on a beach; a couple parts of the message are unreadable. You examine one sentence: "I hath buried the treasure north of the —— by the beach." Needless to say, you are highly motivated to figure out what the missing word might be.

It probably does not occur to you that the word might be "hamster" or "donut" or "toupee." There are a few reasons for this. You have some common sense you can apply: hamsters are restless, donuts are biodegradable, and toupees blow in the wind—none of them reliable landmarks for long-term treasure wayfinding. Anyone hiding loot over a span of presumably months to years, you reason, would need something stable, unlikely to disintegrate or move.

Now imagine you're a computer with a complete lack of such common sense—let alone the ability to put yourself in the shoes of a prospective treasure burier—but what you do have is an extremely large sample (a "corpus") of real-world texts to scan for patterns. How good a job could you do at predicting the missing word *purely* based on the statistics of the language itself?

Constructing these kinds of predictive models has long been a grail for computational linguists.[54] (Indeed, Claude Shannon founded information theory in the 1940s on a mathematical analysis of this very sort, noticing that some missing words are more predictable than others, and attempting to quantify by how much.[55]) Early methods involved what are known as "*n*-grams," which meant simply *counting up* every single chain of, say, two words in a row that appeared in a particular corpus— "appeared in," "in a," "a particular," "particular corpus"—and tallying them in a huge database.[56] Then it was simple enough, given a missing word, to look at the preceding word and find which *n*-gram in the database beginning with that preceding word had appeared most often. That would then be your best guess as to what was missing. Of course, additional context beyond just the immediately preceding word could offer you additional clues, but incorporating it was far from straightforward. Going from storing a list of all possible two-word phrases in your language ("bigrams") to all the three-word phrases ("trigrams"), or four-word phrases or more, meant growing your database to an absurd and untenable size. Moreover, these databases became incredibly sparsely populated, with the vast majority of possible phrases never appearing at all, and much of the rest appearing only once or twice.

Ideally, we'd also want to be able to make reasonable guesses even if a particular phrase had *never* appeared verbatim before in the corpus. Such counting-based methods are no help. In the sentence "I sipped at a jaundiced ———," we might imagine "chardonnay" is more likely than "charcoal," even if neither word has ever been preceded by "jaundiced" in the history of the language. Relying on a count simply won't help in cases like this—and, again, the problem gets worse the more context we try to add, because the longer the phrases we consider, the more likely we are never to have seen something before.

This set of issues is known as the "curse of dimensionality," and has plagued this linguistic approach from the very beginning.[57]

Was there a better way?

There was, and it came in the form of what are called "distributed representations."[58] The idea was to try to represent words by points in some kind of abstract "space," in which related words appear "nearer" to one another. A number of techniques emerged over the 1990s and 2000s for doing this,[59] but one in particular in the past decade has shown exceptional promise: neural networks.[60]

The hypothesis here, the big bet on which the model rests, is simply this: Words will tend to be found near words that are "similar" to themselves. And these similarities can be captured numerically. The neural network model works by transforming ("embedding") every word into a set (a "vector") of numbers that represent its "coordinates" in that space. This set of coordinate numbers is known as the word's *representation*. (In the case of word2vec, it's three hundred decimal numbers between –1.0 and 1.0.) This enables a direct measure of how "similar" any word is to any other: How far away are those coordinates?[61]

All we have to do is—*somehow*—arrange the words in this space to make them do as good a job of predicting these missing words as possible. (At least, we'll have done as good a job as this particular model architecture allows.)

How are we going to arrive at these representations? Why, of course, stochastic gradient descent! We will simply scatter our words *randomly* throughout space to begin with. Then we'll pick a randomly selected phrase from our corpus, hide a word, and ask the system what it expects might fill in that blank.

When our model guesses wrong, we'll adjust the coordinates of our word representations to slightly nudge the correct word *toward* the context words in our mathematical space and slightly nudge any incorrect guesses *away*. After we make this tiny adjustment, we'll pick another phrase at random and go through this process again. And again. And again. And again. And again.[62]

"At this point," explains Stanford computational linguist Christopher Manning, "sort of a miracle occurs."

In his words:

> It's sort of surprising—but true—that you can do no more than set up this kind of prediction objective, make it the job of every word's word vectors to be such that they're good at predicting the words that appear in their context or vice-versa—you just have that very simple goal—and you say *nothing* else about how this is going to be achieved—but you just pray and depend on the magic of deep

learning. . . . And this miracle happens. And out come these word vectors that are just amazingly powerful at representing the meaning of words and are useful for all sorts of things.[63]

In fact, one could argue that these embeddings actually manage to capture *too much* of the nuance of our language. Indeed, they capture with startling clarity the parts we ourselves prefer not to see.

THE DARK SIDE OF EMBEDDINGS

Out of the crooked timber of humanity no truly straight thing was ever made.

—IMMANUEL KANT[64]

Word-embedding models like these, including Google's word2vec and Stanford's GloVe, subsequently became the de facto standard for computational linguistics, undergirding since roughly 2013 almost every application that involves computer use of language, be it ranking search results, translating passages from one language to another, or analyzing consumer sentiment in written reviews.[65]

Indeed, the embeddings, simple as they are—just a row of numbers for each word, based on predicting nearby missing words in a text— seemed to capture a *staggering* amount of real-world information.

You could, for instance, simply add two vectors together to get a new vector, and search for the nearest word. The results, as we have seen, often made a shocking amount of sense:

Czech + currency = koruna
Vietnam + capital = Hanoi
German + airlines = Lufthansa
French + actress = Juliette Binoche*

* With second place going to **Vanessa Paradis** and third to **Charlotte Gainsbourg**.

And you could *subtract* words, too. This meant—incredibly—you could produce "analogies" by getting the "difference" between two words and then "adding" it to a third.[66]

These analogies suggested that the embeddings had captured geography:

$$\texttt{Berlin - Germany + Japan = Tokyo}$$

And grammar:

$$\texttt{bigger - big + cold = colder}$$

And cuisine:

$$\texttt{sushi - Japan + Germany = bratwurst}$$

And science:

$$\texttt{Cu - copper + gold = Au}$$

And tech:

$$\texttt{Windows - Microsoft + Google = Android}$$

And sports:

$$\texttt{Montreal Canadiens - Montreal + Toronto =}$$
$$\texttt{Toronto Maple Leafs}^{67}$$

Unfortunately, as we've seen, that wasn't all the vectors captured. They contained stunning gender biases. For every clever or apt analogy for `man:woman`, like `fella:babe`, or `prostate cancer:ovarian cancer`, there was a host of others that seemed to be reflecting mere stereotypes, like `carpentry:sewing`, or `architect:interior designer`, or `doctor:nurse`.

We are only now coming to a full appreciation of the problem. "There have been hundreds of papers written about word embeddings and their applications, from Web search to parsing Curricula Vitae," as Tolga Bolukbasi, Adam Kalai, and their collaborators write. "However, none of these papers have recognized how blatantly sexist the embeddings are and hence risk introducing biases of various types into real-world systems."[68]

Machine-learning systems like this not only *demonstrate* bias but may silently, subtly *perpetuate* it. Consider an employer, doing a search for "software engineer" candidates. The search will rank millions of possible résumés by some measure of "relevance" and present just the top handful.[69] A system naïvely using word2vec, or something like it, might well observe that John is a word more typical of engineer résumés than Mary. And so, all things being equal, a résumé belonging to John will rank higher in "relevance" than an otherwise identical résumé belonging to Mary. Such examples are more than hypothetical. When one of the clients of Mark J. Girouard, an employment attorney, was vetting a résumé-screening tool from a potential vendor, the audit revealed that one of the two most positively weighted factors in the entire model was the name "Jared." The client did not purchase the résumé-screening tool—but presumably others have.[70]

We already know, of course, that job candidates' names exert influence over real human employers. In 2001 and 2002, the economists Marianne Bertrand and Sendhil Mullainathan mailed out nearly five thousand résumés with randomly assigned names designed to sound either White (Emily Walsh, Greg Baker) or African-American (Lakisha Washington, Jamal Jones). They found a stunning 50% gap in the callback rates, despite the résumés themselves being identical.[71]

Word2vec maps proper names to racial and gender axes just like it does with any other words, putting Sarah – Matthew on a gender axis and Sarah – Kiesha on a race axis. Given that it puts professions on these axes as well, it's not hard to imagine a system inadvertently using such racial or gender dimensions—in effect, stereotypes—to uprank or downrank candidates for "relevance" to a given job opening. In other words, we have reason to be every bit as concerned if it is a machine sifting through these résumés, and not a person.[72]

The obvious solution in the human case—removing the names—will not work. In 1952, the Boston Symphony Orchestra began holding its auditions with a screen placed between the performer and the judge, and most other orchestras followed suit in the 1970s and '80s. The screen, however, was not enough. The orchestras realized that they also needed to instruct auditioners, before walking out onto the wood floor of the audition hall, to remove their *shoes*.[73]

The problem with machine-learning systems is that they are designed precisely to infer hidden correlations in data. To the degree that, say, men and women tend toward different writing styles in general—subtle differences in diction or syntax—word2vec will find a host of minute and indirect correlations between `software engineer` and *all* phrasing typical of males.[74] It might be as noticeable as `football` listed among their hobbies, rather than `softball`, as subtle as the names of certain universities or hometowns, or as nearly invisible as a slight grammatical preference for one preposition or synonym over another. A system of this nature cannot, in other words, ever be successfully blindfolded. It will always hear the shoes.

In 2018, Reuters reported that Amazon engineers had, starting in 2014, been developing a machine-learning tool to sift through online résumés and rank possible job candidates from one to five stars—just like Amazon products themselves—based on how promising they seemed, and that Amazon recruiters would focus their efforts accordingly.[75] "They literally wanted it to be an engine where I'm going to give you a hundred résumés, it will spit out the top five, and we'll hire those," as one source told reporters. The measure of that star rating? Similarity—using a word representation model—to résumés of previous Amazon hires in the preceding ten years.

By 2015, however, Amazon began noticing problems. Most of those previous engineering hires were men. The model, they realized, was assigning a negative score to the word "women's"—for instance, in describing extracurriculars. They edited the model to remove this bias.

They also noticed, however, that it was assigning a negative score to the names of all-women's colleges. They edited the model to remove this bias.

Still, the model found a way to hear the shoes. Engineers noticed that the model was assigning positive scores to seemingly *all* vocabulary choices—for instance, words like "executed" and "captured"—more prevalent among male résumés than female ones.[76]

By 2017, Amazon had scrapped the project and disbanded the team that made it.[77]

DEBIASING WORD EMBEDDINGS

For Tolga Bolukbasi and Adam Kalai, along with their BU and Microsoft collaborators, the question was, of course, not simply the *discovery* of these biases but what to *do* about them.

One option was to find the axis in this high-dimensional vector "space" that captured the concept of gender and *delete* it. But deleting the gender dimension altogether would mean losing useful analogies like `king:queen` and `aunt:uncle`. So the challenge, as they put it, is "to reduce gender biases in the word embedding while preserving the useful properties of the embedding."[78]

As it happens, even identifying the proper gender "dimension" is hard. You could define it, for instance, as the vector "difference" of `woman - man`. But there's more going on here than just gender—you also have idiomatic uses like "man oh man" and the verb form, as in "all hands, man your battle stations." The team decided to take a number of different word pairs of this type—`woman - man`, but also `she - he`, `gal - guy`, and so forth—and then use a technique called principal component analysis (PCA) to isolate the axis that explained the greatest amount of the difference in these pairs: presumably, gender.[79]

Then their task was to try to determine, for words that differ on this gender dimension, whether that gender difference is appropriate or inappropriate. Let's say `king` and `queen` are appropriately separated by gender, and ditto `father` and `mother`, but maybe we *don't* want to regard—as word2vec by default does—`Home Depot` as the gender-flipped version of `JC Penney`; or `alcoholism` and `eating disorders`; or `pilot` and `flight attendant`.

How, then, to tease apart the problematic from the unproblematic gender associations for not just a handful but for hundreds of thousands of different words? How to know which analogies should be kept, which should be adjusted, and which should be purged entirely?

The team of five computer scientists found themselves doing, in effect, social science. Indeed, part of the project ended up requiring consultation beyond their normal disciplinary lines. "We're a bunch of machine-learning researchers," says Kalai. "I work in a lab that includes a bunch of social scientists, and just from listening to them talk about various issues that come up in sociology and social sciences, we were aware of these possible concerns that the machine-learning algorithms might discriminate, but none of the five of us—we're all five guys—had ever worked or read much about gender bias."

The group, perhaps naïvely, asked the sociologists how they should encode a formal definition of which analogies were acceptable and which were not. The sociologists rapidly disabused them of the idea that a simple formal definition of this kind was possible. "We're thinking, how do we define the best thing?" says Bolukbasi. "They said, 'Sociologists can't define what is good.' As an engineer you want to say, 'Okay, this is the ideal, so this is my target, so I'm just going to make my algorithm until I reach that target.' Because it's involved so much with people and culture and everything, you don't know what's optimal. You can't optimize for something. It's very hard actually in that sense."

The group decided to identify a set of words that they felt were appropriate to consider as gendered in some essential or fundamental way: words like "he" and "she," "brother" and "sister," as well as anatomical words like "womb" and "sperm," and social words like "convent" and "monastery" or "sorority" and "fraternity." Some of these required complicated decisions—like the word "nurse," for instance. As a noun, it was a profession that had no intrinsic gender aspect, but as a verb, it could be something that only women were capable of doing. What about a word like "rabbi"? Whether the word had an intrinsic gender dimension depended on whether the Jewish denomination in question was, say, Orthodox or Reform. The team did their best, identifying 218 such gender-specific words in a subset of their model's dictionary, and let their system extrapolate to the rest of the dictionary. "Note that the choice of words is subjective," they wrote, "and ideally should be customized to the application at hand."[80] For all words outside this set, they set the gender component of the word's representation to zero. Then they adjusted the representations of all the gender-related words such that pairs of equivalent terms—say, "brother" and "sister"—were "centered" around this zero point. Said another way, they were adjusted so that neither term was represented in the model as more "gender-specific" or more "gender-neutral" than the other.

Was this new, debiased model an improvement? The team took a methodological page from the social sciences and simply *asked people*. They used American workers on the Amazon Mechanical Turk platform to categorize a number of the model's analogies as either "stereotypes" or not. Even here, the input of the sociologists was critical. The wording of exactly what they asked was going to be important. "We had to talk to them because when we were designing these experiments on Mechanical Turk, the way you ask questions makes a difference, actually," says Bolukbasi. "It's such a sensitive topic."[81]

The results were encouraging. Where the default model was returning `doctor - man + woman` as `nurse`, the system now said `physician`. The Mechanical Turk workers had reported that 19% of the original model's gender analogies reflected gender stereotypes; of the new, debiased model's analogies, only 6% were judged to reflect stereotypes.[82]

This neutralization came at a small cost—the model now, for instance, thought it just as likely that someone could be "grandmothered in" as "grandfathered in" to a legal exemption.[83] But maybe this was a price worth paying—and you could always decide how much prediction error you were willing to trade for how much debiasing and set an appropriate tradeoff.

As the team wrote, "One perspective on bias in word embeddings is that it merely reflects bias in society, and therefore one should attempt to debias society rather than word embeddings. However, . . . in a small way debiased word embeddings can hopefully contribute to reducing gender bias in society. At the very least, machine learning should not be used to inadvertently amplify these biases, as we have seen can naturally happen."[84]

It was an encouraging proof of concept that we might be able to build our systems atop models of language that don't just capture the status quo world as it happens to be, but a model of a *better* world—a model of the world we want.

Still, there is more to the story. In 2019 Bar Ilan University computer scientists Hila Gonen and Yoav Goldberg published an exploration of these "debiased" representations and showed that the debiasing was potentially only, as they put it, "lipstick on a pig."[85] Yes, it removed links from, say, professions like "nurse" or "receptionist" to *explicitly* gendered terms like "woman" and "she." But an implicit connection among these stereotypically "feminine" professions themselves—between "nurse" and "receptionist"—remained. In fact, such only partial debiasing may actually make the problem worse, they argue, in the sense that it leaves the majority of these stereotypical associations intact while removing the ones that are the most visible and easiest to measure.[86]

Bolukbasi, now at Google, and his colleagues are continuing to work on the issue, affirming that there are cases where a debiased model, used in a recruiting context, may in fact actually be worse than the original model.[87] In such cases, it is possible that a system that deletes the gender dimension entirely—even for such fundamentally gendered terms as "he" and "she"—may result in fairer outcomes. The story is not a simple one, and the work continues.

SELF-PORTRAIT IN A STATISTICAL MIRROR

In a recruiting application these biases may simply be dangers to be mitigated, but taken on their own they raise a host of questions. For instance, where do they come from? Are they an artifact of the statistical technique used, or do they reflect something deeper: namely, the bias in our own heads and the bias in the world at large?

A classic test of unconscious bias in humans used in the social sciences is the "implicit association test," where subjects will see a sequence of words and are asked to press a button anytime the word belongs to *either* of two different categories: for instance, a flower (e.g., "iris") *or* something pleasant (e.g., "laughter"). It sounds simple enough, and it is; the story is not in *accuracy*, but in *reaction time*. Asking people to press a button if the word is a flower or something pleasant results in quick reaction times, but asking them to press the button if it's a flower or something *un*pleasant takes longer. This suggests, then, that there is some degree of overlap between the mental categories of "floral" and "pleasant," or that they reflect concepts that are somehow linked.[88]

The group that invented this test famously demonstrated that a group of White undergraduates were quick to identify if a word was either a prevalently White name ("Meredith," "Heather") or a pleasant word ("lucky," "gift"). They were also quick to identify if a word was either a prevalently Black name ("Latonya," "Shavonn") or an *unpleasant* word ("poison," "grief"). But when asked to press a button if the word was a White name or an unpleasant word, they were slow; likewise, they were slow when they had to press a button if the word was a Black name or a pleasant word.

A team of computer scientists at Princeton—postdoc Aylin Caliskan and professors Joanna Bryson and Arvind Narayanan—found that the distance between embeddings in word2vec and other widely used word-embedding models uncannily mirrors this human reaction-time data. The slower people are to identify any two groups of words, the farther away those word vectors were in the model.[89] The model's biases, in other words, are, for better or worse, very much our own.

Beyond these implicit associations, the Princeton team also wanted to know if models like word2vec capture some of what they call the "veridical" bias in the world. Certain names really are more commonly given to women than to men, and certain jobs really are more commonly held by women than by men. To the extent that certain names get representations that skew more male than others or more female than others, does that reflect to some degree this objective reality? And to the extent that certain professions fall at different places along the model's gender axis, might that reflect to some degree the fact that certain professions—nurse, librarian, carpenter, mechanic—really *do* happen to be unevenly balanced? The Princeton team consulted the US Census and the Bureau of Labor Statistics, respectively; in both cases, they found, the answer was yes.

The more strongly a word representation for a profession skews in a gender direction, the more overrepresented that gender tends to be within that profession. "Word embeddings," they write, "correlate strongly with the percentage of women in 50 occupations in the United States."[90] Looking at names, they found the same thing, with a correlation only slightly less strong; then again, the latest census data they had access to was from 1990, and so perhaps the gender distribution of names had in fact changed slightly since then.

The fact that the embeddings that emerge from this "magical" optimization process are so uncannily and discomfitingly useful as a mirror for society means that we have, in effect, added a diagnostic tool to the arsenal of social science. We can use these embeddings to quantify something in precise detail about society at a given snapshot in time. And regardless of causation—whether it's changes in the objective reality that change the way we speak, or vice versa, or whether both are driven by some deeper cause—we can use these snapshots to *watch society change.*

This is exactly what an interdisciplinary group from Stanford, led by Nikhil Garg and James Zou, set out to do. Garg, a PhD candidate in electrical engineering, and Zou, an assistant professor of biomedical data science, joined with historian Londa Schiebinger and linguist Dan Jurafsky to look at word embeddings using not just a corpus of *contemporary* text but samples across the past hundred years.[91]

What emerged was a rich and detailed history of the changing winds of culture. As they put it, "The temporal dynamics of the embedding helps to quantify changes in stereotypes and attitudes toward women and ethnic minorities in the 20th and 21st centuries in the United States."

The Stanford group corroborated the Princeton group's findings about the connection between the representations of profession words and gender, adding that there appears to be something of a "male baseline": namely, professions that we know from census data to be evenly split fifty-fifty between men and women nonetheless have a slight bias in their word embeddings in the "male" direction. As the authors explain, "Common language is more biased than one would expect based on external, objective metrics." Baseline aside, however, there is a consistent trend across time that shows the gender bias in word embeddings for professions moving in lockstep with the change in the workforce itself.

By looking at texts across time, they found a wealth of narratives reflecting social change. The data show that gender bias has broadly decreased over time and, in particular, that the "women's movement in the 1960s and 1970s especially had a systemic and drastic effect in women's portrayals in literature and culture."

The embeddings also show a detailed history of the shift in racial attitudes. In 1910, for instance, the top ten words most strongly associated with Asians relative to Whites included "barbaric," "monstrous," "hateful," and "bizarre." By 1980 the story could not be more different, with the top ten words topped by "inhibited" and "passive" and ending with "sensitive" and "hearty": stereotypes in their own right, of course, but ones that reflect an unmistakable cultural change.

More recent cultural shifts are also visible in the embeddings—for instance, the association between words related to Islam and words related to terrorism goes up sharply both in 1993 (the year of the World Trade Center bombing) and 2001 (the year of 9/11).

One might even imagine using such an approach to look not retrospectively but prospectively: Do the data of the last six months, say, suggest that these biases are getting better or getting worse? One might imagine a kind of real-time dashboard of whether society itself—or, at the very least, our public discourse—appears to be getting more or less biased: a bellwether for the shifts underway, and a glimpse of the world to come.

REPRESENTATIONS AND REPRESENTATION

There are several takeaways here, of which the first is principally, though not purely, methodological. Computer scientists are reaching out to the social sciences as they begin to think more broadly about what goes into the models they build. Likewise, social scientists are reaching out to the machine-learning community and are finding they now have a powerful new microscope at their disposal. As the Stanford authors write, "In standard quantitative social science, machine learning is used as a tool to analyze data. Our work shows how the artifacts of machine learning (word embeddings here) can themselves be interesting objects of sociological analysis. We believe this paradigm shift can lead to many fruitful studies."

Second is that biases and connotations—while they *seem* gossamer, ethereal, ineffable—are *real*. They are measurable, in detail and with precision. They emerge, spontaneously and reliably, from models built to do nothing but predict missing words, and they are measurable and quantifiable and dynamic. They track ground-truth data about labor participation as well as subjective measures of attitude and stereotypes. All this and more are present in models that ostensibly just predict missing words from context: the story of our language is the story of our culture.

Third: These models should absolutely be used with caution, particularly when used for anything other than their initial purpose of predicting missing words. Says Adam Kalai, "I've talked to people who said that after reading our paper . . . [they] are more cautious about using these word embeddings—or at least thinking twice before they use them in their own applications. So, that's one positive consequence." This note of caution is echoed by the Princeton team: "Certainly," they write, "caution must be used in incorporating modules constructed via unsupervised machine learning into decision-making systems."[92] Very few Amazon executives would have been likely to explicitly declare a policy of "hire the people whom, had they applied ten years ago, would have most strongly resembled the people we *did* hire back then." But using language models to filter résumés for "relevance" is making just such a leap.

We find ourselves at a fragile moment in history—where the power and flexibility of these models have made them irresistibly useful for a large number of commercial and public applications, and yet our standards and norms around how to use them appropriately are still nascent. It is exactly in this period that we should be most cautious and conservative—all the more so because many of these models are unlikely to be substantially changed once deployed into real-world use. As Princeton's Arvind Narayanan puts it: "Contrary to the 'tech moves too fast for society to keep up' cliché, commercial deployments of tech often move glacially—just look at the banking and airline mainframes still running. ML [machine-learning] models being trained today might still be in production in 50 years, and that's terrifying."[93]

Modeling the world as it is is one thing. But as soon as you begin *using* that model, you are *changing* the world, in ways large and small. There is a broad assumption underlying many machine-learning models that the model itself will not *change* the reality it's modeling. In almost all cases, this is false.

Indeed, uncareful deployment of these models might produce a feedback loop from which recovery becomes ever more difficult or requires ever greater interventions. If, say, a résumé-search system detects a gender skew with a given position, and upranks (say) male applicants in a way that *exaggerates* that skew, then this may well be the next batch of training data on which the model learns. And it will only learn a more extreme version of its existing bias. The easiest moment in which to intervene is, of course, *as soon as possible.*

Lastly, these models offer us a digital sextant as we look ahead as a society. From this work, we get a portrait not only of our history but of our up-to-the-minute present. As long as new text is being published online each day, there will be a new set of data to sample.

If used wisely—and *de*scriptively rather than *pre*scriptively—the very systems capable of reinforcing and perpetuating the biases latent in society can be used, instead, to make them visible, inarguable. They offer us a gauge on what might have seemed diffuse or formless.[94] That's a start.

Yonatan Zunger, no longer at Google, argues that people sometimes forget the degree to which engineering is inextricable from human society, human norms, and human values. "Essentially, engineering is all about cooperation, collaboration, and empathy for both your colleagues and your customers," he writes. "If someone told you that engineering was a field where you could get away with not dealing with people or feelings, then I'm very sorry to tell you that you have been lied to."[95]

As for Jacky Alciné, who now runs his own software consultancy and still keeps in touch with Zunger, he agrees that the problem neither begins nor ends with tech. "This is actually partly why I want to go into history teaching," he tells me, smiling and at least half serious: "When I turn 35 I'll just stop everything, retire, and transition to history."

2

FAIRNESS

While mankind has been wandering the American continent since the
retreat of the glaciers and possibly before the ice age, it remains somewhat
of a sad commentary on his evolution that one of the problems science has
just undertaken is the question of an accurate prediction of what a man
will do when released from prison on parole.

—*CHICAGO TRIBUNE*, JANUARY 1936[1]

Our law punishes people for what they do, not who they are. Dispensing
punishment on the basis of an immutable characteristic flatly contravenes
this guiding principle.

—SUPREME COURT CHIEF JUSTICE JOHN ROBERTS[2]

As we're on the cusp of using machine learning for rendering basically all
kinds of consequential decisions about human beings in domains such
as education, employment, advertising, health care and policing, it is
important to understand why *machine learning is not, by default, fair or
just in any meaningful way.*

—MORITZ HARDT[3]

The idea that society can be made more consistent, more accurate,
and more fair by replacing idiosyncratic human judgment with
numerical models is hardly a new one. In fact, their use even in criminal
justice is nearly a century old.

In 1927, the new chairman of the Parole Board of Illinois, Hinton Clabaugh, commissions a study on the workings of the parole system in the
state. He is motivated by what he perceives to be a kind of innovation
gap: "Although our industrial and governmental machines are far from
perfect, we probably are the most ingenious and efficient nation indus-
trially," Clabaugh writes. "Can we truthfully say this of our law enforce-

ment?"[4] Despite Illinois having been one of the very first states to enact a parole law, public opinion has soured. As Clabaugh observes, public sentiment is that "the pendulum of justice and mercy has swung to the extreme in favor of criminals." His own view is not far off: that the entire concept of parole may amount to little more than undue leniency, and that the parole system should perhaps be scrapped entirely. But he reasons that, under US law, an individual is entitled to a defense—and so the parole system should probably be entitled to one as well.

Clabaugh asks his state's most prestigious schools—the University of Illinois, Northwestern University, and the University of Chicago—to join forces and prepare a comprehensive report on the parole system to be delivered to him in a year's time. University of Illinois law school dean Albert Harno will report on the workings of the parole board; judge Andrew Bruce of Northwestern will review the history of the penal system in Illinois (including an eye-opening look at the "abolition of the lash" in the nineteenth century); and Chicago sociologist Ernest Burgess is given the intriguing challenge to see whether any factors predict the "success or failure" of a given parolee.

As Burgess writes:

> Two widely divergent pictures of the paroled man are, at present, in the minds of the people of Illinois. One picture is that of a hardened, vicious, and desperate criminal who returns from prison, unrepentant, intent only upon wreaking revenge upon society for the punishment he has sullenly endured. The other picture is that of a youth, perhaps the only son of a widowed mother, who on impulse, in a moment of weakness, yielded to the evil suggestion of wayward companions, and who now returns to society from the reformatory, determined to make good if only given a chance.[5]

The question, of course, is whether it might be possible to anticipate which potential parolees were which.

Burgess gathers data on some three thousand different Illinois parolees and does his best to categorize them into one of four groups: "first offenders," "occasional offenders," "habitual offenders," and "professional offenders." From a twenty-first-century perspective, some of his work seems strikingly dated: for instance, the eight possible "social types" into which he divides people are "Hobo," "Ne'er-do-well," "Mean citizen," "Drunkard," "Gangster," "Recent immigrant," "Farm boy," and "Drug addict." Nonetheless, Burgess's work appears impressively thorough, particularly for the time—examining criminal histories, work histories, residential histories, type of crime, length of sentence, time served, psychiatric diagnoses, and more. He undertakes a study of whether the data reveals undue political influence on behalf of certain inmates, and questions whether the public's jaded view of the justice system is warranted. And in his final chapter, he directly tackles the question that will launch something of a movement in criminal justice, one that will continue well into our present century: "Can scientific methods be applied to parole administration?"

He writes, "Many will be frankly skeptical of the feasibility of introducing scientific methods into any field of human behavior. They will dismiss the proposal with the assertion that human nature is too variable for making any prediction about it. But in the analysis of factors determining success and failure on parole some striking contrasts have already been found."

Burgess notes, for instance, that parole violation among those with a strong work record, high intelligence, and a farming background who had served one year or less was *half* the state average. On the other hand, parole violation among those living "in the criminal underworld," whose prosecutors or judges had argued against leniency, and who had served five or more years was *twice* the state average. "Do not these striking differences, which correspond with what we already know about the conditions that mould the life of the person," he asks, "suggest that they be taken more seriously and objectively into account than previously?"

"It would be entirely feasible and should be helpful to the Parole Board," he says, "to devise a summary sheet for each man about to be paroled in order for its members to tell at a glance the violation rate for each significant factor. . . . The prediction would not be absolute in any given case, but, according to the law of averages, would apply to any considerable number of cases."

His conclusion is that rehabilitation is, in many cases, eminently possible. What's more, it does appear to be at least somewhat predictable in which cases it will succeed. Wouldn't a system built on this statistical foundation be better than the status quo of subjective, inconsistent, and idiosyncratic human decisions made by judges on the fly? "There can be no doubt of the feasibility of determining the factors governing the success or the failure of the man on parole," Burgess writes. "Human behavior seems to be subject to some degree of predictability. Are these recorded facts the basis on which a prisoner receives his parole? Or does the Parole Board depend on the impressions favorable or unfavorable which the man makes upon its members at the time of the hearing?"

The report's ultimate conclusion is both clear and firm. Parole and "indeterminate sentencing" should indeed be continued in the state of Illinois—and, just as critically, "their administration can and should be improved both by the placing of the work of the parole board on a scientific and professional basis and by further safeguards against the constant pressure of political influence."

Chairman Clabaugh reads the report, and he completely changes his mind about the system he oversees. "My first impression was that indeterminate sentences and parole laws were an asset to criminals," he admits. "Proof to the contrary is overwhelming, and I now believe these laws, properly administered, to be decidedly beneficial to society, as well as to the individual." Parole, put simply, is a good thing, he writes—"even where the method of administration is faulty. And of course, no machinery is more efficient than the human element that operates it."[6]

SCIENTIFIC PAROLE IN PRACTICE

In the early 1930s, buoyed by Burgess's enthusiasm and Clabaugh's approval, a predictive parole system goes into use in Illinois, and by 1951 a "Manual of Parole Prediction" is being published, reporting back on what is now twenty years of both study and practice.

The tone is upbeat. Burgess, in the book's Introduction, writes, "In the past two decades social scientists have made significant progress in their efforts to find out which prisoners succeed on parole and which fail, and under what conditions success or failure occurs. Out of their research has grown a conviction that, notwithstanding the difficulties involved, it is possible to predict to some extent how prisoners will behave on parole."[7]

The book notes several areas for future study and possible improvement, for instance additional factors—like evaluations by prison staff—that might help improve predictions. A particularly prescient section at the end of the book, titled "Scoring by Machine Methods," considers the use of punch-card machines to automate and streamline the process of collecting the data, building the models, and outputting individual predictions.

Despite this early optimism, and the seeming success story of Illinois, the adoption of parole prediction instruments would be surprisingly slow. By 1970, for instance, just two states were using such tools. But that was about to change.

In 1969 a Scottish-born statistician named Tim Brennan was working for Unilever in London, building statistical models of a different sort: he was the company's leading expert on market segmentation. One project, for instance, segmented buyers of bathroom soap into those for whom glamour is paramount and those who prioritize gentleness to the skin. He was good at it, and enjoyed the work—but something wasn't right. "I had a values crisis," Brennan tells me. A report crossed his desk, and he noticed that Unilever, over the past year, had spent more money studying the packaging for its "Sqezy" line of squeezable liquid dish soap—the wording and colors that would draw the eye on a supermarket shelf, all backed by the latest in perceptual psychology—than the entire British government had spent on literacy.[8]

"This was in the late sixties," he says, "and you know how the hippie phenomenon was at the time. Anyway, I couldn't see working on developing packs for Sqezy. So I resigned my job and I applied for graduate school." He ended up at Lancaster University, applying his statistics for market segmentation to the problem of education: identifying students with different learning styles and different needs in the classroom.[9]

Following his girlfriend to the United States, Brennan found himself at the University of Colorado, working for Delbert Elliott (who would later become the president of the American Society of Criminology), and then ultimately starting his own research firm, working with the National Institute of Corrections and the Law Enforcement Assistance Administration. His techniques for classification had found a third use: bringing a more consistent and rigorous approach to jails and prisons, each of which has to organize inmates among bunks and wards according to their risks and needs—both for the safety of others and for their own rehabilitation. This was typically done either at random or by instinct; Brennan's math said there was a better way.

Brennan took a trip up to Traverse City, Michigan, where he'd heard there was some pioneering machine-learning work on using so-called "decision-tree" models to classify inmates and—in response to overcrowding—identify whom to release. There Brennan met "this young guy . . . with straggly hair and a beard" who had invented the model. The young guy's name was Dave Wells. "He was on fire about reforming the criminal justice system," says Brennan. "It was a job of passion for Dave." Brennan and Wells decided to team up.[10] They called their company Northpointe.

As the era of the personal computer dawned, the use of statistical models at all points in the criminal justice system, in jurisdictions large and small, exploded. In 1980, only four states were using statistical models to assist in parole decisions. By 1990, it was twelve states, and by 2000, it was twenty-six.[11] Suddenly it began to seem strange *not* to use such models; as the Association of Paroling Authorities International's 2003 *Handbook for New Parole Board Members* put it, "In this day and age, making parole decisions without benefit of a good, research-based risk assessment instrument clearly falls short of accepted best practice."[12]

One of the most widely used tools of this new era had been developed by Brennan and Wells in 1998; they called it Correctional Offender Management Profiling for Alternative Sanctions—or COMPAS.[13] COMPAS uses a simple statistical model based on a weighted linear combination of things like age, age at first arrest, and criminal history to predict whether an inmate, if released, would commit a violent or nonviolent crime within approximately one to three years.[14] It also includes a broad set of survey questions to identify a defendant's particular issues and needs—things like chemical dependency, lack of family support, and depression. In 2001, the state of New York began a pilot program using COMPAS to inform probation decisions. By the end of 2007, all fifty-seven of its counties outside of New York City had adopted the use of the COMPAS tool in their probation departments. And by 2011, state law had been amended to *require* the use of risks and needs assessments like COMPAS when making parole decisions.[15]

But, from the perspective of the *New York Times* editorial board, there was a problem: the state wasn't using them *enough*. The tools, even where their use was mandated, still were not always given appropriate consideration. The *Times* urged wider acceptance of risk-assessment tools in parole, writing in 2014, "Programs like COMPAS have been proved to work."[16] In 2015, as a parole reform case went before the state's highest court, the editorial board published a second opinion piece, arguing again that statistical risk-assessment tools like COMPAS offered a significant improvement over the status quo. They alleged that the New York parole board "clings stubbornly to the past, routinely denying parole to long-serving inmates based on subjective, often unreviewable judgments." Adoption of COMPAS, they wrote, would "drag the board into the 21st century."[17]

Then—abruptly—the tone changed.

Nine months later, in June 2016, the paper ran an article titled "In Wisconsin, a Backlash Against Using Data to Foretell Defendants' Futures," which closed by quoting the director of the ACLU's Criminal Law Reform Project saying, "I think we are kind of rushing into the world of tomorrow with big-data risk assessment."[18]

From there the coverage in 2017 only got bleaker—in May, "Sent to Prison by a Software Program's Secret Algorithms"; in June, "When a

Computer Program Keeps You in Jail"; and in October, "When an Algorithm Helps Send You to Prison."

What had happened?

What had happened was—in a word—ProPublica.

GETTING THE DATA

Julia Angwin grew up in Silicon Valley in the 1970s and '80s, the child of two programmers and a neighbor of Steve Jobs. She assumed from an early age she'd be a programmer for life. Along the way, however, she discovered journalism and fell in love with it. By 2000 she was a technology reporter for the *Wall Street Journal*. "It was hilarious," she recounts. "They were like, 'You know computers? We'll hire you to cover the internet!' And I was like, 'Well, anything in particular about the internet?' And they're like, 'No—*everything*!' "[19]

Angwin stayed at the paper for fourteen years—from the dot-com crash through the rise of social networks and smartphones—writing not only about technology itself but also the social questions it often left in its wake; she authored a long-running series on privacy-related issues called "What They Know." She went on leave from the *Journal*, writing a book about privacy, in 2013.[20]

Angwin never came back to the paper from her book sabbatical, and instead joined ProPublica, a nonprofit news outlet started by former *WSJ* managing editor Paul Steiger. If the theme of her earlier work was "What They Know," the answer inevitably raised another question: *What are they going to do with it?* "And so I thought," she says, "I need to move to data usage. That is the next story. What are they going to do? . . . What kind of judgments are they going to make about you?"[21]

Angwin set about trying to find the most consequential, and overlooked, decisions being made based on data. She landed at criminal justice. Statistical risk assessments, COMPAS and others, were rapidly being adopted in hundreds of jurisdictions: not just for parole, but for pretrial detention, bail, and even sentencing. "I was shocked, actually," she says. "I realized that our whole country was using this software. . . .

And then what I found even more shocking was that none of them had independently been validated."

New York State, for instance, which had been using COMPAS since 2001, did its first formal evaluation of the software in 2012—after *eleven* years of use. (New York ultimately found that it "was both effective and predictively accurate."[22])

Stories like this are stunningly common. Minnesota's Fourth Judicial District, which includes Minneapolis and handles 40% of all cases filed in the state, had developed their own pretrial risk-assessment tool in 1992. "The report that was written at that time suggested that this new scale should be validated within the first few years of use," the state's official evaluation begins. "As it turned out, it has been closer to 14 years."[23]

The long-overdue evaluation, in 2006, found that four of the variables in their model—whether the defendant had lived in Minnesota for more than three months, whether they lived alone, their age at the time of booking, and whether a weapon was involved in their charge—had virtually no bearing whatsoever on their actual risk of committing a new offense while waiting for their trial date or of missing a required court appearance. And yet the model had been recommending pretrial detention on those grounds. Worse, three of those four factors strongly correlated with race. The district scrapped their model and started over from scratch.[24]

The more Angwin learned about risk-assessment models, the more concerned she became. Angwin knew she had found her next story: "From a journalist's perspective, it's the perfect storm. It's something that's never been audited; it has extremely high human stakes; and people who are very smart are like, 'These are all proxies for race.' And so I was like, 'I'm going to test this.'"[25]

She decided to focus on COMPAS in particular, which was being used not just in New York but in California, Wisconsin, Florida—some two hundred different jurisdictions in total. It was ubiquitous, at that time not well studied, and—being a closed-source, proprietary tool, though its basic design was available in white papers—something of a black box. In April 2015 she submitted a Freedom of Information Act request to

Broward County, Florida. After five months of legal scuffling, the data at last came through: all eighteen thousand COMPAS scores given in Broward County during 2013 and 2014.

Angwin's team began doing some exploratory data analysis. Immediately something looked strange. The risk scores—which range from 1, meaning lowest risk, to 10, meaning highest risk—were more or less evenly distributed for Black defendants, with roughly ten percent of defendants in each of the ten buckets. For White defendants, they saw a totally different pattern: *vastly* more defendants in the very lowest-risk bucket, and vastly fewer in the highest-risk buckets.

Angwin was tempted to publish a story then and there. But she realized that the wildly different distributions weren't necessarily evidence of bias—perhaps it was just *actually* how risky those defendants happened to be. So how risky *did* those defendants turn out to be? There was only one way to find out. "I had the sad realization," Angwin recounts, "that we had to look up the criminal records of every one of those eighteen thousand people. Which we did. And it sucked."[26] To link the set of COMPAS scores to the set of criminal records—what data scientists call a "join"—would take Angwin, and her team, and the county staff, almost an entire additional year of work.

"We used, obviously, a lot of automated scraping of the criminal records," she explains. "And then we had to match them on name and date of birth, which is the most terrible thing you could possibly ever imagine. There's so many typos, so many spelling errors. I might have cried every day. It was so awful trying to do that join. The data was so messy. Broward County actually themselves had never done the join."[27] County staff pitched in to help ProPublica clean the data and make sense of it.

What resulted was the piece that Angwin and her team published in May 2016. Titled "Machine Bias," it ran with the logline "There's software used across the country to predict future criminals. And it's biased against blacks."[28]

Brennan and his Northpointe colleagues, in early July, published an official rebuttal to ProPublica's findings.[29] In their words, "When the correct classification statistics are used, the data do not substantiate the ProPublica claim of racial bias towards blacks."[30] Namely, they said, COMPAS met two essential criteria for fairness.

First, its predictions were just as *accurate* for Black defendants as they were for White defendants. Second, its 1-to-10 risk scores had the same *meaning*, regardless of the defendant's race, a property known as "calibration." Defendants rated with a risk score of 7 out of 10, say, for committing a violent reoffense went on to reoffend the same percentage of the time, regardless of race; ditto for a risk score of 2, 3, and so forth. A 1 was a 1, a 5 was a 5, and a 10 was a 10, regardless of race. COMPAS had both of these properties—equal accuracy and calibration—and because of this, Northpointe argued, it was essentially *mathematically impossible* for the tool to be biased.

By the end of July, ProPublica responded in turn.[31] Northpointe's claims, they wrote, were true. COMPAS really was calibrated, and it was equally accurate across both groups: predicting with 61% accuracy for Black and White defendants alike whether they would go on to reoffend ("recidivate") and be re-arrested. However, the 39% of the time it was wrong, it was wrong in *strikingly* different ways.

Looking at the defendants whom the model misjudged revealed a startling disparity: "Black defendants were twice as likely to be rated as higher risk but not re-offend. And white defendants were twice as likely to be charged with new crimes after being classed as lower risk."[32]

The question of whether the tool was "fair" in its predictions had sharpened: into the question of which statistical measures were the "correct" ones by which to define and measure fairness in the first place.

The conversation was about to take a new turn—and it would come from another community entirely, one that had also begun to slowly but surely train its attention on the question of fairness.

WHAT FAIRNESS IS NOT

Harvard computer scientist Cynthia Dwork is arguably best known for developing a principle called "differential privacy," which enables companies to collect data about a population of users while maintaining the privacy of the individual users themselves. A web browser company might want to understand user behavior, but without knowing which sites *you* personally went to; or a smartphone company might want to learn how to improve their spelling-correction or text suggestions without knowing the details of your personal conversations. Differential privacy made this possible. It would come to be almost ubiquitous in major tech companies starting around 2014, and would win Dwork the Gödel Prize, one of computing's highest honors.[33] But in the summer of 2010, however, she saw her theoretical work as done—and she was on the hunt for a new problem.

"I had started working in privacy in 2000, 2001," she explains. "Differential privacy was differential privacy by 2006. I had in my mind one last set of questions I wanted to look at—and I did that work—and then I said, Okay, I want to think about something else."[34]

Dwork, then at Microsoft Research, came up to Berkeley to meet with fellow computer scientist Amos Fiat. They spent the entire day talking. By lunchtime, as they sat down together in beloved local restaurant Chez Panisse, they had landed on the topic of *fairness*. Dwork recalls: "In order that the people around us wouldn't be disconcerted by our discussions of racism and sexism, we were using terms like 'purple ties' and 'striped shirts,' and stuff like that. But by lunchtime we were . . . we were on to this."

The term "fairness" in theoretical computer science comes up in a number of contexts, from game-theoretic mechanisms for cutting a cake (or dividing an inheritance) so that everyone gets a proper share, to scheduling algorithms that ensure that every process on a CPU gets to run for an appropriate amount of time. But there was something more to the idea of fairness, Dwork thought, that the field had yet to truly reckon with.

As it happens, Dwork had read one of Julia Angwin's "What They Know" columns in the *Wall Street Journal,* on the topic of online advertising. It showed that as early as 2010, if not before, companies could discern nearly the *exact* personal identity of every user visiting their websites—narrowing an ostensibly anonymous user down to one of just several dozen possible people—and were making split-second decisions about, say, which types of credit cards to suggest to whom.[35]

Having thought about the privacy side for the previous decade, Dwork too began to shift her thinking—from the question of "what they know" to the question of "what they're doing with it."

Dwork came back to her lab at Microsoft and said, "I found our problem."

One of her lab members was then–PhD student Moritz Hardt, interning that summer from Princeton. Hardt hadn't started off wanting to work on real-world problems. He was interested in theory: complexity, intractability, randomness. The more abstract, the better. "No applications," he jokes. "Old-school."[36]

"I learned a lot," he says, "but I found the range of problems I could address in that space . . . it didn't get at the questions about the world I was curious about. And I quickly got attracted by some of the more social problems that [computer science] was touching on." This had started with privacy-preserving data analysis, working with Dwork. She pulled him into the fairness project.

As Hardt recalls, "Cynthia had the hunch that sometimes when people were asking for privacy, they were actually worried about somebody using their data in the wrong way. It wasn't so much about hiding the data at all costs but preventing harm from the way that data was used. . . . That was a pretty accurate hunch. Over time, the public discussion shifted from privacy to fairness, and everything that used to look like a privacy problem suddenly looked like a fairness problem."

What they and their colleagues began to find was that not only were there enormous complexities in translating our philosophical and legal ideas about fairness into hard mathematical constraints but, in fact, much of the leading thought and practice, some of it decades old, was deeply misguided—and had the potential to be downright harmful.

US antidiscrimination law, for instance, defines a number of "protected attributes"—things like race, gender, and disability status—and it is typically understood that one should strictly prohibit the use of these variables in machine-learning models that might affect people in settings like hiring, criminal detention, and so on. If we hear in the press that a model "uses race" (or gender, etc.) as an attribute, we are led to believe something has already gone deeply wrong; conversely, the company or organization behind a model typically defends its model by showing that it "doesn't use race as an attribute," or is "race-blind." This seems intuitive enough—how can something be discriminatory toward a particular group if it doesn't know who is in that group to begin with?

This is a mistake, for several reasons.

Simply *removing* the "protected attribute" is insufficient. As long as the model takes in features that are *correlated* with, say, gender or race, avoiding explicitly mentioning it will do little good.

As we discussed in the case of the Boston Symphony Orchestra and, moreover, in the résumé-screening language models, simply omitting the variable you're concerned about (gender, in that case) may not be enough, particularly if there are other factors that *correlate* with it. This is known as the concept of "redundant encodings." The gender attribute is *redundantly encoded* across other variables.

In a criminal justice context, a history of treating one demographic group differently creates redundant encodings all over the place. Policing a minority neighborhood more aggressively than others, for instance, means that all of a sudden something as seemingly neutral as length-of-rap-sheet, i.e., number of previous convictions, can become a redundant encoding of race.[37]

Because of redundant encodings, it's not enough to simply be *blind* to the sensitive attribute. In fact, one of the perverse upshots of redundant encodings is that being blind to these attributes may make things *worse*. It may be the case, for instance, that the maker of some model wants to measure the degree to which some variable is correlated with race. They can't do that without knowing what the race attribute actually is! One engineer I spoke with complained that his management repeatedly stressed the importance of making sure that models aren't skewed by sensitive attributes like gender and race—but his company's privacy policy prevents him and the other machine-learning engineers from *access-*

ing the protected attributes of the records they're working with. So, at the end of the day, they have no idea if the models are biased or not.

Omitting the protected attribute makes it impossible not only to measure this bias but also to mitigate it. For instance, a machine-learning model used in a recruiting context might penalize a candidate for not having had a job in the prior year. We might not want this penalty applied to pregnant women or recent mothers, however—but this will be difficult if the model must be "gender-blind" and can't include gender itself, nor something so strongly connected to it as pregnancy.[38]

"The most robust fact in the research area," Hardt says, "is that fairness through blindness doesn't work. That's the most established and most robust fact in the entire research area."[39]

It will take time for this idea to percolate from the computer scientists through to legal scholars, policy makers, and the public at large, but it has begun making its way. "There may be cases where allowing an algorithm to consider protected class status can actually make outcomes fairer," as a recent *University of Pennsylvania Law Review* article put it. "This may require a doctrinal shift, as, in many cases, consideration of protected status in a decision is presumptively a legal harm."[40]

The summer Moritz Hardt had spent working with Cynthia Dwork had borne fruit, in the form of a paper that established some of these early results.[41] More than the results themselves, the paper was a beacon, signaling to their fellow theoreticians that there was something here worth looking into: something with both meaty open theoretical questions and undeniable real-world importance.

Back at Princeton and finishing his PhD, Hardt found himself set up on a kind of academic blind date. Dwork had been talking with Helen Nissenbaum, herself a pioneer in thinking about ethical issues in computing.[42] Nissenbaum had a fellow graduate student in the area named Solon Barocas, and she and Dwork realized their respective mentees might be kindred spirits.[43]

The first meeting got off to an awkward start.

Hardt was sitting at a table at Sakura Express sushi, on Witherspoon Street in Princeton. In walked Barocas. "He sits down," Hardt recalls, "and he pulls out the paper, and to my dismay, he had, with yellow marker, like, carefully underlined passages in the paper. That's not how you're supposed to read a computer science paper!" He laughs. "Till that point I had always thought of words in a paper as a filling for the math. . . . And it was very embarrassing, because what I'd written made no sense."

The two got to talking. Bumpy start aside, the resonance their mentors had suspected was clearly there. Ultimately, the two decided to submit a proposal to the 2013 Neural Information Processing Systems (NeurIPS) conference to present a workshop on fairness. NeurIPS said no. "The organizers felt there wasn't enough work or material on this," Hardt tells me. "Then in 2014, Solon and I got together and said, 'Okay, let's try one more time before we give up.'" They gave it a new name, and a broader mandate: "Fairness, Accountability, and Transparency in Machine Learning," or "FATML" for short. This time NeurIPS said yes. A day-long workshop was held at that year's conference in Montreal, with Barocas and Hardt introducing the proceedings and Dwork giving the first talk.

Hardt graduated and went from Princeton to IBM Research. He kept chipping away at fairness questions in the background. "I always had to have another thing to continue to exist as a computer scientist. In terms of my career, this was always like a side project." IBM, to its credit, gave him a lot of freedom, even if his passion wasn't widely shared. "My group at IBM wasn't particularly interested in that topic," he says, "but neither was anybody else."

The two came back and repeated the conference again in 2015. "The room was full," Hardt recalls, "people attended, but it didn't spark a revolution at that point, it's safe to say." He kept spending the lion's share of his time on more conventional computer science. Another year went by, and he and Solon reprised their FATML workshop for a third time.

This time, something was different; 2016 "was the year where it got out of hand," Hardt says, "the year where somehow everybody started working on this." Mathematician and blogger Cathy O'Neil, who'd presented at the original 2014 conference, had published the best-selling book *Weapons of Math Destruction*, about social problems that can stem from the careless (or worse) use of big data. A series of shock-ing election results, defying the consensus of pollsters worldwide, had shaken confidence in the trustworthiness of predictive models; mean-while, the work of data-driven political firms like Cambridge Analytica had raised questions about machine learning being wielded to directly influence politics. Platforms like Facebook and Twitter were caught in the crossfire over how—and whether—to use machine learning to fil-ter the information being shown to their billions of users. And a group of reporters at ProPublica had, after a year of tireless data cleaning and analysis, gone public with their findings about one of the country's most widely used risk-assessment tools.

Fairness was no longer just a problem. It was becoming a movement.

THE IMPOSSIBILITY OF FAIRNESS

Since 2012, Cornell computer scientist Jon Kleinberg and University of Chicago economist Sendhil Mullainathan had been working on a proj-ect using machine learning to analyze pretrial detention decisions, com-paring human judges against predictive machine-learning models. "Part of that is thinking through what are the concerns that people have about algorithmic tools in the context of racial inequities," Kleinberg says. "I think we had done a fair amount of thinking about that. Then the Pro-Publica article appeared, and just . . . our social media channels just got completely filled with reshares of this article. It was just really capturing people's attention. We felt like, 'They put their finger on something. . . . Let's really dig into it and figure out how does this relate to the themes that we've been thinking about.'"[44]

In Pittsburgh, Carnegie Mellon statistician Alexandra Chouldechova had been working since the spring of 2015 with the Pennsylvania Com-mission on Sentencing, developing a visual dashboard to explore various mathematical properties of risk-assessment tools. "I started thinking more and more about these issues . . . to understand different notions of fairness in terms of classification metrics and how those relate," she says. "All of that was happening while I was still focused on other projects. . . .

I'd been talking to my senior colleague about maybe writing a paper about some ideas on the validation of risk-assessment instruments—and we'd done our literature review and everything like that—and the Pro-Publica piece hits. I think that really accelerated a lot of people's thinking in the area."[45]

An almost identical narrative was playing out on the other side of the country. Stanford doctoral student Sam Corbett-Davies had started his degree "wanting to do robotics, just like straight computer science." A year in, he found that he simply wasn't enjoying it—"and at the same time was obsessively reading about US public policy just as a hobby. I was like, 'I've got to find a way to sort of combine the technical skills I have with something that's a bit more policy focused.'"[46] An assistant professor named Sharad Goel had recently joined the faculty, working on computational and statistical approaches to public policy.[47] A collaboration seemed obvious, and in short order the two began working on a number of projects examining bias in *human* decisions, including things like a close look at traffic stops in North Carolina. They found, for instance, that police appeared to apply a lower standard when searching Black and Hispanic drivers than White drivers: searching them more often, *and* finding contraband less of the time.[48] "We're working on this," Corbett-Davies explains. "We're sort of thinking about these criminal justice issues, and what it means for discrimination to occur. Then . . . the ProPublica article came out."

ProPublica emphasized the *types* of errors COMPAS made, and highlighted the fact that it appeared to consistently overestimate the risk of Black defendants who didn't reoffend and underestimate the risk of White defendants who did reoffend. Northpointe highlighted the *rate*, rather than the kind, of errors and emphasized that the model was equally *accurate* in its predictions of both Black and White defendants and that, moreover, at every risk score from 1 to 10, COMPAS was "calibrated": a defendant with that score is equally likely to reoffend, whatever their race. The question of whether COMPAS was "fair" appeared, then, to boil down to a conflict between two different mathematical definitions of fairness.

As sometimes happens in science, the time is so ripe for a certain idea or insight that a group of people have it almost in unison.

A trio of papers emerge.[49] All three efforts turn up similar, complementary takes. The news is not good.

Kleinberg says, having identified as the crux of the debate ProPublica's definition of fairness—the propensity for Black defendants who would *not* go on to reoffend being almost twice as likely as their White counterparts to be miscategorized as high-risk—"we could then basically line it up with other definitions people had been paying more attention to, and ask, To what extent are these compatible?

"The answer was, they aren't compatible."

Only in a world in which Black and White defendants happened to have equal "base rates"—that is, happened to actually reoffend exactly as often as one another—would it be possible to satisfy ProPublica's and Northpointe's criteria at the same time. Otherwise, it is simply impossible.

This has nothing to do with machine learning. Nothing to do with criminal justice as such. "It is simply," Kleinberg and his colleagues write, "a fact about risk estimates when the base rates differ between two groups."[50]

Chouldechova's analysis lands exactly in the same place: A tool that is calibrated, she writes, "cannot have equal false positive and negative rates across groups, when the recidivism prevalence differs across those groups."

"So you just can't have it all," she says. "It's a general principle, but in this case it leads you to interesting conclusions . . . that [have] implications for risk assessment in the real world."[51]

One of those implications is that if a set of equally desirable criteria are impossible for *any* model to satisfy, then any exposé of any risk-assessment instrument whatsoever is *guaranteed* to find something headline-worthy to dislike.

As Sam Corbett-Davies explains, "There isn't a world in which Pro-Publica couldn't have found some number that was different that they could call bias. There's no possible algorithm—there's no possible version of COMPAS—where that article wouldn't have been written."[52]

(Ironically, in his analysis of the Broward County data, Corbett-Davies discovered that COMPAS is *not* calibrated with respect to gender. "A woman with a risk score of 5 reoffends about as often as a man with a risk score of 3," he says.[53] "[ProPublica] could have written *that* article.")[54]

The brute mathematical fact of this impossibility also means that these problems affect not just machine-learning models of risk-assessment but *any* means of classification, human *or* machine. As Kleinberg writes: "Any assignment of risk scores can in principle be subject to natural criticisms on the grounds of bias. This is equally true whether the risk score is determined by an algorithm or by a system of human decision-makers."[55]

Having established, then, the answers to the question of how these different, seemingly equally intuitive, equally desirable measures of fairness can be reconciled—they can't—and the question of whether good old-fashioned human judgment is any better in this regard—it isn't—a further question suggests itself.

Now what?

AFTER IMPOSSIBILITY

I ask Jon Kleinberg what he makes of his own impossibility result—and what he thinks it suggests we *ought* to do. "I don't have anything particularly controversial to say on this," he says. "I think it depends. . . . I think they're both important definitions, and which one carries more weight depends on the domain you're working in."

It's true that critical aspects of these tradeoffs change radically from one domain to another. Consider, for instance, the domain of *lending*, which differs from criminal justice in a number of important ways. Denying a loan to someone who *would have* paid it back not only means lost interest income for the lender, but potentially serious consequences for the borrower: there is a home they can't purchase, a business they can't start. On the other hand, making the opposite mistake—loaning money to someone who *doesn't* pay it back—might only be a monetary loss for the lender. Maybe this asymmetry changes our sense of what fairness means in that setting: maybe we want to ensure, for instance, that all creditworthy borrowers from two different groups have the same chance of getting a loan—even if, for instance, the math tells us that not only will profit for lenders be reduced, but also that more *unworthy* people will get loans in one group than another.[56]

In a criminal justice setting, however, and particularly in the context of violent crime, both false positives ("high-risk" people who don't reoffend) *and* false negatives ("low-risk" people who do reoffend) have a serious human toll. The tradeoffs we seek are likely to be quite different as a result.

Sam Corbett-Davies has argued, for instance, that equalizing false positive rates in a risk-assessment context—ensuring that the defendants who *won't* reoffend, be they Black or White, are no more likely to be improperly detained—would, so long as the actual rates of offense between the groups are different, entail applying different standards to defendants of different races. For instance, it might mean detaining every Black defendant who is a risk of 7 or greater but detaining every White defendant who is a 6 or greater. Such an approach, he says, "probably violates the Equal Protection Clause of the Fourteenth Amendment."[57]

It's actually even worse than that, he says. "Because we're detaining low-risk defendants while releasing some relatively higher-risk defendants, there's going to be an increase in violent crime committed by released defendants. And because we know that crime generally occurs *within* communities as opposed to between communities, this crime will be concentrated in minority neighborhoods, and the victims will also bring a Fourteenth Amendment case."[58]

The impossibility proofs also show that equalizing the false positive and false negative rates means giving up on calibration—namely, the guarantee that for every numerical risk level, the chance of a defendant reoffending is the same regardless of gender or race. "Without calibration," says Corbett-Davies, "it's unclear what it means to have a risk score.

So if you ask me, 'How risky is this defendant?' and I say, 'They're a 2,' and you say, 'What does that mean?' and I say, 'Well, if they're male, it means there's a 50% chance they'll reoffend; if they're female it means there's a 20% chance they'll reoffend'—you can see how '2' loses its meaning."[59]

Tim Brennan shares the view that calibration is paramount: "If a Black guy and a White guy both score 7, does that have the exact same rate of failure for the two of them? The arrest rate? . . . The standard method is to make sure—damn sure—that your calibration is not racially biased and your accuracy level is about the same. . . . And we've got both. And if we did something else, we would be breaking the law."[60]

However, even those who emphasize the importance of calibration think that it alone isn't enough. As Corbett-Davies says, "Calibration, though generally desirable, provides little guarantee that decisions are equitable."[61]

Just because these properties can't be satisfied all at once in their entirety, however, doesn't mean we can't look for certain tradeoffs between them that might be better than others. Other researchers have been exploring the space of these possible tradeoffs, and the work continues.[62]

I ask Julia Angwin what she herself makes of the storm of theoretical results that her article prompted, and of the ultimate impossibility of doing what her team seemed to demand—namely, to make a tool both equally calibrated *and* with an equal balance of false positives and false negatives.

"So what I feel about that," she says, "is that that's a policy question. And that's a moral question. But what I'm really happy about is no one knew that that was a question until we came up with it. To me, that is like the greatest thing as a journalist. . . . I do feel that defining the problem so accurately allows it to be solved."[63]

Not everyone welcomes the public dialogue that ProPublica's work spurred. A group of criminal justice scholars, led by Anthony Flores, published a "rejoinder" to the initial ProPublica report, not only defending COMPAS as a calibrated model (and calibration as the appropriate measure of fairness) but, furthermore, lamenting the damaging effects of the controversy itself.[64] "We are at a unique time in history," they write. "We are being presented with the chance of a generation, and perhaps a lifetime, to reform sentencing and unwind mass incarceration in a scientific way and that opportunity is slipping away because of misinformation and misunderstanding about [statistical risk-assessment models]. Poorly conducted research or misleading statements can lead to confusion and/or paralysis for those charged with making policy."

Tim Brennan shares much of this concern, and, of course, sharply disagrees with ProPublica's characterization of COMPAS. But when I ask him if there is a silver lining to the controversy, in particular the cross-disciplinary movement it has generated within and beyond computer scientists, he agrees: "I think unpacking fairness into its various different coefficients—alerting people to the meanings, benefits, costs, and the impossibility of using certain coefficients— Once you understand a problem or name a problem, solutions become, certainly, more sought after than if you don't even know the problem existed."[65]

Most in the computer science community agree that these things are better discussed openly than not.

"I mean, so first of all, the math is the math," says Cynthia Dwork. "So the best we can hope for is what the math will permit. And we are much, much better off knowing what the limits are than not."[66]

For Moritz Hardt, the initial wave of academic and technical work following ProPublica's study has legitimated ethical concepts like fairness as valid topics for research, not only in academic departments but also in the corporate world. Hardt himself was at Google in 2016 when this wave of work—including his own—appeared, and recounts the impact he observed. "Before that, people weren't even so sure, is this something we should even be touching? It was a hot potato," he says. From 2016 onward there was an explosion of interest and research, both within Google and beyond. "So that's something I'm proud of," Hardt says. "It actually made a difference for the culture."[67]

For Jon Kleinberg, the role of computer science in questions like these is to furnish the stakeholders with the tools to articulate the issue. "Our point is not to tell you which one is right and which one is wrong but to give you the language to have that discussion. . . . I certainly view this as part of my mandate as a computer scientist: to take things that have always existed informally and qualitatively, and try and think about, Can we actually talk about them rigorously and precisely? Because the world is moving in that direction."

The world certainly is. In the late summer of 2018, California passed SB 10, a monumental justice reform bill that would eliminate cash bail altogether—leaving the state with the choice simply to detain or release a defendant awaiting trial—and, furthermore, *mandated* the use of algorithmic risk-assessment tools in informing these determinations.[68] Just before the end of that year, the US Congress, in a strikingly bipartisan vote, passed the First Step Act, a sweeping criminal justice reform bill that, among its many reforms, required the Department of Justice to develop a statistical model that will be used to assess the recidivism risk of all federal prisoners, as well as to determine their course of rehabilitation.[69]

Though the sudden and widespread adoption of these tools is only increasing, the race between their adoption and the wisdom with which they are built and used at least appears to be drawing somewhat closer to even. We are coming, haltingly, to a clearer understanding of what it means to make "good" predictions of this kind.

A bigger question lurks behind this discussion, however—which is whether "prediction" is really even what we want to be doing at all.

After the ProPublica story broke, a *Washington Post* reporter called Chouldechova to ask for her response. The reporter had essentially two questions. The first was about the tension between ProPublica's claims and Northpointe's defense.

"There I had an answer prepared," says Chouldechova. "If the prevalence of reoffense differs across populations, then you can't have everything hold all at once. What I didn't have a ready canned answer for was when he asked, *What actually matters?*"

"And so, that was . . . That's really the question that prompted me to think about things from a different perspective."[70]

BEYOND PREDICTION

Although prediction is feasible on the basis of data now accessible, exclusive reliance should not be placed on this method.

—ERNEST BURGESS[71]

Your scientists were so preoccupied with whether or not they could . . . that they didn't stop to think if they *should.*

— JEFF GOLDBLUM AS IAN MALCOLM, *JURASSIC PARK*

One of the most important things in any prediction is to make sure that you're actually predicting what you think you're predicting. This is harder than it sounds.

In the ImageNet competition, for instance—in which AlexNet did so well in 2012—the goal is to train machines to identify what images depict. But this isn't what the training data captures. The training data captures what human volunteers on Mechanical Turk *said* the image depicted. If a baby lion, let's say, were repeatedly misidentified by human volunteers as a cat, it would become part of a system's training data *as a cat*—and any system labeling it as a lion would be docked points and would have to adjust its parameters to correct this "error."

The moral of the story: Sometimes the "ground truth" is not the ground truth.

Gaps of this kind are even more significant in the case of criminal justice predictions. One often talks in shorthand of predicting recidivism *itself,* but that's not what the training data captures. What the training data captures is not re*offense,* but rather re*arrest* and re*conviction.* This is a potentially crucial distinction.

Kristian Lum, lead statistician at the Human Rights Data Analysis Group, and William Isaac of the Department of Political Science at Michigan State articulated this in a 2016 paper on the use of predictive models in policing:

> Because this data is collected as a by-product of police activity, pre-
> dictions made on the basis of patterns learned from this data do
> not pertain to future instances of crime on the whole. They per-
> tain to future instances of *crime that becomes known to police*. In
> this sense, predictive policing is aptly named: it is predicting future
> policing, not future crime.[72]

As it happens, this critique was being made all the way back in
the 1930s. When the parole reform proposed by Ernest Burgess was
deployed in Illinois, critics—particularly those skeptical of parolees'
ability to return productively to society—argued that the official figures underestimated the rate of reoffense. As Elmer J. Schnackenberg,
the Republican minority leader in the Illinois House of Representatives,
complained in 1937, "Because a parolee isn't caught at crime during the
first or second years of his parole, he is listed during those two years as
having made good."[73]

This gap, between what we intend for our tool to measure and what
the data actually captures, should worry conservatives and progressives
alike. Criminals who successfully evade arrest get treated by the system
as "low-risk"—prompting recommendations for the release of other similar criminals. And the overpoliced, and wrongfully convicted, become
part of the alleged ground-truth profile of "high-risk" individuals—
prompting the system to recommend detention for others like them.

This is particularly worrisome in the context of predictive policing,
where this training data is used to *determine* the very police activity
that, in turn, generates arrest data—setting up a potential long-term
feedback loop.[74]

A person who commits crimes in an area that is less aggressively
policed, or who has an easier time getting their charges dropped, will be
tagged by the system as someone who *did not recidivate*. And even *fewer*
police will be dispatched to that neighborhood. Any preexisting disparity in policing between two otherwise similar neighborhoods is likely
only to grow. As Lum and Isaac put it, "The model becomes increasingly
confident that the locations most likely to experience further criminal

activity are exactly the locations they had previously believed to be high in crime: selection bias meets confirmation bias."[75] The system begins to sculpt the very reality it is meant to predict. This feedback loop, in turn, *further* biases its training data.

Lum and Isaac conclude not only that "drug crimes known to police are not a representative sample of all drug crimes" but, what's more, "rather than correcting for the apparent biases in the police data, the model reinforces these biases. The locations that are flagged for targeted policing are those that were, by our estimates, already over-represented in the historical police data."[76]

There is reason to believe we already have significant disparities in the present day. As Alexandra Chouldechova notes, "Self-reported marijuana use rates among young Black males and young White males are roughly the same. But the arrest rates for marijuana-related crimes are two and a half to five times higher among Black young males."[77] A 2013 report by the ACLU found that on average a Black American is four times more likely to be arrested for marijuana possession than a White person; in Iowa and Washington, D.C., for instance, the disparity is more than eightfold.[78] A 2018 investigation by the *New York Times* found that Black residents of Manhattan were *fifteen* times more likely than White residents to be arrested on marijuana charges, despite similar rates of use.[79]

There are options here to mitigate these issues in a model. The COMPAS model, for instance, makes three distinct predictions of "risk": violent reoffense, reoffense, and failure to appear in court. Violent crimes—for instance, homicides—are much more consistently reported to police than nonviolent crimes, and the police, in turn, are more consistent about enforcement and arrest. And *all* failures of a defendant to appear in court by *definition* become known to the court system, leaving virtually no room for biased sampling or differential enforcement. A wise use of risk-assessment tools, then, might emphasize the violent reoffense and failure to appear predictions over the nonviolent reoffense prediction, on the grounds that the model's training data is more trustworthy in those cases—which is exactly what a number of jurisdictions are beginning to do.[80]

Other mitigations include somehow building the model to take into account vast disparities in enforcement of certain offenses—treating, say, a Black Manhattanite with several marijuana arrests no differently than a White Manhattanite with just one. (Of course this would require the model to use the defendant's race as an input.) A third mitigation in the case of marijuana, in particular, is to simply cut the Gordian knot by legalizing or at least decriminalizing it—which then makes downstream machine-learning questions moot.

A second, and equally serious, concern is whether—even if a predictive model measured exactly what it claimed to—we are in practice using it for its intended purpose, or for something else.

For instance, some states are using the COMPAS tool to inform *sentencing* decisions, something that many regard as an awkward if not inappropriate use for it. Says Christine Remington, Wisconsin assistant attorney general, "We don't want courts to say, this person in front of me is a 10 on COMPAS as far as risk, and therefore I'm going to give him the maximum sentence." But COMPAS *has* been used to inform sentencing decisions—including in Wisconsin. When Wisconsinite Paul Zilly was given a longer than expected sentence in part due to his COMPAS score, Zilly's public defender called none other than Tim Brennan himself as a witness for the defense. Brennan testified that COMPAS was not designed to be used for sentencing.[81] At a minimum, it seems clear that we should know exactly what it is that our predictive tools are designed to predict—and we should be very cautious about using them outside of those parameters. "USE ONLY AS DIRECTED," as the label reads on prescription medications. Such reminders are just as necessary in machine learning.

Zooming out even further, however, we might challenge one of the most fundamental unspoken premises of the entire enterprise: that better predictions lead to better public safety.

At first it seems crazy to think otherwise. But there are reasons to pause, and to challenge some of the assumptions that underlie that view.

Columbia Law professor Bernard Harcourt, in his book *Against Pre-*

diction (which Angwin cites as an inspiration for her own work), raises several such objections. As he argues, the link between better predictions and less crime isn't as straightforward or foolproof as it may seem. For instance, imagine a predictive tool that identifies that most reckless drivers are male. It may be the case that aggressively pulling over male drivers doesn't substantially reduce their recklessness on average—but *does* cause female drivers, who recognize that they are less likely to be pulled over, to drive *more* recklessly. In that case, roads may become *less* safe on average—precisely because of the use of this predictive policy. "In other words," as Harcourt puts it, "profiling on higher past, present or future offending may be entirely counterproductive with regard to the central aim of law enforcement—to minimize crime."[82] These sorts of backfiring scenarios are not as far-fetched or rare as they may seem, he argues: they may well describe exactly what is currently happening, for instance, with drug use and tax fraud. If differential enforcement emboldens the overlooked group more than it deters the scrutinized group, it may only make the problem worse.

Predictions also might fail at the ultimate goal of making society safer if they can't be meaningfully turned into *actions*. In 2013, the city of Chicago piloted a program designed to reduce gun violence by creating a "Strategic Subjects List" (informally a "heat list") of people at high risk of being victims of gun violence. Those people as a group turned out to be 233 times more likely than the average Chicagoan to become homicide victims. In this sense, the list's predictive power appeared to be spot-on. Homicides are so rare, though, that even among those on the "heat list," only 0.7% were victimized, while 99.3% were not. What do you do, then, with that predictive information? What kind of intervention for a thousand people would help the seven that will actually go on to be victims?

"By leveraging advanced analytics, police departments may be able to more effectively identify future crime targets for preemptive intervention," noted a 2016 RAND Corporation report on predictive policing in Chicago.[83] However, "improvements in the accuracy of predictions alone may not result in a reduction in crime . . . perhaps more importantly, law enforcement needs better information about what to *do* with the predictions" (emphasis mine).[84]

Predictions are not an end in themselves. What is better: a world in which we can be 99% sure where a crime will occur and when, or a world in which there is simply 99% less crime? In a narrow-minded pursuit of predictive accuracy—or, for that matter, fairness—in a particular prediction tool, we may be missing something larger.[85]

In pretrial release—for which COMPAS is designed—there is, similarly, a gap from predictions to interventions that needs to be considered more broadly. Predicting that someone won't make it to their court date doesn't necessarily mean that *jailing* them between now and then is the correct intervention.[86]

Alexandra Chouldechova explains: "If you think about it from that perspective, then you're saying, Okay this particular population, maybe they're less able to provide for themselves: they actually maybe have lower *risk*, but higher *needs*." Maybe they need day care for their children on their court date or a ride to court—not detention. As it turns out, simply *reminding* people about their court date can significantly improve their rate of appearance.[87] Unfortunately, many risk assessment tools, unlike COMPAS, conflate a prediction of failure to appear with a prediction of criminal reoffense.[88] This is an enormous problem if the solution to one risk is incarceration while the solution to another is a text message.

This is a point that particularly resonates with Tim Brennan. The COMPAS information sheet about a defendant that gets shown to a judge was designed so that the risk assessment was in red but the *needs* assessment was in green. "In green," he says, "you know, because these are the things that you want to cultivate and help the guy."[89] The whole point was to steer as many defendants away from incarceration as possible, toward treatment, community supervision, and the like—the AS in COMPAS stands for "Alternative Sanctions," after all. But some judges simply view the "needs" scores—addiction issues, homelessness, lack of close community—not as a road map for rehabilitation but as all the

more reason to lock someone up.[90] Of course, the ability to assign someone to such alternative sanctions or to treatment programs, classes,
counseling, and the like requires such services to actually exist. If they
don't, then there is a problem that no statistical model, and indeed no
judge, can solve.

"So this brings me to my main point," Moritz Hardt tells me.[91] A
machine-learning model, trained by data, "is by definition a tool to predict the future, given that it looks like the past. . . . That's why it's fundamentally the wrong tool for a lot of domains, where you're trying to
design interventions and mechanisms to change the world."[92]

He elaborates: "Reducing crime and incarceration rates is a really,
really hard problem that I would like to leave to experts in criminal justice. I feel like prediction offers a bit of a dystopian perspective on the
topic, which is 'Let's assume that we're not gonna structurally reduce
crime. We're going to predict where it's gonna happen and go and try to
catch people before it's happening.' It doesn't really offer, to me, a mechanism to structurally reduce crime. And that's what I find dystopian
about it. I don't want to know how to predict where crime's going to happen. I guess that's useful, but, much rather I would have a mechanism
to reduce crime structurally. I, as a computer scientist, have nothing
to offer on that topic, absolutely nothing. I can't tell you the first thing
about this. It would take me years to get to a point where I could."

The importance of stepping back to take a wider, more macroscopic
view of the criminal justice system was not lost on the earliest pioneers
in the field.

Ernest Burgess, writing in 1937—after his initial report on the parole
system had prompted a risk-assessment model that went into practice
statewide—felt that it was high time to move on to something more
comprehensive. "The time has arrived in Illinois, in my judgment," he
wrote, "to stop tinkering with parole as an isolated part of our penal
problem. What is required is a major operation which involves a complete reorganization of the prison system of the state."[93]

Eighty-some years have passed since then. It's still true.

3

TRANSPARENCY

The rules are supposed to be clear, uniformly expressed, and accessible to all. As we all know, this is rarely actually the case.

—DAVID GRAEBER[1]

Providing overwhelming amounts of information without adequate structure or documentation is not transparency.

—RICHARD BERK[2]

In the mid-1990s, Microsoft's Rich Caruana was a graduate student at Carnegie Mellon working on neural networks, when his advisor, Tom Mitchell, approached him for help with something.

Mitchell was working on an ambitious, interdisciplinary, multi-institutional project—bringing together biostatisticians, computer scientists, philosophers, and doctors—to better understand pneumonia. When a patient is first diagnosed, the hospital needs to make one critical decision fairly early on, which is whether to treat them as an inpatient or an outpatient—that is, whether to keep them in the hospital overnight for monitoring or to send them home. Pneumonia was at the time the sixth leading cause of death in the United States, with about 10% of pneumonia patients ultimately dying—and so correctly identifying which patients were at greatest risk would translate fairly straightforwardly into lives saved.

The group had been given a dataset of about fifteen thousand pneumonia patients and was tasked with building a machine-learning model to predict patient survival rates that could help the hospital triage new patients. The result was a head-to-head competition of sorts among a menagerie of different machine-learning models: logistic regression, a rule-learning model, a Bayesian classifier, a decision tree, a nearest-neighbor classifier, a neural network—you name it.[3]

Caruana worked on the neural network (in the '90s, the state of the art was *wide* networks, rather than deep), and it was not without some pride that he digested the results. His neural network had won—handily. It was the best of all the complex models, and outperformed more traditional statistical methods like logistic regression by a significant margin.[4]

As the group would write in their reports, "Even small improvements in predictive performance for prevalent and costly diseases, such as [pneumonia], are likely to result in significant improvements in the quality and efficiency of healthcare delivery. Therefore, seeking models with the highest possible level of predictive performance is important."[5]

So naturally the Pittsburgh hospitals that were partnering on the study decided to deploy the highest-performing model. Right?

"We started talking about that—whether it was safe to use a neural net on patients," says Caruana.[6]

"And I said, *Hell no, we will not use this neural net on patients.*"

They deployed one of the simpler models that his neural net had so handily beaten. Here's why.

THE WRONG RULES

One of the other researchers on the project, Richard Ambrosino, had been training a much different, "rule-based" model on the same dataset. Rule-based models are among the most easily interpreted machine-learning systems; they typically take the form of a list of "if x then y" rules. You simply read your way down the list, from top to bottom, and as soon as a rule applies, you're done. Imagine a flow chart that doesn't branch and simply looks like a single vine running from top to bottom: "Does this rule apply? If so, here's the answer and you're done. If not, keep reading." In this way, rule-based models resemble "conditionals" or "switch statements" in traditional software programming; they also sound a lot like the way humans think and write. (More complex models use "sets" rather than "lists," in which multiple rules can be applied at once.)[7]

Ambrosino was building a rule-based model using the pneumonia data. One night, as he was training the model, he noticed it had learned a rule that seemed very strange. The rule was "If the patient has a history of asthma, then they are low-risk and you should treat them as an outpatient."

Ambrosino didn't know what to make of it. He showed it to Caruana. As Caruana recounts, "He's like, 'Rich, what do you think this means? It doesn't make any sense.' You don't have to be a doctor to question whether asthma is good for you if you've got pneumonia." The pair attended the next group meeting, where a number of doctors were present; maybe the MDs had an insight that had eluded the computer scientists. "They said, 'You know, it's probably a real pattern in the data.' They said, 'We consider asthma such a serious risk factor for pneumonia patients that we not only put them right in the hospital . . . we probably put them right in the ICU and critical care.'"

The correlation that the rule-based system had learned, in other words, was real. Asthmatics really *were*, on average, less likely to die from pneumonia than the general population. But this was precisely *because* of the elevated level of care they received. "So the very care that the asthmatics are receiving that is making them low-risk is what the model would deny from those patients," Caruana explains. "I think you can see the problem here." A model that was recommending outpatient status for asthmatics wasn't just wrong; it was life-threateningly dangerous.[8]

What Caruana immediately understood, looking at the bizarre logic that the rule-based system had found, was that his neural network must have captured the same logic, too—it just wasn't as obvious.

Amending or editing the rule-based system was fairly straightforward; the neural network was harder to "correct" in this way, though not impossible. "I don't know where it is in the neural net, but one way or another I can solve the problem," Caruana recounts. "I can publish more papers in the process of doing it—so this is good—and we'll make that problem go away. I said, the reason why we're not going to deploy the neural net is not actually because of *asthma*, because I already know about that.

"I said, what I'm worried about is things that the neural net has learned that are just as risky as asthma but the rule-based system *didn't* learn." Because the neural net is more powerful, more flexible, it was capable of learning things that the rule-based system didn't. This, after all, is the advantage of neural networks—and the reason Caruana's neural net had won the group's internal contest. "I said it's *those* things that will make us not use this model. Because we don't know what's in it that we would need to fix. So it's this transparency problem with the neural net that ultimately caused me to say we're not going to use it. This has bothered me for a long, long time. Because the most accurate machine-learning models are often not transparent like this. And I'm a machine-learning person. So I want to use accurate models, but I also want to use them safely."

It's often observed in the field that the most powerful models are on the whole the least intelligible, and the most intelligible are among the least accurate. "This really pissed me off, by the way," he tells me. "I want to do machine learning for health care. Neural nets are really good, they're accurate; but they're completely opaque and unintelligible, and I think that's dangerous now. It's like, maybe I shouldn't be doing machine learning for health care."[9] Instead, though, Caruana decided to spend the next twenty years developing models that attempt to have the best of both worlds—models that are, ideally, as powerful as neural networks but as transparent and legible as a list of rules.

One of his favorites is an architecture called "generalized additive models," first pioneered by statisticians Trevor Hastie and Robert Tibshirani in 1986.[10] A generalized additive model is a collection of graphs, each of which represents the influence of a single variable. For instance, one graph might show risk as a function of age, another would show risk as a function of blood pressure, a third would show risk as a function of temperature or heart rate, and so forth. These graphs can be linear, or curved, or incredibly complex—but all of that complexity can be immediately apprehended *visually*, simply by looking at the graph. These individual one-variable risks are then simply added up to produce the final prognosis. In this way it is more complex by far than, say, a linear regression but also much more interpretable than a neural net. You can visualize, on a plain old two-dimensional graph, every factor going into the model. Any strange patterns should immediately stand out.

Many years after the original pneumonia study, Caruana revisited the dataset and built a generalized additive model to explore it. The generalized additive model turns out to be just as accurate as his old neural net, and far more transparent. He plotted the pneumonia mortality risk, for example, as a function of age. It was mostly what one would expect: it's good to be young or middle-aged if you have pneumonia, and more dangerous to be older. But something in particular stood out: an abrupt, sharp jump beginning at age 65. It seemed unusual that a *particular* birthday would trigger a sudden increase in risk. What was going on? Caruana realized that the model had managed to learn the impact of *retirement*. "It's really annoying that it's dangerous, right? You would hope that the risk goes *down* when you retire; sadly, it goes up."[11]

More importantly however, the closer he looked, the greater the number of troubling connections he saw. He had feared that his old neural network had learned not just the problematic asthma correlation but others like it—though the simple rule-based models at the time weren't powerful enough to show him what else might be lurking in the neural network. Now, twenty years later, he *had* powerful interpretable models. It was like having a stronger microscope and suddenly seeing the mites in your pillow, the bacteria on your skin.

"I looked at it, and I was just like, 'Oh my— I can't believe it.' It thinks chest pain is good for you. It thinks heart disease is good for you. It thinks being over 100 is good for you. . . . It thinks all these things are good for you that are just obviously not good for you."[12]

None of them made any more medical sense than asthma; the correlations were just as real, but again it was precisely the fact that these patients were prioritized for more intensive care that made them as likely to survive as they were.

"Thank God," he says, "we didn't ship the neural net."

Today Caruana says he finds himself in a different position than most researchers. He continues to work on developing model architectures that promise neural-network levels of predictive accuracy while remaining easily understood. But instead of evangelizing any particular solution of his own invention—for instance, an updated version of generalized additive models[13]—he is evangelizing the problem itself. "Everyone is committing these mistakes," he says, "just like I have committed them for decades, and didn't know I was doing it."

"My goal right now," he tells me, "is to scare people. To terrify them. I feel I've succeeded if they've stopped and they've thought, *Oh shit. We really got a problem here.*"

THE PROBLEM OF THE BLACK BOX

Nature conceals her secrets because she is sublime, not because she is a trickster.

—ALBERT EINSTEIN

The act of giving a reason is the antithesis of authority. When the voice of authority fails, the voice of reason emerges. Or vice versa.

—FREDERICK SCHAUER[14]

Rich Caruana is far from the only one who in recent years has had some version of the thought *Oh shit. We really got a problem here.* As machine-learning models proliferate throughout the decision-making infrastructure of the world, many are finding themselves uncomfortable with how little they know about what's actually going on inside those models.

Caruana found himself particularly uncomfortable with the use of large neural networks because of their longstanding reputation as "black boxes." With the breathtaking rise in neural networks in everything from industry to the military to medicine, a growing number are feeling that same unease.

In 2014, United States Defense Advanced Research Projects Agency (DARPA) program manager Dave Gunning was talking to Dan Kaufman, director of DARPA's Information Innovation Office. "We were just try-ing to kick around different ideas on what to do in AI," Gunning tells me.[15] "They had had a whole effort where they had sent a whole group of data scientists to Afghanistan to analyze data, try to find patterns that would be useful to the war fighters. And they were already beginning to see that these machine-learning techniques were learning interesting patterns, but the users often didn't get an explanation for why." A rap-idly evolving set of tools was able to take in financial records, movement records, cell phone logs, and more to determine whether some group of people might be planning to strike. "And they might get some pat-tern that would be suspicious," says Gunning, "but now they'd want an explanation of why." There was no such explanation on offer.

Around that same time, Gunning attended a meeting sponsored by the intelligence community at the Lab for Analytical Sciences at North Carolina State University. The workshop brought together machine-learning researchers and data visualization experts. "We had a govern-ment intel analyst in the room listening to us talk about what all the machine-learning technology could do," Gunning remembers. "And this one analyst just was really kind of adamant that her problem was she already has these big-data algorithms giving her recommendations, but she has to put her name on the recommendation that goes forward. And she gets scored if you will—or worse—based on whether that recom-mendation is correct. But she didn't understand the rationale for the rec-ommendation she was getting from the learning algorithm." Should she sign her name to it, or not? And on what basis, exactly, should she decide?

As computing technology progresses, the defense community has found itself increasingly thinking about what an automated battlefield might look like—what risks and questions surround the idea of ever more autonomous weapons. But many of those questions and problems remain—for the moment, anyway—theoretical.

"With the intel analysis problem, those systems are already there; people are using them," says Gunning. "You know what I mean? This problem is already there. And they want help."

Over the next two years, Gunning would find himself the program manager of a multi-year DARPA program in an attempt to meet that problem head-on. He would call it XAI: Explainable Artificial Intelligence.

Across the Atlantic Ocean, the European Union was getting ready to pass an omnibus law called the General Data Protection Regulation, or GDPR. The GDPR would go into effect in 2018 and would sharply change how companies collect, store, share, and use data online. The regulations—all 260 pages worth—amounted to one of the most important documents in the history of data privacy. They also included something a bit more curious and intriguing, and perhaps just as profound.

At the Oxford Internet Institute in the fall of 2015, Bryce Goodman was leafing through the draft legislation when something caught his attention. "I'd learned a bit about machine learning, and how some of the best methods don't really lend themselves to being transparent or interpretable," he says, "and then I came across this. In the earlier drafts of the GDPR, it was much more explicit. . . . They said people should have the right to ask for an explanation of algorithmically made decisions."[16]

"I find it really interesting," he says, "when there are these pieces of legislation or these things that somebody just sort of puts a stake in the ground and says, *This thing exists now.*"

Goodman approached his Oxford colleague Seth Flaxman, who had just finished a PhD in machine learning and public policy. "Hey, I read this thing in the GDPR," Goodman said. "This seems like it would be a problem."

"He was like, 'Yeah, that seems like it would be.'"

Whether it was getting rejected for a loan, being turned down for a credit card, being detained pending trial or denied parole, if a machine-learning system was behind it, you had a right to know not just what happened but *why.*

The following spring, in 2016, the GDPR was officially adopted by the European Parliament, and the collective hair went up on the back of the necks of executives across the entire tech sector. Attorneys described sitting down with EU regulators. "You realize," they said, "that getting intelligible explanations out of a deep neural network is an unsolved scientific problem, right?" As Goodman and Flaxman had written, it "could require a complete overhaul of standard and widely used algorithmic techniques." The regulator was not moved by this: "That's why you have until 2018."

As one researcher put it: "They decided to give us two years' notice on a massive research problem."

The GDPR has now gone into effect, though the exact details of what EU regulators expect, and what constitutes a sufficient explanation—to whom, and in what context—are still being worked out. Meanwhile transparency—the ability to understand what's going on inside a machine-learning model and why it behaves as it does—has emerged as one of the field's most clear and crucial priorities. Work on this problem continues, full steam, to this day—but recent progress has been made on a number of fronts. Here is what we are coming to know.

CLINICAL VERSUS STATISTICAL PREDICTION

> There is a prevalent myth that the expert judge of men succeeds by some mystery of divination. Of course, this is nonsense. He succeeds because he makes smaller errors in the facts or in the way he weights them. Sufficient insight and investigation should enable us to secure all the advantages of the impressionistic judgment (except its speed and convenience) without any of its defects.
>
> —EDWARD THORNDIKE[17]

The first thing we need to consider as we confront the problem of transparency—particularly in large, complex models—is whether, in fact, we ought to be using large or complex models in the first place. For that matter, should we be using models at all?

These are questions that reach deeply into the history of statistics and machine learning, ones that are as relevant today as they have ever been, and ones whose answers are rather surprising indeed.

In 1954, Robyn Dawes was an undergraduate philosophy major at Harvard, specializing in ethics. His thesis—"A Look at Analysis"—investigated whether, and to what degree, moral judgments were rooted in emotion.

Not only did Dawes think these were important questions, he thought *"empirical* work might be important. But how would you do empirical work?" With this question, he realized that his interests lay more with psychology than philosophy. He applied to psychology graduate programs, and the best one in the country, the University of Michigan, let him in off the waiting list on the final day of the admissions window. Dumbstruck, Dawes broke the good news to his advisor, who told him, "Wire them back immediately before they realize they made a clerical error."[18]

Dawes went to Michigan and began his training in clinical psychology. At the time—the late 1950s and early '60s—this meant a heavy emphasis on Rorschach tests. Dawes found himself growing skeptical of just how useful the Rorschach was as a clinical tool. "It was all intuitive and it made intuitive sense. But then I started reading and found out that in fact, empirically, a lot of this stuff doesn't work."

During his dawning skepticism, Dawes was working as a resident in a psychiatric ward. "There was a client who had this delusion," he recounts, "and the delusion was that he was growing breasts. And he was in my group—on a locked ward by the way, because he had this delusion that must be schizophrenia. And uh, why did he think he was growing breasts? Well, earlier that week, one of his parents had committed suicide before he developed the delusion. Okay, makes perfect sense, right? . . . They'd never asked this guy to take off his shirt even. He'd just been sent to the psychiatric ward—six weeks on a locked ward in my group. And when they asked him to take off his shirt, it was in fact true: he was growing breasts." The man had Klinefelter syndrome: a genetic condition caused by an additional X chromosome, with symptoms including lack of body hair and the development of breast tissue. Dawes was incensed. "Okay, well that's six weeks of his life just thrown out," he says, "because people were so convinced about, *Oh this is a fascinating delusion.*"

This led to a major pivot in Dawes's career, away from clinical practice and toward what was at the time being called "mathematical psychology." He had a paper he was intending to submit for publication, comparing expert clinical judgments against simple mathematical models, and he showed the paper to a friend. The friend's reaction caught Dawes off guard. "He sort of looked at me funny and said, you know, in effect, 'Are you sure you didn't plagiarize?'"

As it happened, Dawes was joining an academic lineage of which he had no idea. The question of pitting expert judgment against simple mathematical models—then called "actuarial" methods—went back to the early 1940s work of a colleague of Ernest Burgess's named Ted Sarbin. Sarbin looked at predictions of academic performance for incoming freshmen at the University of Minnesota. The "actuarial" model was a simple linear regression to predict their college GPA from just two data points: their high school class rank and their college aptitude test scores. The human predictions were made by experienced clinical psychologists who had access to these two data points, plus additional tests, an eight-page dossier, notes from a colleague's interview, and their own firsthand impression of the student.

Sarbin found no measurable difference between the two predictions. If anything, the actuarial model was more accurate, though not markedly so. From Sarbin's perspective, it seemed incredible that the additional information available to the clinicians appeared to add nothing in predictive accuracy. "What of strivings, habits of work and play, special aptitudes, emotional patterns, systematic distractions, and the hundreds of other conditions," he wrote, "which seem to be related to this complex form of social psychological behavior known as academic achievement?"[19] Ironically, Sarbin found that the human counselors didn't place much emphasis on these things themselves and, in fact, made their predictions chiefly on the basis of class rank and test scores—the very same data used in the regression model. They just weren't as consistent or finely tuned in how they weighted it.

Sarbin's conclusion was that the time-intensive effort spent in conducting interviews was a waste. And he cautioned: "Unless checked by statistical studies, the case-study method in the social sciences will become intellectually bankrupt."

Sarbin's findings had caught the attention, in turn, of a young psychologist named Paul Meehl. Inspired by Sarbin's provocation, and unsure of where he himself stood, Meehl began an investigation that would turn into an entire book on the subject: 1954's *Clinical Versus Statistical Prediction*. Though much if not most of the book was dedicated to understanding what goes into clinical judgments and how clinicians make decisions, by far its biggest impact stemmed from the chapter in which Meehl pitted clinical and actuarial judgments head to head. The human experts, Meehl discovered, didn't stand a chance. Out of almost a hundred different domains, in only a half dozen did there seem to be even a slight edge for the human decision makers. "I am told," Meehl recounts, "that half the clinical faculty at one large Freudian oriented midwest university were plunged into a six-month reactive depression as a result of my little book."[20]

Dawes, having found himself to be an unwitting plagiarist, of course became keenly interested in Meehl's work—but his advisors seemed to disapprove of the influence. "I was told by my psychoanalytic mentors that, well, Meehl is a genius, and everyone knows he's a genius, but what he does has nothing to do with what we do. And I started worrying about if what he does has nothing to do with what we do, maybe I don't *want* to do what we do."

By the mid-1970s, Dawes, by then at the Oregon Research Institute, authored another concussive paper in the lineage. As he summarized, Sarbin and Meehl had opened the question of how purely statistical analysis might compare to expert human judgment. "The statistical analysis was thought to provide a floor to which the judgment of the experienced clinician could be compared," Dawes wrote. "The floor turned out to be a ceiling."[21]

Thirty years after Sarbin's original paper, and many dozens of studies later, he concluded, "A search of the literature fails to reveal *any* studies in which clinical judgment has been shown to be superior to statistical prediction when both are based on the same codable input variables" (emphasis mine).[22]

The picture was humbling indeed. Even in cases where the human decision makers were given the statistical prediction as yet another piece of data on which to make their decision, their decisions were *still* worse than just using the prediction itself.[23] Other researchers tried the reverse tack: feeding the expert human judgments into a statistical model as input. They didn't appear to add much.[24]

Conclusions like these, which have been supported by numerous studies since, should give us pause.[25] For one, they seem to suggest that, whatever myriad issues we face in turning decision-making over to statistical models, human judgment alone is not a viable alternative. At the same time, perhaps complex, elaborate models really aren't necessary to match or exceed this human baseline.

A tantalizing question lurks, however: Namely, what explains this surprising verdict? Is human judgment really that bad? Are simple linear models of a handful of variables really that good? Or . . . a third possibility: Has human expertise somehow managed to enter into the simple models where we least expect it? Were we looking for it in the wrong place?

IMPROPER MODELS: KNOWING WHAT TO LOOK AT

The fact that some real problems are hard does not imply that all real problems are hard.

—ROBERT HOLTE[26]

The fact that man cannot combine information as efficiently as a computer, for example, does not imply that man can be replaced by machines. It does imply that the necessity for a man-computer system is at hand.

—HILLEL J. EINHORN[27]

Dawes, with his colleagues at the Oregon Research Institute, wanted to get to the bottom of the shocking effectiveness of simple, linear models in decision-making.

One hypothesis was that perhaps the models were outperforming the experts because of a kind of "wisdom of the crowd" effect: that a single model was aggregating the judgments of a whole group of experts, so of course it would outperform any single expert in isolation. It sounded plausible—but it wasn't true. Amazingly, even if a model was trained only to mimic a *single* expert's judgments, it still outperformed the expert themselves![28]

Maybe there was something about the way that the linear models were optimized, the way they were tuned to have the optimal coefficients that weighted each variable. With his collaborator Bernard Corrigan, Dawes compared the judges against what they called "improper" linear models, ones whose weights were not optimized. They tried *equal* weights, and they tried *random* weights.

The results were stunning. Models with random weights—as long as they were constrained to be positive—were as accurate or more than the judges themselves. Models with equal weights were even better.[29]

The practical possibilities for such quick-and-easy improper models seemed endless. In the mid-1970s, Dawes was on a panel with a medical physician; afterward, sharing drinks at the bar, the doctor asked him, "Could you, for example, use one of your improper linear models to predict how well my wife and I get along together?"[30]

Dawes immediately imagined the simplest model he could think of: a good marriage, he figured, should probably involve more sex than fights. With a colleague's dataset, he did the math. He added up the number of times a couple had sex ("defined as genital union with or without orgasm") over a period of weeks or months, and subtracted the number of times they had a fight ("situations where at least one party became uncooperative") in that same time. "The linear prediction is the quintessence of simplicity: subtract the rate of arguments from the rate of sexual intercourse. A positive difference predicts happiness, a negative one unhappiness."[31] The data, from couples in the Kansas City, Missouri, area, showed that indeed, sex-minus-fights was positive for 28 out of the 30 couples that self-identified as "happy," while sex-minus-fights was negative for all 12 couples that self-identified as "unhappy."[32] Study after study—in Oregon, in Missouri, in Texas—confirmed this correlation.

In fact, in many cases the equal-weighted models were better even than the optimal regressions—a fact that seems impossible, as the opti-

mal weights are chosen precisely to be, well, optimal. But they were optimal for a particular context and a particular set of training data, and
that context didn't always transfer: optimal weights for predicting academic performance at the University of Minnesota, for instance, weren't
necessarily optimal for predicting academic performance at Carnegie
Mellon. In practice, equal weights seemed to hold up better and be more
robust across contexts.[33]

Dawes was fascinated by this. Given the complexity of the world,
why on earth should such dead-simple models—a simple tally of equally
weighted attributes—not only work but work *better* than both human
experts and optimal regressions alike?

He came up with several answers. First, despite the enormous complexity of the real world, many high-level relationships are what is known
as "conditionally monotone"—they don't interact with one another in
particularly complex ways. Regardless of whatever else might be happening with a person's health, it's almost *always* better if that person
is, say, in their late twenties rather than their late thirties. Regardless of
whatever else might be happening with a person's intellect, motivation,
and work ethic, it's almost *always* better if that person's standardized
test scores are ten points higher than ten points lower. Regardless of
whatever else might be happening with a person's criminal history, self-
control, and so forth, it's almost *always* better if they have one fewer
arrest on their record than one more.

Second, there is almost always error in any measurement. For theoretical as well as intuitive reasons, the more error-prone a measurement
is, the more appropriate it is to use that measurement in a linear fashion.

Perhaps most provocatively from the perspective of alignment, Dawes
argued that these equally weighted models surpass their "proper," opti-
mally weighted counterparts because those weights have to be tuned, as
we've seen, with respect to some kind of objective function. In real life,
often we either can't define how exactly we plan to measure success, or
we don't have time to wait for this ground-truth to come so we can tune
our model. "For example," he writes, "when deciding which students to
admit to graduate school, we would like to predict some future long-

term variable that might be termed 'professional self-actualization.' We have some idea what we mean by this concept, but no good, precise definition as yet. (Even if we had one, it would be impossible to conduct the study using records from current students, because that variable could not be assessed until at least 20 years after the students had completed their doctoral work.)" Even here, Dawes argues, when we don't know what exactly we want and have no data at all, improper models should serve us at least as well if not better than raw intuition.[34]

One might, however, object that Dawes and Corrigan's comparisons, particularly against human experts, don't seem entirely fair. The models were not random linear combinations of random properties; they were random linear combinations of *precisely* the things that humans have established, through decades if not generations of best practices, to be the most relevant and predictive things to consider.

It would be tempting to argue, perhaps, that all of this "pre-processing" activity—deciding which two or five or ten things out of the infinite information available is most pertinent to the decision at hand—reflects the real wisdom and insight into the problem: in effect, that we've done all of the hard work already before passing it off to the linear model, which then gets all the credit. This, indeed, was exactly Dawes's point. As he wrote: "The linear model cannot replace the expert in deciding such things as 'what to look for,' but it is precisely this knowledge of what to look for in reaching the decision that is the special expertise people have."[35]

It was Dawes's conclusion that human expertise is characterized by knowing what to look for—and not by knowing the best way to integrate that information. One of the clearest demonstrations of this idea came from a study by decision theorist Hillel Einhorn in 1972.[36] Einhorn looked at physicians' judgments of biopsy slides for patients diagnosed with Hodgkin's lymphoma. The physicians were asked to specify the factors they considered important when looking at a slide, and then to score each slide for those factors. The physicians then gave an overall rating about the severity of the patient. It so happened that their overall severity ratings had *zero* correlation with patient survival. However, in a result that is becoming something of a refrain, a simple model using the experts' individual factor scores was a powerful predictor for patient mortality.

Put differently, we've been looking for human wisdom in the wrong place. Perhaps it is not in the mind of the human decider, but embodied in the standards and practices that determined exactly which pieces of information to put on their desk. The rest is just math—or, at any rate, should be.

Dawes put it a third way, in arguably the most famous sentence of his storied career. "The whole trick is to know what variables to look at," he wrote, "and then to know how to add."[37]

OPTIMAL SIMPLICITY

Simple can be harder than complex: You have to work hard to get your thinking clean to make it simple. But it's worth it in the end because once you get there, you can move mountains.

—STEVE JOBS[38]

The only simplicity for which I would give a straw is that which is on the other side of the complex—not that which never has divined it.

—OLIVER WENDELL HOLMES JR.[39]

Perhaps no one so carries the spirit of Dawes forward into the twenty-first century like Duke University computer scientist Cynthia Rudin. Rudin has made simplicity one of the central drives of her research; she is interested in not only arguing against the use of overly complex models, but in pushing the envelope of what simple models can do. In the criminal justice domain, for instance, Rudin and her colleagues published a paper in 2018 showing that they could make a recidivism-prediction model as accurate as COMPAS that could fit into a single *sentence*: "If the person has more than three prior offenses, or is an 18-to-20-year-old male, or is 21-to-23 years old *and* has two or more priors, predict they will be rearrested; otherwise, not."[40]

For Rudin, Dawes's research is both an inspiration and a kind of challenge, like a gauntlet being thrown down. So simple models, made from hand-selected high-level variables, perform about as well as more complex models—sometimes better—and consistently as well as or better than human experts. But even that leaves a lot of questions—and avenues for research open. How, for instance, might one build not just *a* simple model from a given dataset but the *best* simple model?

Surprisingly, answers have come only in the last several years.

Rudin looks at the simple models currently being used in twenty-first century health care, and takes a different and much less sanguine view than Dawes. Instead of seeing present-day models as a superior alternative to clinical intuition, she sees models overly shaped *by* clinical intuition. And ones with a lot of room for improvement.

She raises the example of the coronary heart disease score sheet for men. "So if you're a guy, you go into the doctor's office, and they will try to compute your ten-year risk of coronary heart disease. And they'll do it by asking you five questions: what's your age, what's your cholesterol level, do you smoke, and so on and so forth. So they ask you these five questions, and then you get points for each answer you give, and then you add up the points and that translates into your ten-year risk of coronary heart disease." Her voice takes a sharp turn. "But where do they get these five questions? And how did they get the points? And the answer is, *They made it up!* This was made up by a team of doctors! This is— This is not how I want to do this. What I want to do is, I want to build something that's that interpretable—but I want to build it from data."[41]

As it happens, finding *optimal* simple rules is not for the faint of heart. In fact, it requires tackling an "intractable," or "NP-hard" problem: a thicket of complexity in which there is no straightforward means of obtaining the guaranteed best answer. Given tens of thousands of patient records, each with dozens or perhaps hundreds of different attributes—age, blood pressure, etc.—how do you find the best simple flow chart for diagnosis? Computer scientists have a toolkit full of ways to make progress here, but Rudin felt that the existing algorithms for building simple rule lists and scoring systems from big data—algorithms like CART, developed in the '80s,[42] and C4.5, developed in the '90s[43]—

just weren't sufficient. There was also something available to computer scientists in the 2010s that wasn't available in the '80s and '90s: about a million-fold speedup in computing power. Instead of using that computational horsepower to train a huge, complex model—like AlexNet, with tens of millions of parameters—why not put it to use, instead, searching the vast space of all possible *simple* models? What might be possible? Her team went back to the drawing board and came up with new approaches—one for rule-based models and another for models based on score sheets—and set about comparing these against the status quo.

In particular, Rudin and her lab set their sights on beating one of the most commonly used models in all of medicine: $CHADS_2$. $CHADS_2$, developed in 2001, and its successor, CHA_2DS_2-VASc, developed in 2010, are designed to predict the risk of stroke in patients with atrial fibrillation.[44] Each was designed by doctors and researchers working closely with a dataset, along with their clinical expertise, to identify what they thought the most relevant factors were. And subsequent studies have confirmed the predictive utility of the tools. Both models, despite being generally accepted as valid instruments and having found incredibly widespread use, remain to a degree "artisanal," handcrafted. Rudin wanted to *computationally* identify the most relevant factors to combine together into a single scoring instrument.

Working with more than six thousand times as much data as was used in the original $CHADS_2$ study, Rudin let her algorithm, called Bayesian Rule Lists, loose on a set of 12,000 patients, to pore over some 4,100 different properties for each—every drug they were taking, every health condition they had reported—to make the best possible scoring system.[45] She then compared her own model to both $CHADS_2$ and CHA_2DS_2-VASc against held-out portions of that same dataset.

The results showed a marked improvement over both $CHADS_2$ and CHA_2DS_2-VASc. More intriguingly, they also showed a marked *decrease* in accuracy from the original $CHADS_2$ to the more recent CHA_2DS_2-VASc. The newer model appeared—at least by this measure, on this data—to be *worse* than the old one. This, as she and her colleagues put it delicately in their paper, "highlights the difficulty in constructing these interpretable models manually."

In a subsequent project, Rudin and her PhD student Berk Ustun worked with Massachusetts General Hospital to develop a scoring system for sleep apnea, a condition that affects tens of millions of Americans, and more than a hundred million people worldwide.[46] Their goal was to create a model that was not only as accurate as possible, but also so simple that it could run quickly and reliably on some decidedly old-school hardware: physicians' notepads.

Because of the constraint that the model would be deployed on paper, Ustun and Rudin had to make their model almost impossibly simple. It would need very few explicit features to consider, and integer coefficients as small as possible.[47] Even into the twenty-first century, it was not uncommon for practitioners to simply come up with an ad hoc model based on their own intuition. This is sometimes derisively referred to as the "BOGSAT method": a bunch of guys sitting around a table. Even in cases where machine learning was used in building the model, it was often more complex models that were simplified manually after the fact.[48] It is still true today that models in current medical practice were designed in this ad hoc fashion, meaning that accuracy—and, therefore, real patients—are suffering.[49] Ustun and Rudin wanted to see if there was a better way.

They developed a model called SLIM ("Supersparse Linear Integer Model") to find not just decent heuristics but provably optimal ways to make decisions under these severe constraints. The upshot of their work was twofold, with concrete benefits in medicine and machine learning alike.

First, the model showed—contrary to received wisdom and current practice—that patient *symptoms* were significantly less useful than their *histories*. When Ustun and Rudin trained a model on patients' medical histories—things like past heart attacks, hypertension, and the like—it was significantly more predictive than one trained on their immediate symptoms: things like snoring, gasping, and poor sleep. What's more, *adding* symptoms to the model based on histories didn't register much of an improvement. The screening of sleep apnea—which in severe untreated form triples one's risk of death[50]—had taken a measurable step forward.

Second, the machine-learning community had scored a methodological victory it could carry through into other collaborations and other domains. "SLIM accuracy was similar to state-of-the-art classification models applied to this dataset," Ustun and Rudin's team report, "but with the benefit of full transparency, allowing for hands-on prediction using yes/no answers to a small number of clinical queries."[51]

As Rudin puts it, "I want to design predictive models with the end user in mind. I want to design these things . . . not just so they're accurate, but so that people can use them, people can make decisions using them. I want to create predictive models that are highly accurate, yet highly interpretable, that we can use for trustworthy decision making. And I'm working under the hypothesis that I believe is true, that many data sets permit predictive models that are also surprisingly small. I'm not the first person to hypothesize this; it was hypothesized many years ago. But now we have the computational ability and new ideas and new techniques that will really allow us to test this hypothesis."

It's an exciting time for researchers working on this set of questions. Simple models are amazingly competitive—and then some—with human expertise. Modern techniques give us ways of deriving *ideal* simple models.

With that said, there are cases where complexity is simply unavoidable; the obvious one is models that don't have the benefit of human experts filtering their inputs to meaningful quantities of manageable size. Some models must, for better or worse, deal not with human abstractions like "GRE score" and "number of prior offenses" but with raw linguistic, audio, or visual data. Some medical diagnostic tools can be fed human inputs, like "mild fever" and "asthmatic," while others might be shown an X-ray or CAT scan directly and must make some sense of it. A self-driving car, of course, must deal with a stream of radar, lidar, and visual data directly. In such cases we have little choice but the kinds of large, multimillion-parameter "black box" neural networks that have such a reputation for inscrutability. But we are not without resources here as well, on the science of transparency's other, wilder frontier.

SALIENCY: THE WHITES OF ITS EYES

Humans, relative to most other species, have distinctly large and visible sclera—the whites of our eyes—and as a result we are uniquely *exposed* in how we direct our attention, or at the very least, our gaze. Evolutionary biologists have argued, via the "cooperative eye hypothesis," that this must be a feature, not a bug: that it must point to the fact that cooperation has been uncommonly important in our survival as a species, to the point that the benefits of shared attention outweigh the loss of a certain degree of privacy or discretion.[52]

It might be understandable, then, for us to want to expect something similar from our machines: to know not only what they think they see but where, in particular, they are looking.

This idea in machine learning goes by the name of "saliency": the idea is that if a system is looking at an image and assigning it to some category, then presumably some parts of the image were more important or more influential than others in making that determination. If we could see a kind of "heat map" that highlighted these critical portions of the image, we might obtain some crucial diagnostic information that we could use as a kind of sanity check to make sure the system is behaving the way we think it should be.[53]

The practice of such saliency methods has been full of surprises that highlight just how unintuitive machine-learning systems can be. Often they latch onto aspects of the training data we did not think were relevant at all, and ignore what we would imagine was the critical information.

In 2013, Portland State University PhD student Will Landecker was working with a neural network trained to distinguish images in which an animal was present from those with no animals present. He was developing methods for looking at which portions of the image were relevant to the ultimate classification, and noticed something bizarre. In many cases, the network was paying more attention to the background of the picture than the foreground. A closer look showed that blurry backgrounds—known in photographer's lingo as "bokeh"—were commonly present in images of an animal, with the face in sharp focus and the background artfully out of focus. Empty landscapes, by contrast, tended to be more uniformly in focus. As it turns out, he hadn't trained an animal detector at all. He'd trained a bokeh detector.[54]

In 2015 and 2016, dermatologists Justin Ko and Roberto Novoa led a collaboration between researchers from Stanford's medical and engineering schools. Novoa had been struck by the progress in computer vision systems' ability to differentiate between hundreds of different dog breeds. "I thought," he says, "if we can do this for dogs, we can do this for skin cancer."[55] They put together the largest dataset ever assembled of benign and malignant skin patterns, 130,000 images spanning two thousand different diseases as well as healthy skin. They took an off-the-shelf open-source vision system, Google's Inception v3, which had been trained on the ImageNet dataset and categories, and they retrained the network to tell the difference, not between Chihuahuas and Labradors, but between acral lentiginous melanoma and amelanotic melanoma, and thousands of other conditions.

They tested their system against a group of twenty-five dermatologists. The system outperformed the humans. This "dermatologist-level" accuracy landed them a widely cited paper in *Nature* in 2017.[56] For Ko—who prides himself on the keenness of his own diagnostic eye, honed over the better part of a decade of training and clinical practice—the result was both deeply humbling and inspiring. "I've been spending years and years and years," says Ko, "and this thing can do it in a few weeks."[57] And yet what such a system promised was the ability to "essentially extend high-quality, low-cost diagnostic abilities to the furthest reaches of the globe."

As it turns out, a system like this is useful not only in places where first-rate diagnosticians are hard to find, but also as a second opinion for trained experts like Ko himself. Ko remembers the date—April 17, 2017—when a patient came in to his clinic with a funny-looking spot on their shoulder. "I was completely on the fence," Ko says. Something about it, he said, just seemed "not quite right. But then when I looked at it with my dermatoscope, I didn't see any features that would suggest to me that this was an early evolving melanoma." Still, it didn't sit right with him.

"And so I said, Okay, this is a perfect time to whip out my iPhone." Ko took a series of photos, with every kind of angle and lighting he could, and fed each of them into the network. "Remarkably," he says, "regardless of the photo, it was pretty stable in the read—and it was pretty adamant that this was a malignant lesion." Ko had it biopsied and talked to the clinic's dermatopathologist. "Lo and behold, she said, 'Hey, you know what? This is really fascinating. This was a really early, subtle example of an evolving melanoma.' So we caught it at a completely treatable stage." The date sticks in Ko's mind. It was the first time the network had made a clinical impact. "And," he says, "I hope it's just the first of many."

The full story, though, is a bit more complicated. Ko, Novoa, and their collaborators submitted a letter the following year to the *Journal of Investigative Dermatology*, urging caution in moving neural-network models too quickly into routine clinical practice.

They felt that extreme care was due before such models are deployed widely into the field, and underscored their point with a cautionary tale from their own experience. The vision system they were using was much more likely to classify any image with a *ruler* in it as cancerous. Why? It just so happened that medical images of malignancies are much more likely to contain a ruler for scale than images of healthy skin. "Thus the algorithm inadvertently 'learned' that rulers are malignant."[58] Saliency-based methods can catch some of these issues. Still, the ultimate solution—whether ensuring that the dataset contains diverse enough variation or that all input images are standardized in some way—is complex. "We must continue addressing the many nuances involved," they conclude, "in bringing these new technologies safely to the bedside."

TELL ME EVERYTHING: MULTITASK NETS

One of the simplest ideas in making complex models more transparent and understandable is simply to have them output *more things*. When

Rich Caruana was working on neural networks to predict medical outcomes, he realized that the network could be used to make not just a single prediction—say, whether the patient would live or die—but potentially dozens: how long they'd stay in the hospital, how large their bill would be, whether they'd need breathing assistance, how many courses of antibiotics they'd require, and so on.

All of this additional information in a dataset was useless in practice as additional *inputs* to the model. Learning to predict a patient's risk of death based on their hospital bill won't actually help you when a new patient arrives, because of course you don't *know* their final bill yet. But rather than serving as additional inputs, this information is useful as additional *outputs*, additional sources of ground truth in training the model. The technique has come to be known as "multitask learning."[59]

"Strangely enough, that can be easier than training it to predict one of those things at a time," Caruana tells me.[60] "Together, you can think of training it on a hundred related things at the same time as providing you more signal. More information."

As an example, he invites me to imagine that I'm admitted to the hospital with some serious condition, perhaps pneumonia. Suppose I live, he says—but I'm in the hospital for a month, and my bill is a half million dollars. "I now know something was terribly wrong," says Caruana. "The shit hit the fan for you. Maybe it didn't qualify as one of our 'dire outcome' things. But I now know that you were very, very sick."

To a system built narrowly to predict mortality risk, my case might serve as training data with a ground-truth mortality risk of zero, seeing as I survived, but something would be missing. My case was, as he puts it, "a pretty high-risk zero." If I indeed had simply been lucky, maybe the system ought to predict, say, an 80% chance of death, not a 0% chance, in future cases like mine. Giving it a wider range of outputs to train on, and simultaneous predictions to make, might nudge the system toward that more accurate assessment.

What Caruana came to realize was that not only were these "multitask" models better in the traditional senses—they trained faster, achieved higher accuracy—but they were also more transparent, in a way that made it easier to identify problems. If a health-care system like the one he built in the '90s were generating predictions only for whether a patient would live or die, you might find it predicting unexpected things—like, for instance, that asthmatics are less likely to die than regular outpatients. On the other hand, if you had a multitask net predicting all sorts of things from the data—not just death but length of hospital stay or dollar cost of treatment—these anomalies would be much more visible. The asthmatics, for instance, might have better-than-average morbidity but astronomical medical bills. It would be much clearer that these were no ordinary "low-risk" patients to be sent home with instructions to take two pills and call back in the morning.

In some cases, these additional output channels can also offer something more significant. A team spanning Google, its life-science spinoff Verily, and the Stanford School of Medicine was working in 2017 and 2018 on similarly adapting Google's Inception v3 network to classify images of the retina.[61] They found encouraging diagnostic results, with the model detecting diseases like diabetic retinopathy as accurately as human experts. The team realized, however, that the dataset they were working with had all sorts of other information about the patient: age, sex, body mass index, whether they smoked, and so on. "So we sort of added those variables to the model," says Google researcher Ryan Poplin.[62] Just as Caruana had done, they thought that since they had all this additional patient data, why not have their model predict *all* of it? What if they treated this trove of ancillary data—age, sex, blood pressure, etc.—not as additional inputs to the model, but as additional *outputs*? It might offer a way to make the model more robust, and might offer some insight into cases where the model's disease predictions were off. "We felt like it was a great sort of control or ground truth that we could add to the model," says Poplin.

They were in for an enormous shock. The network could almost perfectly tell a patient's age and sex from nothing but an image of their retina.

The doctors on the team didn't believe the results were genuine. "You show that to someone," says Poplin, "and they say to you, 'You must have a bug in your model. 'Cause there's no way you can predict that with such high accuracy.' . . . As we dug more and more into it, we discovered that this wasn't a bug in the model. It was actually a real prediction."

The team used saliency methods to reveal, if not exactly *how* the network was doing it, then at least what the relevant features were. Age, as it happened, was determined by the model looking mostly at the blood vessels; sex, in contrast, by looking at the macula and the optic disc.

At first, says Poplin, when showing these results to a doctor, "They kind of laugh at you. They don't believe it. But then, when you show them that heat map and show that it's focusing on the optic disc or maybe features around the optic disc, then they say, 'Oh yeah, of course we knew that, of course you can see that.' By showing where in the image the model is using to make its prediction, it really does provide a level of trust and also, you know, a level of validity to the results."

More than achieving mere predictive accuracy, the model suggested an intriguing path forward for medical science itself. The combination of multitask learning and saliency techniques showed the field that there were sex differences in the retina that had been overlooked. Not only that; it showed where to find them.

These methods of explanation, in other words, don't just make for better medicine. They might also make for better doctors.

POPPING THE HOOD: FEATURE VISUALIZATION

We've seen how multitask networks, with additional outputs, can give us important context for what a network is predicting. Saliency methods offer context for the network's input, and can offer us information about, in effect, where a model is *looking*. But neither tells us much about what's going on inside the black box—that is, what the model is actually *seeing*.

The signature breakthrough in machine learning since AlexNet in 2012 have been neural network models that learn from the messy welter of raw sensory perception: millions of colored pixels. The models have *tens of millions,* not dozens, of parameters, and those parameters represent fairly ineffable things: summed thresholds of earlier layers, which

are themselves summed thresholds of earlier layers, all the way back to
the millions of raw pixels. This is not the raw grist from which intelligible explanations are made.

So what to do?

At NYU, PhD student Matthew Zeiler and his advisor, Rob Fergus,
were fixated on this question. The success of these enormous, bewildering models, they argue, is undeniable. "However," they write, "there is no
clear understanding of why they perform so well, or how they might be
improved. . . . From a scientific standpoint, this is deeply unsatisfactory."[63]
In other words, the results are impressive, but as Zeiler puts it, "With all
these nice results, it's not clear what these models are learning."[64]

People knew that the bottommost layer of a convolutional network
represented basic things: vertical edges, horizontal edges, diagonal edges,
a strong single color, or a simple gradient. And it was known that the
final output of these networks was a category label: cat, dog, car, and so
forth. But it wasn't really known how to interpret the layers in between.

Zeiler and Fergus developed a visualization technique they called
"deconvolution," which was a way to turn intermediate-level activations
of the network back into images.[65]

For the first time they were *seeing* the second layer. It was a menagerie of shapes. "Parallel lines, curves, circles, t-junctions, gradient patterns, colorful blobs: a huge variety of structure is present already at
the second layer." The third layer was even more complex, beginning
to represent portions of objects: things that looked like parts of faces,
eyeballs, textures, repeated patterns. It was already detecting things like
the white fluff of a cloud, the multicolor stripes of a bookshelf, or the
green comb of grass. By the fourth layer, the network was responding
to configurations of eyes and nose, to tile floors, to the radial geometry
of a starfish or a spider, to the petals of a flower or keys on a typewriter.
By the fifth layer, the ultimate categories into which objects were being
assigned seemed to exert a strong influence.

The effect was dramatic, insightful. But was it *useful*? Zeiler popped
the hood of the AlexNet model that had won the ImageNet competition
in 2012 and started digging around, inspecting it using deconvolution.

He noticed a bunch of flaws. Some low-level parts of the network had normalized incorrectly, like an overexposed photograph. Other filters had gone "dead" and weren't detecting anything. Zeiler hypothesized that they weren't correctly sized for the types of patterns they were trying to match. As astoundingly successful as AlexNet had been, it was carrying some dead weight. It could be improved—and the visualization showed *where*.

In the fall of 2013, within a matter of frantic months, Zeiler finished his PhD, left NYU, started his own company—Clarifai—and entered that year's ImageNet competition. He won.[66]

Other groups, subsequently and in parallel, have explored further ways of directly visualizing a neural network. In 2015, Google engineers Alexander Mordvintsev, Christopher Olah, and Mike Tyka experimented with a method of starting from an image of random static, and then tweaking its pixels to maximize the probability that the network assigns it a particular label—say, "banana" or "fork."[67] This deceptively simple method has proven to be remarkably powerful. It results in fascinating, memorable, often psychedelic, and occasionally grotesque images. Start optimizing static for "dog," for instance, and you're likely to get some unholy amalgam of dozens of eyes and ears, growing fractally, one on another, at a variety of different scales.

It's fertile ground for artistic mischief and almost a novel visual aesthetic unto itself. The Google engineers had a further idea: rather than starting from static and manually specifying a category label, they would start from a real image—clouds, say, or leaves—and simply adjust the image to amplify whatever neurons in the network already happened to be most active. As they write, "This creates a feedback loop: if a cloud looks a little bit like a bird, the network will make it look more like a bird. This in turn will make the network recognize the bird even more strongly on the next pass and so forth, until a highly detailed bird appears, seemingly out of nowhere." They dubbed this rather hallucinatory method "DeepDream."[68]

Other, more bizarre possibilities exist as well. When Yahoo's vision team open-sourced a model used to detect whether an uploaded image was pornographic, UC Davis PhD student Gabriel Goh used this generative method to tune static into shapes the network regarded as *maximally* "not safe for work." The result was like pornography by Salvador Dalí. If you optimize for some *combination* of the obscenity filter and normal ImageNet category labels—for instance, volcanoes—you get, in this case, obscene geography: what look like giant granite phalluses, ejaculating clouds of volcanic ash. Such images are, for better or worse, not easily forgotten.[69]

On a more philosophical note, such techniques suggest that, at least as far as neural networks are concerned, the line between critic and artist can be a thin one. A network trained to *recognize,* say, lakes or cathedrals can be made, in this way, to endlessly churn out images of lake after lake after lake, or cathedral after cathedral after cathedral, that it has never seen before. It's a beautiful, if painstaking, prescription for artistic practice: anyone who can tell good art from bad can be a creator. All you need is good taste, random variations, and plenty of time.

Techniques like these not only open up a vast space of aesthetic possibility but have important diagnostic uses as well. For example, Mordvintsev, Olah, and Tyka used their start-from-static technique to have an image classification system "generate" images that would maximally resemble all of its different categories. "In some cases," they write, "this reveals that the neural net isn't quite looking for the thing we thought it was." For example, pictures that maximized the "dumbbell" categorization included surreal, flesh-colored, disembodied arms. "There are dumbbells in there alright," they write, "but it seems no picture of a dumbbell is complete without a muscular weightlifter there to lift them. In this case, the network failed to completely distill the essence of a dumbbell. Maybe it's never been shown a dumbbell without an arm holding it. Visualization can help us correct these kinds of training mishaps."[70] This is also a useful technique for exploring issues of bias and representation. If starting from random static and fine-tuning hundreds of images to maximize the "face" category produces a set of faces that are, say, exclusively white and male, then that's a pretty good indication that the network won't recognize other types of faces as readily.

Since the initial DeepDream research, Tyka has gone on to cofound the Artists and Machine Intelligence program at Google, and continues to explore the aesthetic possibilities of machine learning. Meanwhile Olah and Mordvintsev, along with their collaborators, have continued to explore the scientific and diagnostic frontier of visualization.[71] Today Olah leads the clarity team at OpenAI. "I've always been kind of fascinated by explaining things," he tells me. "My goal, which I think some people think is kind of crazy, is just to completely reverse engineer neural networks."[72] This work has pushed the boundaries not only of science but of *publishing;* Olah found that traditional scientific journals just weren't suited for the kinds of rich, interactive, full-color and high-resolution visualizations he was making. So he launched a new one.[73]

Overall, the group expresses a measured but contagious optimism about the work underway. "We have a lot of work left ahead of us to build powerful and trustworthy interfaces for interpretability," they write. "But, if we succeed, interpretability promises to be a powerful tool in enabling meaningful human oversight and in building fair, safe, and aligned AI systems."[74]

Much progress continues to be made in the visual domain especially— but what about understanding a network like this not at a visual level but a *conceptual* one? Might there be a way to make sense of the interior of the network through words, for instance? This is one of explanation's latest frontiers.

DEEP NETWORKS AND HUMAN CONCEPTS

In the fall of 2012, MIT graduate student Been Kim had found herself, at the beginning of her PhD, homing in on the project that would shape the next years of her life. She had previously been working in robotics— including going so far as getting a forklift driver's license in order to better understand the industrial robotics environment—but decided in the end that it wasn't the right fit. "I realized that robotics had a hardware limitation," she tells me. "My thoughts go faster than what physical stuff can do."[75]

Increasingly Kim believed that interpretability might well become her dissertation topic, if not her life's work. Her first time at the NeurIPS conference, that December, she found herself talking with an older colleague. "I was telling a faculty member that I work on interpretability, and he was like, 'Oh, why? Neural networks will solve everything! Why do you care?' And I was like, 'Well . . . !'" She chuckles at the recollection. "Imagine if you walk into a doctor's office and he says, 'Oh, I'm going to open you up and remove maybe a couple of things.' And you ask, 'Oh, why?' He says, 'Oh, I don't know. This machine says that that's the best option for you. 99.9%.'"

"What would you ask?" she says. Her question is both rhetorical and more than rhetorical. "What would you say to the doctor? It was so abundantly obvious to me that this is an area we need to solve. What I wasn't sure is how soon the time will come that people realize this is important." Late-afternoon sunlight filters through the glass of the conference room at Google Brain, where Kim has worked since 2017. "I think that time has come."

Kim's belief is that there is a dimension to explanation and interpretation that is inherently *human*—and so there is an inherent messiness, an inherent interdisciplinarity to the field. "A large portion of this—still underexplored—is thinking about the human side," she says. "I always emphasize HCI [human-computer interaction], cognitive science. . . . Without that, we can't solve this problem." Recognizing the ineluctably human aspect of interpretability means that things don't always translate neatly into the familiar language of computer science.

"Some folks think that you have to put down a mathematical definition of what explanation must be. I don't think that's a realizable direction," she says. "Something that is not quantifiable makes computer scientists uncomfortable—inherently very uncomfortable."

Almost all of Kim's research papers contain something fairly unusual in computer science, but typical outside of it: namely, actual studies using human subjects. "Iterating with the users is critical," she says. "Because if your sole reasoning for existence is for human consumption . . . we have to *show* that it is good for human consumption." This iteration is critical because often what designers *think* is useful to actual human users simply isn't. If you're designing explanations or interpretable models to be used by real people, then the process should be every bit as iterative as designing, say, cockpit controls or a software user interface.[76] Not to let that kind of empirical feedback guide the process is simply hubris.

Such studies add some complexity to the story we explored earlier, which celebrated the use of simple models. Kim notes that it's ultimately an empirical question whether this is the best approach to interpretability. "In some cases, in very limited cases where you can sufficiently identify and empirically validate that *if* I have a small number of features, that it's interpretable for this particular task—in that case, yes, you can write down what it means to be most interpretable, given a problem, and you can optimize it." But real human studies—Kim's and others'—show that things are rarely so straightforward in practice.

For instance, in 2017, Microsoft Research's Jenn Wortman Vaughan and her colleagues studied the way that human users interacted with a machine-learning model of home values, which predicted price from features like square footage, bathrooms, and so on. Users were better at anticipating the model's predictions when it used fewer factors and was made more "transparent" to the user. But neither the simplicity nor the transparency actually affected the level of *trust* that people reported toward the model. And, in fact, people were *less* likely to realize that the model had made a mistake when the model was more transparent.[77]

One of Kim's beliefs is that "humans think and communicate using concepts," not numbers.[78] We communicate—and, for the most part, think—*verbally*, leveraging high-level concepts; we don't talk about the raw minutiae of sensory experience. For this reason, Kim thinks that many saliency-based methods don't go far enough. She and her collaborators have been working on something they call "testing with concept activation vectors," or TCAV, which offers a way to use such human concepts to understand the internal workings of the network.

For instance, imagine a model correctly identifying a picture of a zebra. Suppose TCAV shows that the network used "stripes," "horse," and "savanna" in its prediction: that seems reasonable. On the other hand, a network trained somewhat naïvely on a set of images of doctors might—if the dataset were biased—assume that the concept of "male" had some predictive value. TCAV would show this, and would offer an indication that we may want to adjust the model or the training data accordingly to remove such a bias.[79]

How exactly might we get this kind of insight? Every image generates from our model not only an ultimate category label output, but a vast pattern of *internal* activity in the network, across however many tens of millions of its artificial neurons. These internal activations may seem like an overwhelming cacophony to human eyes, but that doesn't mean that you can't show them to a *machine*. The basic idea behind TCAV is that for any concept you're interested in—say, "males," in the doctor case—you present the network with a bunch of images of males, and then a bunch of images of random other things: women, animals, plants, cars, furniture, and so on. You feed the internal state of your network (say, at a particular layer) to a second system, a simple linear model that learns to tell the difference between the activations typical of your category ("males" in this case) and the activations from random images.[80] You can then look to see whether this activation pattern is present, and to what degree it contributes, when the network categorizes an image as a "doctor."

"I think this method offers a unique benefit," she says, "which is that explanation now speaks your users' language. Your users don't have to come and learn machine learning. . . . We can provide an explanation in the language that *they* speak, and answer the hypothesis that *they* have, using their own terms."[81]

When Kim used TCAV to look at two popular image recognition models, Inception v3 and GoogLeNet, she found a number of such issues. The concept of "red," for instance, was critically important to the concept of "fire truck." "So that makes sense," she says, "*if* you're from an area where fire engines are red."[82] It's almost always true in the United States, for instance, but not in Australia, where, depending on the district, fire trucks can sometimes be white—or, in Canberra, neon yellow. This would suggest, say, that a self-driving-car model developed on a

US-centric dataset might need modification before it was safe to deploy down under.

Kim also found that the concept of "arms" was important to identifying "dumbbells," corroborating the earlier visual findings of the Deep-Dream group from Google, and hinting that the network might struggle to identify a dumbbell on a rack or on the ground.[83] The concept "East Asian" was important to "ping-pong ball," and the concept "Caucasian" was important to "rugby." Some of these may reflect patterns that, while not inaccurate—according to the International Table Tennis Foundation, seven of the top ten male players, and all ten of the top ten female players in the world are from East Asian countries[84]—may nonetheless still signal aspects of the model that deserve scrutiny, from the perspective of both accuracy and bias. We wouldn't be happy, on either front, with a system that—as Facebook AI Research's Pierre Stock and Moustapha Cisse discovered—categorized a portrait of Chinese president Xi Jinping as a "ping-pong ball."[85] TCAV offers a way to explicitly quantify such issues—and ideally to nip them in the bud.

Kim's work on TCAV was featured by Google CEO Sundar Pichai during his keynote address at Google's 2019 I/O conference. "It's not enough to know if a model works," said Pichai. "We need to know *how* it works."

In 2012, it had felt to Kim like there was hardly an audience for these concerns, and that only she, along with Cynthia Rudin and Julie Shah, her mentors at MIT, were even "vaguely interested in this topic." By 2017, there were entire symposia at the field's largest conferences devoted to interpretability and explanation. By 2019, the CEO of Google was proudly describing her work on the company's biggest stage.

I ask Kim if that feels like vindication.

"We have such a long way to go," she says—seemingly unwilling to allow herself a moment's satisfaction, or a feeling of I-told-you-so. Instead, she feels mostly responsibility: to make sure that these issues are addressed faster than the already rapid progress and deployment of the technology itself. She has to somehow get out ahead of that wave, in order to make sure there aren't any high-stakes problems.

As we wrap up our conversation and start to walk out of the conference room, I thank Kim for her time and ask her if there's anything else she wants to add that we haven't already touched on. She pauses thoughtfully for about ten seconds, then suddenly lights up.

"By the way, this faculty that I was talking about?" The one who, in 2012, scoffed at her when she, as a first-year PhD student, mentioned that her PhD work was on interpretability, and tried to steer her away from what he saw as a dead end. She bears no grudge, though: "I think genuinely, he was giving me advice. . . . It's a really reasonable, sweet thing."

Kim grins. "That person is now working on interpretability."

PART II

Agency

4

REINFORCEMENT

The role of reinforcement in human affairs has received steadily increasing attention—not through any changing fashion in learning theory but as the result of the discovery of facts and practices which have increased our power to predict and control behavior and in doing so have left no doubt of their reality and importance. The scope of reinforcement is still not fully grasped, even by those who have done most to demonstrate it, and elsewhere among psychologists cultural inertia is evident. This is understandable because the change has been little short of revolutionary: scarcely anything in traditional learning theory is left in recognizable form.

—B. F. SKINNER[1]

The trouble with modern theories of behaviorism is not that they are wrong but that they could become true.

—HANNAH ARENDT[2]

Gertrude Stein, in the spring of 1896, enrolled in a psychology seminar at Harvard with the great William James. There she would study "motor automatism," or the ability to write words on paper without thinking deliberately about them. The peer-reviewed article that resulted would be her very first appearance in print; more significantly, what began as undergraduate psychology research would lead her directly to the trademark modernist "stream of consciousness" prose style for which she would be famous.[3] Stein described her classmates in James's seminar as a "funny bunch," and one in particular was a bit of a character: he was preoccupied with incubating chickens.[4]

The student was Edward Thorndike, and, to be fair, turning his apartment into a makeshift chicken coop wasn't his first idea. He *wanted* to study the mechanisms of learning in human children—and to debunk the idea of extrasensory perception while he was at it. Harvard didn't approve the project. So the incubator full of chirping chicks just sort of . . . happened as a fallback plan.[5]

For an aspiring psychologist to be tending to a flock of chicks was indeed strange at the time. Animal research had yet to come into its twentieth-century vogue, and Thorndike's own classmates thought he was a bit odd. It may have raised the eyebrows of his peers, and the hackles of his landlady, but tending his flock was—despite being a logistical nightmare—not without its perks. When traveling in Massachusetts, for instance, Thorndike was often forced to stop at the house of some family friends—the Moultons—to warm the chicks on their stove before continuing on to Cambridge. This may have been a research necessity, though it becomes clear from his letters that Thorndike was motivated in no small part by the chance to flirt with their younger daughter, Bess. She later became his wife.

Thorndike's landlady in Cambridge finally declared his incubator a fire hazard, and gave him an ultimatum: the chicks had to go. James attempted to secure him a laboratory space on the Harvard campus, but the university wasn't any more enthusiastic. As a last resort, James— over the protestations of his wife—let Thorndike move the chicks, incubator and all, into the basement of his own home. (James's children, at least, appear to have been delighted.)

In 1897, Thorndike graduates from Harvard and moves to New York City for a fellowship at Columbia, this time working with all manner of animals. The year is frequently bleak. Two of his cats run away, he can't get any dogs, and "my monkey is so wild I can't touch him." To make matters worse, not all of the animals in Thorndike's urban menagerie are there for science; a number are simply pests. In a letter to Bess on February 14, 1898, he writes, "A mouse just ran across my foot. A lot of rats are gnawing the bureau; three chicks are sleeping within a yard of me; the floor of my room is all over tobacco and cigarette ends and newspapers and books and coal and a chicken pen and the cats' milk dish and old shoes, and a kerosene can and a broom which seems rather out of place. It's a desolate hole, this flat of mine." Still, he promises he'll clean it, and—he's writing on Valentine's Day, after all—he invites her to pay him a visit.

Out of that squalor will come some of the most original and consequential research of the nineteenth century. Thorndike has built a set of contraptions he calls "puzzle boxes," full of latches and levers and buttons. He'll drop an animal inside the box, place some food outside, and observe the way that the animals—chicks, cats, and dogs—manage to find their way out. "You'd like to see the kittens open thumb latches, and buttons on doors, and pull strings, and beg for food like dogs and other such stunts," Thorndike writes to Bess, "me in the meantime eating apples and smoking cigarettes."

Thorndike is interested in learning, and a theory is taking shape in his mind. What Thorndike observes—while eating those apples and smoking those cigarettes—is that the animals, who have no idea how the box actually works, initially behave almost purely randomly: biting things, nudging things. Once one of these random actions gets them out of the box, they very quickly learn to repeat that action, and become more and more capable of quickly escaping that same box in the future. "The one impulse, out of many accidental ones, which leads to pleasure," he observes, "becomes strengthened and stamped in thereby."[6]

Thorndike sees here the makings of a bigger, more general law of nature. As he puts it, the results of our actions are either "satisfying" or "annoying." When the result of an action is "satisfying," we tend to do it more. When on the other hand the outcome is "annoying," we'll do it less. The more clear the connection between action and outcome, the stronger the resulting change. Thorndike calls this idea, perhaps the most famous and durable of his career, "the law of effect."

As he puts it:

> The Law of Effect is: When a modifiable connection between a situation and a response is made and is accompanied or followed by a satisfying state of affairs, that connection's strength is increased: When made and accompanied or followed by an

annoying state of affairs its strength is decreased. The strengthening effect of satisfyingness (or the weakening effect of annoyingness) upon a bond varies with the closeness of the connection between it and the bond.[7]

From this seemingly modest and intuitive idea will be built much of twentieth-century psychology.[8] In 1927, Ivan Pavlov's *Conditioned Reflexes* appears in English, translated by his student Gleb von Anrep, who uses the term "reinforcement" to refer to this effect. By 1933, Thorndike himself is using the term "reinforcement"—as is a young Harvard postdoc by the name of Burrhus Frederic (B. F.) Skinner (whom we will meet more properly in Chapter 5).[9] The language of "reinforcement"— along with the idea of using animals to understand the mechanisms of trial-and-error learning and, thereby, the *human* mind—sets the frame in which much of psychology will operate for decades to come.

As one of the leading psychologists in the next generation, Edward Tolman, will write: "The psychology of animal learning—not to mention that of child learning—has been and still is primarily a matter of agreeing or disagreeing with Thorndike."[10]

If Thorndike ultimately accepted his place in the canon of psychology, he was uncommonly modest about it. When Thorndike's textbook *The Elements of Psychology* came out in 1905, it threatened to unseat the previous mainstream textbook, *Psychology: Briefer Course,* written by none other than his own Harvard mentor, William James. Thorndike mailed him a check for a hundred dollars to offset the lost royalties. James replied, "Seriously, Thorndike, you're a freak of nature. When the first law of nature is to kill all one's rivals, (especially in the schoolbook line) you feed them with the proceeds!" James refused to cash the check, and sent it back.[11] The torch had, in a manner, been passed—and twentieth-century psychology was on its way.

DIGITAL TRIAL AND ERROR

> What sort of force acting through what sort of process or mechanism can
> be and do what the confirming reaction is and does? The answer which
> seems to me to fit all or nearly all the facts is . . . the force and mechanism
> of reinforcement, applied to a connection.
>
> —EDWARD THORNDIKE[12]

If the animal researchers following Thorndike were, like he was, ultimately interested in the psychology of the human child, they were not
alone; computer scientists—the very first ones—were too. Alan Turing's most famous paper, "Computing Machinery and Intelligence,"
in 1950, explicitly framed the project of artificial intelligence in these
terms. "Instead of trying to produce a programme to simulate the adult
mind," he wrote, "why not rather try to produce one which simulates
the child's? If this were then subjected to an appropriate course of education one would obtain the adult brain." Turing imagined these ersatz
child minds as what he called "unorganized machines," beginning in a
random configuration that would then be modified by the quality of the
outcomes of their (initially random) actions.

A road map to artificial intelligence, then, was already taking shape.
The "unorganized machines" would borrow directly from what was
known about the nervous system, and the "course of education" would
borrow directly from what the behaviorists were discovering about how
animals (and children) learned. Warren McCulloch and Walter Pitts had
by the early '40s shown that a large assembly of artificial "neurons," suitably wired, was capable of computing just about *anything*. Turing had
begun to sketch out ways that such a network might be trained through
trial and error. Indeed, this was precisely the process of "stamping in" that
Thorndike had described fifty years before, in his smoky apartment zoo;
Turing's description resembles, nearly verbatim, Thorndike's law of effect:

> When a configuration is reached for which the action is undeter-
> mined, a random choice for the missing data is made and the appro-

priate entry is made in the description, tentatively, and is applied. When a pain stimulus occurs all tentative entries are cancelled, and when a pleasure stimulus occurs they are all made permanent.[13]

By the end of the 1950s, IBM's Arthur Samuel, working out of the company's Poughkeepsie laboratory, had built a program to play checkers that, in a crude and early way, adjusted its own parameters on the basis of won and lost games. It wasn't long before Samuel was losing matches to his own creation. As the *New Yorker* reported in 1959, "Dr. Samuel has thus become probably the first scientist in history to concede defeat by an adversary of his own devising."[14] He published a report called "Some Studies in Machine Learning Using the Game of Checkers," embracing the term "machine learning" to describe this methodology. Samuel wrote:

> The studies reported here have been concerned with the programming of a digital computer to behave in a way which, if done by human beings or animals, would be described as involving the process of learning. . . . We have at our command computers with adequate data-handling ability and with sufficient computational speed to make use of machine-learning techniques, but our knowledge of the basic principles of these techniques is still rudimentary. Lacking such knowledge, it is necessary to specify methods of problem solution in minute and exact detail, a time-consuming and costly procedure. Programming computers to learn from experience should eventually eliminate the need for much of this detailed programming effort.[15]

In plainer English, he explained, "It's one of the most satisfying things that have ever happened to me. . . . As far as I know, nobody else has ever got a digital computer to improve of its own accord. You see, the kind of mental activity that computers can simulate has always been severely limited, because we've had to tell them exactly what to do and exactly how to do it."

Continuing to develop machines that could learn, in other words—by human instruction or their own experience—would alleviate the need for *programming*. Moreover it would enable computers to do things we didn't know *how* to program them to do.

The unveiling of Samuel's research became the stuff of computer science legend. Fellow AI pioneer John McCarthy recounts that when Samuel was getting ready to demonstrate his checkers program on national television, "Thomas J. Watson Sr., the founder and President of IBM, remarked that the demonstration would raise the price of IBM stock 15 points. It did."[16]

THE HEDONISTIC NEURON

The struggle of the nerve cells, each for its own life, parallels the struggle of our wants for satisfaction.

—EDWARD THORNDIKE[17]

In 1972, Harry Klopf—a researcher working for the United States Air Force out of Wright-Patterson Air Force Base in Dayton, Ohio—published a provocative report titled "Brain Function and Adaptive Systems: A Heterostatic Theory." Klopf argued that "the neuron is a hedonist": one that works to maximize some approximate, local notion of "pleasure" and minimize some notion of "pain." The whole complexity of human and animal behavior, Klopf believed, arises as a result of these individual cellular "hedonists" wired up into systems of increasing complexity.

The generation before Klopf, the so-called cybernetics movement of the '40s through the '60s, had framed intelligent behavior specifically in terms of what they called "negative feedback." Organisms were motivated, they argued, largely by homeostasis or equilibrium. They strove to maintain a comfortable temperature. They ate to quell their hunger. They mated to quell their desire. They slept to quell their fatigue. Everything seemed to be about getting back to baseline.

Indeed, the seminal 1943 cybernetics paper "Behavior, Purpose and

Teleology"—which, incidentally, coined the term "feedback" in its now common sense of "information used for adjustment"—set out to distinguish purposeful from purposeless (or random) behavior.[18] For the cyberneticists, purpose was tantamount to a goal that could be arrived at as a place of rest. For arch-cyberneticist Norbert Wiener, one of the canonical "intrinsically purposeful" machines is a thermostat: when the temperature is too low, it turns on the heat, and when the temperature gets high enough, it shuts it off. He thought also of the "governor" of an engine—more than coincidentally, the term is an etymological kin to the word "cybernetics" itself, coming from the same Greek root *kybernetes*.[19] (Thus "cybernetics," for all its exotic sci-fi flavor, could well have been the much blander and more bureaucratic-sounding field of "governetics.") A mechanical "governor" opens valves when the engine is running too fast, and closes them when it's running too slow, helping it to maintain equilibrium. "Notice that the feedback tends to oppose what the system is already doing, and is thus negative," Wiener writes.[20] This was an essential part of any goal-oriented system, in the cybernetic view. "All purposeful behavior," the cyberneticists wrote, "may be considered to require negative feed-back."[21]

Klopf was having none of it. For him, organisms were *maximizers*, not minimizers. Life was about growth, reproduction, endless and boundless and insatiable forward progress in any number of senses. For Klopf, the goal was not homeostasis at all, but the opposite. "Living adaptive systems seek, as their primary goal, a maximal condition (heterostasis), rather than . . . a steady-state condition (homeostasis)." More poetically, he espoused the virtues of positive, rather than negative feedback: "Both positive and negative feedback are essential to life's processes. However, it is positive feedback that is the dominant force—it provides the 'spark of life.'" This notion ran all the way through from single cells to organisms to societies. Klopf was not modest about what he saw as the implications of this idea. "It appears to be the first theory capable of providing a single unifying framework within which the neurophysiological, psychological, and sociological properties of living adaptive systems can be understood," he wrote. "Neurons, nervous systems, and nations are heterostats."[22]

Neurons were insatiable maximizers? And this explained the behavior of *nations*? It was an ambitious, unorthodox, and quite possibly absurd idea. The air force offered him funding to find or assemble a lab of researchers to look into it. He found his team at UMass Amherst. There he hired a postdoctoral researcher named Andrew Barto to determine—as Barto puts it—"the scientific significance of this idea: was it a crazy, crackpot idea, or did it have some scientific merit?"[23]

Klopf had, perhaps unusually, been keeping a running correspondence during this time with a bright undergraduate psychology major from Stanford. When Barto came onboard, Klopf encouraged him: "There's this really bright kid, you should get him into this project."[24] The bright kid was named Richard Sutton; still a teenager, he would join Barto in Amherst and become Barto's very first graduate student.

I ask Barto if there was any foreshadowing of what was to come. "We had no idea," he tells me. "We had no idea." Sutton and Barto would, with Klopf's Air Force grant, begin a forty-five-year collaboration that would essentially found a new field. The field, which would cross neuroscience, behaviorist psychology, engineering, and mathematics, was dubbed "reinforcement learning"; and their names, forever linked in bibliographies of AI—"Barto & Sutton," "Sutton & Barto"—would become synonymous with the definitive textbook of the field they coauthored and, indeed, almost a shorthand for the field of reinforcement learning itself.[25]

I meet up with Barto on the University of Massachusetts campus, where he's now emeritus. "Happily retired," he says. "Frankly, glad to be out of the maelstrom of hype and excitement about AI and reinforcement."

I tell him I'm excited to talk about the history of reinforcement learning as a field—"RL" as it is often called—and in particular its implications for safety, for decision making, for human cognition— He interrupts me.

"So, how much time have you got?"

Most of the day, I tell him. That's what we take.

THE REWARD HYPOTHESIS

Barto and Sutton took Harry Klopf's idea of organisms as maximizers and gave it a concrete, mathematical form. Imagine you are in an environment that contains rewards of some numerical value, which are available to you if you take certain actions and reach certain states. Your job is to get the most reward—the highest "score"—that you can.

The environment could be a maze, your actions could be moving north, south, east, or west, and your reward could come from reaching the exit (perhaps minus a small time penalty). Or the environment could be a chessboard, your actions the moving of the pieces, and your reward one point for checkmating your opponent (and half a point for reaching a draw). Or perhaps the environment is the stock market, your actions are buying and selling, and your reward is measured in the dollar value of your portfolio. There is virtually no limit to the complexity of these scenarios—the environment could be a national economy, the actions could be issuing words of legislation or diplomacy, and the reward could be long-term GDP growth—as long as the rewards are what is known as *scalar:* they are commensurate, fungible, of a common currency.

The reinforcement-learning framework has proven to be *so* general, in fact, that it has led to an idea known as the "reward hypothesis": "That *all* of what we mean by goals and purposes can be well thought of as the maximization of the cumulative sum of a received scalar reward."[26]

"This is almost a philosophical thing," says Richard Sutton. "But you kind of buy into this."[27]

Of course, not everyone buys into this premise so readily. In sports, in board games and video games, in finance, there may indeed be such a scalar number, a single currency (literal or figurative) for measuring all outcomes—but can that really be applied to something as complex and multiform as human, or even animal, life? We often have to make decisions whose outcomes seem like apples and oranges. Do we work late, improving our standing with our boss but testing the patience of our spouse? Do we prioritize a life of achievement, a life of adventure, a life of human connection, or a life of spiritual growth? Oxford University philosopher Ruth Chang, for instance, has spent decades arguing that nothing so characterizes the human condition as the *incommensurability* of the various motives and goals we have. We *can't* simply put major life choices on a scale and weigh out which one comes out best—or else there would be nothing to morality but pure reason, no opportunity to make meaning, to forge an identity.[28]

Sutton himself concedes that the reward hypothesis is "probably ultimately wrong, but so simple we have to disprove it before considering anything more complicated."

Even if we accept—provisionally—the framework of reward maximization, and the commensurability of scalar rewards, we find that being a heterostatic maximizer is harder than it sounds. Indeed, the problem of reinforcement learning is fraught with difficulty both philosophical and mathematical.

The first challenge is that our decisions are *connected*. Here reinforcement learning is subtly—but importantly—different from both unsupervised learning (which builds the vector word representations we explored in Chapter 1) and supervised learning (which is used for everything from the ImageNet contest to face recognition to recidivism risk assessment). In those settings, every decision stands alone. Your system is shown a picture—say, a chanterelle mushroom—and asked to categorize it. The system may get the answer right, it may get it wrong, its parameters may be fine-tuned a little bit in the event of a mistake, but in any case you simply pull another photo at random from your collection and move on. The data coming in as input are what statisticians call "i.i.d.": independent and identically distributed. There is no causal connection between what we see, what we do, and what we see next.

In reinforcement learning—in a maze, in a chess game, indeed in life—we don't have the luxury of making our decisions in a vacuum. Every decision we make sets the context in which our next decision will be made—and, in fact, it may change that context *permanently*. If we develop our chess pieces a certain way, we are strongly constraining the types of positions we will encounter and the types of strategies that may be useful going forward. In a spatial world, whether virtual or real, it is intrinsic to locomotion that the actions we take—move north in the maze, turn to face our lover, spend the winter in Florida—sculpt the future input we receive, either temporarily or forever. If we sacrifice our queen, it is unavailable to us for the rest of the game. If we jump off the roof, we may never jump again. If we are unkind to someone, we may forever alter how that person behaves toward us, and we may never know how they might have behaved had we been kinder.

The second challenge of reinforcement learning, relative to supervised and unsupervised learning, is that the rewards or punishments we get from the environment—owing to their very scalar quality—are *terse*. A language model trained to predict a missing word is told, after each guess, what the correct word was. An image classifier attempting to classify an image is immediately given the "correct" label for that image. It then updates itself explicitly in the direction of that correct answer. In contrast, a reinforcement-learning system, trying its best to maximize some quantity in some environment, eventually comes to learn what score it achieved, but it may never know, win or lose, what the "correct" or "best" actions should have been. When a rocket blows up, or a bridge collapses, or a pile of plates you're trying to carry topples, or the ball you kick fails to go into the net, the world is crystal clear about the result. But it's mum on exactly what you should have done differently.

As Andrew Barto puts it, reinforcement learning is less like learning with a *teacher* than learning with a *critic*.[29] The critic may be every bit as wise, but is far less helpful. The teacher would look over your shoulder as you work, correcting your mistakes instantly, and telling you or showing you exactly what you *should* be doing instead. The critic waits until your labor is done, then shouts "Boo!" from the back of a dark auditorium, leaving you clueless as to what exactly they didn't like, or what they'd have preferred. The critic may not even be withholding anything by way of insight or constructive feedback; they may not themselves be left with anything other than their degree of—in Thorndike's terms—"satisfaction or annoyance."

Third, not only is feedback terse and not especially constructive, it's *delayed*. We may make an unrecoverable blunder on the fifth move of a game, for instance, in which the coup de grâce comes a hundred moves later. The idiom that comes most to mind when someone experiences a failure or defeat—the disappointed parent, bankrupt business owner, apprehended thief—is "Where did I go wrong?" In reinforcement learning, this is known as the "credit-assignment problem," and had been vexing researchers since the midcentury. MIT's Marvin Minsky, for instance, wrote in his famous 1961 paper "Steps Toward Artificial Intelligence," "In playing a complex game such as chess or checkers, or in writing a computer program, one has a definite success criterion—the game is won or lost. But in the course of play, each ultimate success (or failure) is associated with a vast number of internal decisions. If the run is successful, how can we assign credit for the success among the multitude of decisions?"

Minsky elaborates, driving home the point: "Suppose that one million decisions are involved in a complex task (such as winning a chess game). Could we assign to each decision . . . one-millionth of the credit for the completed task?"[30]

After a car accident, several days into a cross-country road trip, we don't retrace our actions all the way back to turning the key in the ignition, and think, "Well, that's the last time I ever turn *left* out of the driveway!" Nor, though, do we imagine, having been checkmated on move 89 of a game of chess, our blunder *must* have happened on move 88.

So how do we determine the right lessons to take away from success and failure both? The reinforcement-learning framework began opening up a vista on the fundamental problem of learning and behavior that would develop into an entire field, and would direct the course of artificial intelligence research through our present decade. It would also, just as Harry Klopf envisioned, give us a new set of questions to ask of the natural intelligence that surrounds us.

THE RIDDLE OF DOPAMINE

If humans and other animals can be thought of as "heterostatic" reward maximizers, it stands to reason that those rewards operate by way of some mechanism in the brain. If there really is some kind of singular, scalar "reward" that humans and animals are designed to maximize, might it be as simple as a chemical or a circuit in the brain? In the 1950s, a pair of researchers at McGill University in Montreal, James Olds and Peter Milner, appeared—tantalizingly—to have found its location.

Olds and Milner were experimenting with placing electrodes in various places in the brains of rats, and giving the rats an opportunity to press a lever that would deliver electrical current through those electrodes to that particular part of the brain. Some areas of the brain, they discovered, seemed to have no effect on the rats' behavior. In other areas, the rats would appear to go out of their way to make sure the lever did *not* get pressed. But there were some areas—the so-called "septal area" in particular—where rats appeared to want almost nothing more than to press the lever that delivered current to that area. The rats pressed the lever as many as 5,000 times an hour, and for as long as 24 hours without rest.[31] "The control exercised over the animal's behavior by means of this reward is extreme," Olds and Milner wrote, "possibly exceeding that exercised by any other reward previously used in animal experimentation."[32] The foundation had been laid for a study, not only of human and animal behavior through the lens of reward maximization, but of the actual molecular mechanisms of reward itself.

At first these areas were dubbed "reinforcing structures," but soon Olds began referring to them as "pleasure centers."[33] Subsequent research showed that not just rats but humans, too, would go out of their way to receive electrical current in similar parts of their brain.

In time, studies began to establish that the areas of the brain in which this electrical stimulation was most compelling were those areas involving neurons that produced a neurotransmitter called 3,4-dihydroxyphenethylamine—better known by its abbreviated nickname: dopamine.[34] These cells were rare—much less than 1% of 1% of the brain—and they clustered in very specific areas.[35] A single one of these dopamine cells could be wired in some cases to *millions* of other neurons, across a wide area of the brain.[36] In fact, they were almost uniquely broadly connected, with the most highly connected cells each having nearly fifteen *feet* of axonal wiring within the brain.[37] At the same time, however, they were fairly limited in the range and complexity of their output. As NYU neuroscientist Paul Glimcher puts it, this means that "they cannot say much to the rest of the brain, but what they say must be widely heard."[38]

It sounded a lot, in other words, like a "reward scalar"—in essence, like points on the brain's scoreboard—dead simple, but widely broadcast and all-important. Dopamine also appeared to be intimately involved in drugs of addiction, including cocaine, heroin, and alcohol. Could it be that dopamine was literally the molecular currency of reward in the brain?

Work in the late 1970s appeared to suggest as much. A study in 1978, for instance, showed that when rats were given the dopamine-blocking drug pimozide before being taught that pressing a lever would give them food pellets, they showed no more interest in pressing the lever than rats who had never learned the food connection at all. Somehow the food reward just didn't seem to have an effect on the rats treated with pimozide. Pimozide, wrote the researchers, "appears to selectively blunt the rewarding impact of food and other hedonic stimuli."[39] As neuroscientist Roy Wise wrote in 1982, it was as if "all of life's pleasures—the pleasures of primary reinforcement and the pleasures of their associated stimuli—lose the ability to arouse the animal."[40] In fact, when given drugs known to block dopamine receptors, the rats stopped delivering electrical current to their brains as well. The drug, in the words of neuroscientist George Fouriezos, took "the jolts out of the volts."[41] It was as if everything, from food and water to sex to self-administered current, had lost its pleasurable effect.

Over the 1980s, electrophysiology began advancing to the point that individual dopamine neurons could be monitored in real time. The German neurophysiologist Wolfram Schultz, at a lab in Fribourg, Switzerland, began studying the behavior of dopamine neurons in monkeys as they reached into a box that was sometimes empty and other times contained tiny pieces of fruit or baked goods. Sure enough—*wham*—a "burst of impulses occurred when the monkey's hand touched a morsel of food inside the box."[42] This seemed to confirm that scientists had, in fact, found the chemical of reward.

But there was something strange.

In cases where there was a visual or auditory cue of some kind that indicated when food had appeared in the box, the *cue* triggered a rush of dopamine activity. Then the monkeys would reach in and grab their food, while Schultz dutifully monitored the readings—and nothing happened. Just the calm background static of their normal baseline of activity. No spike. The dopamine neurons, he wrote, "discharged a burst of impulses in response to [the cue] but entirely failed to respond to the touch of food."[43]

What in the world was going on?

"We couldn't figure out what it was," says Schultz.[44] He rifled through several hypotheses. Maybe the monkeys were simply satiated, and didn't really want any more food. He tried starving them. It didn't work. They ate the food ravenously. But there was no dopamine spike.

Schultz and his collaborators spent the late '80s and early '90s trying to find a plausible story for what they were seeing.[45] Over the course of repetition, the dopamine spike had moved from the food to the cue—but what did it *mean*? If the food had somehow stopped being "rewarding," then why were the monkeys always so quick to grab and eat it? That didn't make sense, and so dopamine couldn't be directly representing reward. They ruled out a connection to working memory. They ruled out a connection to motion, and a connection to touch.

"We could not pin it to something," Schultz tells me. "You start to pin it to incentive, motivation, as a response to a stimulus that makes you move. . . . We thought that initially, but then we found the concept far too fuzzy." His lab started to home in on the idea that it had something to do with surprise, or unpredictability. There was an idea from psychology called the Rescorla-Wagner model, which suggested that learning is critically dependent on surprise.[46] Perhaps dopamine had something to do with that link: perhaps it somehow represented either surprise itself or the learning process that surprise induced. That would explain why the food generated a dopamine spike when it was unexpected but not when it was cued—and why, in those cases, it was the unexpected *cue* that prompted the spike. "And that was okay," says Schultz. "But that would not explain some of the data we had."[47]

In particular, there was one further phenomenon Schultz observed, one that was even more mysterious than the first. He was doing a follow-up study, using a similar setup but this time with levers and juice instead of boxes and food. Once the monkey learned that the cue reliably preceded the juice, however, Schultz tried something new: he gave it a false alarm. He triggered the cue; as usual, the monkey's dopamine activity spiked above the neurons' normal baseline hum. The monkey pressed the lever. No juice came. The monkey's dopamine neurons fell—briefly, but unmistakably—*silent*. "Then I said, *Well. That is something different than surprise*."[48]

Dopamine was a riddle. It seemed so clearly, at first, to be the brain's currency of reward. It was clearly measuring *something*. If not reward, if not attention, not novelty, not surprise—then *what*?

THE POLICY AND THE VALUE FUNCTION

It's hard to be so smart that the next minute can't fool you.

—LEO STEIN[49]

Over that same period in the 1980s while Schultz was beginning his experiments on the dopamine system, on the other side of the Atlantic, Barto and Sutton were starting to make mathematical progress on the reinforcement-learning problem.

The first big step was to decompose the problem. Learning how to take actions within an environment to maximize reward, they realized, involved two related but potentially independent subproblems: action and estimation. As you began to understand and ultimately master a domain, you learned two important things: how to take the right action in a given situation, and how to estimate the likely future rewards that a state of affairs might hold. A chess player looking at the board, as Barto and Sutton came to realize, has *two* distinct hunches. They have a certain

instinct for which moves suggest themselves to be played next (and what
the opponent is likely to do in response). And they have another instinct
for which player in this position is likely to win. These two dimensions of
the reinforcement-learning problem have come to be known by the technical terms of the *policy*—what to do when—and the *value function*—
what rewards or punishments to expect.

Either of these alone was, at least in theory, enough to solve the problem. If a chess player *always* knew the right move to play, then it didn't
matter if they were lousy at predicting who would win. The reverse, however, was also true: if a chess player, say, *always* knew exactly who stood
better in a given position, then it didn't matter if they weren't sure what
to do. Given enough time, they could simply weigh the consequences of
each move, and choose the one that led to the most promising future.

Policy-based approaches led to a system—be it animal, human, or
machine—with highly trained "muscle memory." The right behavior just
flowed effortlessly. Value-based approaches, by contrast, led to a system with a highly trained "spider-sense." It could tell right away if a situation was threatening or promising. Either, alone, if fully developed,
was enough.

In practice, however, policy-based approaches and value-based
approaches went hand in hand. Barto and Sutton began to elaborate an
idea known as the "actor-critic" architecture, where the "actor" half of
the system would learn to take good actions, and the "critic" half would
learn to predict future rewards.[50] In rough terms, the actor-critic architecture also neatly described their collaboration. Barto was first and
foremost interested in behavior. Sutton was first and foremost interested
in prediction.[51]

Going back all the way to his undergraduate thesis in 1978, Sutton had
been preoccupied with creating, as he put it, "A Unified Theory of Expectation."[52] This fascination with how organisms form and refine their
expectations followed him to Amherst in his work with Klopf and Barto.

As Sutton reasoned, developing good expectations—a good value
function—meant reconciling your moment-to-moment expectations
with the ultimate verdict that came from reality: the final score of the

game, the end-of-quarter report, the successful moon landing, the late-career encomia from admiring colleagues, the standing ovation, the smiling grandchildren. But if you *actually* had to wait until the end of a game to learn from it, then the credit-assignment problem would indeed be virtually impossible. The logic, he says, is threefold.

First, it may be impractical or impossible to *remember* everything we were thinking and doing. At the end of a ninety-minute soccer game in which we lost soundly, do we really gather in the locker room and work backward through every single play? Every shift in our hopes and fears?

Second, we want to be able to learn even without a final verdict. A chess game interrupted before its conclusion should still offer us *something*. If we were about to be checkmated or the position felt hopeless, then that still meant we needed to do something different—regardless of whether the loss or the punishment we felt we had coming ever actually arrived. Likewise, a game decided by a freak, improbable, unforeseeable event in the final seconds doesn't necessarily mean that our earlier expectations were necessarily wrong. Maybe we really *were* on track to victory. Judging entirely by the final outcome doesn't necessarily make sense, especially in situations affected to some degree by chance.

Third, we ideally want to be learning not just after the fact but *as we go along*. This is of particular and critical importance for human lives. Many of the most crucial junctures and most important goals of life—getting into college, say, or raising children, or retiring comfortably—we typically only get one shot at. And many mistakes—a falling grade point average, an expanding waistline, a deteriorating relationship—are such that we can realize things are heading off course and adjust long before it's too late. If it is trial and error by which we learn, it mercifully does not require the *whole* trial—nor the whole error—for us to do so.

Each of these factors, as Sutton explains, makes a theory of expectations more tricky. "So obviously we can try to ignore all these things, think of them as nuisances, but I think of them as clues," he says. "These are hints from nature, about how we should proceed."[53]

It was Sutton himself, as it turned out, who would be first to take those hints.

The breakthrough that Sutton made in thinking about predictions is this. As you move forward toward an uncertain future, you maintain a kind of "running expectation" of how promising things seem. In a chess game, it might be the odds you give yourself to win the game. In a video game, it might be how much progress you expect to make or how many points you expect to rack up in total. These guesses fluctuate over time, and in general they get more accurate the closer you are to whatever it is you're trying to predict. (The weekend weather forecast is almost always more accurate on Thursday than it was on Monday. Your ETA as you drive home from the airport will tend to be more accurate the closer you are to home.) This means that, in general, as our expectation fluctuates, we get differences between our successive expectations, each of which is a learning opportunity; Sutton called these *temporal differences*, or *TD errors*. When one of these temporal differences happens, the later of the two estimates is the one more likely to be correct.

And so maybe we don't *need* to wait until we get the eventual ground truth in order to learn something. Maybe we can learn from these fluctuations themselves. *Any* time our expectation changes can be treated as an error in our previous estimate, and, hence, an opportunity to learn: not from the ultimate truth, which has yet to arrive, but from the new estimate, made by our *very* slightly older and wiser self. As Sutton puts it: "We are learning a guess from a guess."[54]

(He adds, as if parenthetically, "Sounds a bit dangerous, doesn't it?")

By the end of the 1980s, this idea—Sutton called it "the method of temporal differences," or "TD learning," for short—had sharpened into an algorithm, which he called "TD(λ)" ("TD-lambda"), for precisely tuning predictions in the light of subsequent predictions.[55] Inspired by Sutton and Barto's ideas, Cambridge PhD student Chris Watkins had devised a TD algorithm called "Q-learning," which would turn these predictions into actions.[56] Promisingly, he showed that Q-learning would always "converge," namely, as long as the system had the opportunity to try every action, from every state, as many times as necessary, it would *always, eventually* develop the perfect value function: the perfect set of expectations for that environment, be it a maze, a chessboard, or something truer to life. It was a significant theoretical milestone for the field—of course, exactly *how* significant depended on whether you were the kind of person who emphasized the *always* or the *eventually*.

The theory looked good, but TD learning had yet to be truly tested in practice. That first real kicking of the tires would happen at IBM Research in New York. A young researcher named Gerald Tesauro— initially interested in computational models of classical conditioning— had been working, over the late '80s and early '90s, on using neural networks to play backgammon. The early results were promising. Then, in 1992, he plugged TD learning into his model, and it took off like a rocket.[57] It was learning guesses from guesses, steadily coming to learn what an advantageous position looked like. "This is apparently the first application of this algorithm to a complex non-trivial task," Tesauro wrote. "It is found that, with zero knowledge built in, the network is able to learn from scratch to play the entire game at a fairly strong inter- mediate level of performance, which is clearly better than conventional commercial programs, and which in fact surpasses comparable net- works trained on a massive human expert dataset. This indicates that TD learning may work better in practice than one would expect based on current theory." By 1994, his program—dubbed "TD-Gammon"— had reached, as he wrote, "a truly staggering level of performance: the latest version of TD-Gammon is now . . . extremely close to the world's best human players."[58]

The field had long known that it was possible for a programmer to write a system that played better than him- or herself; Arthur Samuel had discovered as much with checkers in the 1950s. But this was some- thing new: a program as good as *any* human player. And it was entirely "self-taught": initialized from a random state, and tuned over countless games against itself. It was an enormous validation for TD learning. And it would offer a direct blueprint for success that would be used—with surprisingly few changes—by twenty-first-century game-mastering software like AlphaZero.

Sutton and others in the burgeoning field of reinforcement learning had taken nature's hints. But now it was their turn to offer something back. In one of the most stunning moments in the field's history, the mathematical framework that had grown initially out of psychology and neuroscience was about to reenter those fields in a big way.

A NEURAL SUBSTRATE OF PREDICTION AND REWARD

In the early 1990s, as Tesauro was fine-tuning his backgammon system, a young cognitive scientist named Peter Dayan, who had spent fertile time in Barto's lab and had worked with Watkins on his convergence proof, found himself moving from one side of the field—and the world— to the other. From studying the mathematics of temporal difference at the University of Edinburgh, he landed with a group of neuroscientists at the Salk Institute in San Diego.

Dayan and a fellow postdoc named Read Montague, working with Salk Institute neuroscientist Terry Sejnowski, had a hunch that not only did the framework of reinforcement learning explain how real human and animal brains might operate—but that it might *literally* be what brains were doing. "We went after the role of a set of systems in your brain that report on value and reinforcement," says Montague. "And we, at the time, just sort of thought, 'These things *should* be implementing a certain kind of algorithm.'"[59] Montague thought that these sorts of learning mechanisms might be one of the bedrock properties of almost every animal that learns. Dayan was coming from his work on temporal-difference learning, which seemed like it might be a candidate for one of these universal algorithms. It worked on the whiteboard—that is, it converged. It worked in machines— TD-Gammon was a stronger player than all but the very best human players alive. Perhaps it worked in brains, too. The two of them began speculating about how a neurological system for learning from temporal differences might work.

"Obviously we knew there must be a link to neuroscience," Dayan explains to me over a conference table at Uber headquarters in San

Francisco, where he's spending his sabbatical year from University College London.

"And then," he says, "we came across these data from Wolfram Schultz."[60]

Suddenly Dayan hops up and begins scrawling dopamine-response diagrams on the whiteboard excitedly. He points to the flat, background static, ho-hum response when the monkey gets the juice reward it fully expects after seeing the cue light. "So *this* part of the signal is very transparent. And that's what you see in the Rescorla-Wagner rule; that's the beautiful aspect of what comes from older psychology."

Then he points to the initial dopamine spike caused by the cue light itself. "But it's *this* part of the signal which is confusing," he says. "Because that is not expected from the psychological perspective. So at various times you can see Wolfram sort of struggling with what that means."

When Dayan and Montague looked at the data from Schultz—the data that Schultz, along with the entire neurophysiology community, found so inscrutable—they knew exactly what it meant.

It was a temporal difference. It was a sudden fluctuation in the monkey's expectation—in its value function, its prediction of how good a state of affairs it was in.

A sudden spike above the brain's dopamine background chatter meant that suddenly the world seemed more *promising* than it had a moment ago. A sudden hush, on the other hand, meant that suddenly things seemed *less* promising than imagined. The normal background static meant that things, however good or bad they were, were as good or bad as *expected*.

A spike in the dopamine system was not reward as such, but it was related to reward; it wasn't uncertainty, or surprise, or attention per se—but it was intimately, and for the first time *legibly,* related to all of them. It was a fluctuation in the monkey's expectation, indicating that an earlier prediction had been in error; it was its brain learning a guess from a guess.

The algorithm that worked so well on paper, and so well in silicon, had just been found in the brain. Temporal-difference learning didn't just *resemble* the function of dopamine. It *was* the function of dopamine.

Schultz, Dayan, and Montague published an explosive paper together in *Science* in 1997, announcing their discovery to the world. They had found, as they put it, "A Neural Substrate of Prediction and Reward."[61]

The effect, across neuroscience and computer science alike, was huge. An idea developed in a pure machine-learning context, inspired by models of classical and operant conditioning from psychology, had suddenly come full circle. It wasn't just a model of how *artificial* intelligence might be structured. It appeared to offer a description of one of the universal principles for intelligence, period.

"It's thought that eyes evolved anywhere from sort of forty to fifty separate times. . . . Biology has discovered eyes over and over again—all kinds of different eyes," Montague explains.[62] "I think it's the same thing in the learning domain. Learning algorithms are so profoundly important to understand how to take experience now, reorganize your internal structure, and produce actions in the future, that . . . you should expect that these algorithms have been stumbled into by biology in many different contexts. And so you see reinforcement learning systems, reward systems, in bees, and sea slugs, and birds learning songs, and human beings, and rodents, etc."[63]

Dayan sees it the same way. "It's no surprise that we come endowed with these sorts of things," he says. "The idea that you would see it quite so transparently in the activity of these dopamine neurons. . . . That was quite a revelation."[64]

Visiting Wolfram Schultz's lab in Cambridge—at age 75, he is still as animated and energized by his work as ever—I ask him if this felt like a revelation at the time. Surprisingly, he says no: the revelation, in his view, is that the humble TD-learning model, which described the activity in the brains of his monkeys as they fumbled for food, is, twenty years later, enabling the current contemporary breakthroughs of AI. That we've been able to take these universal ideas from nature and implement them synthetically is what blows his mind.

"The revelation came," he says, "when I realized that the TD model goes into what we now see, the Go programming, and this artificial intelligence, machine-learning thing. That was a revelation, where I said, *My God, what have I done?* You know, understanding my data came from the Rescorla-Wagner model, from a prediction error, but the consequence, like, Jesus Christ! I mean, we knew Tesauro was programming backgammon with a TD model. Okay, I'm not a backgammon player, but I'm a Go player. Not a very good one, but . . . I said, dammit, if they can program Go, which has been a problem before, it's a *really* good model. And I saw that before they programmed Go."

I tell him I think it's remarkable to see such a clear and powerful synthesis of work across neurophysiology and machine learning, with such powerful implications for both fields.

"Absolutely," he says. "Absolutely. That's the charm of it. You get everything together. Just makes sense."[65]

The effect on neuroscience has been transformative.[66] As Princeton's Yael Niv puts it, "The potential advantages of understanding learning and action selection at the level of dopamine-dependent function of the basal ganglia can not be exaggerated: dopamine is implicated in a huge variety of disorders ranging from Parkinson's disease, through schizophrenia, major depression, attentional deficit hyperactive disorder etc., and ending in decision-making aberrations such as substance abuse and addiction."[67]

It's true that there is much to be worked out, and what now seems like the canonical story may be overturned or complicated in time.[68] But it is clear, says Niv, that "reinforcement learning has already left its permanent mark on the study of decision making in the brain."[69]

She recounts her first time attending the annual Society for Neuroscience meeting, a gathering of some 30,000 researchers across the field. "I remember looking up reinforcement learning the first time I went, which was probably 2003, 2004. There were maybe five posters in the whole meeting. It's a whole week of meetings; every day has two poster sessions, so one in the morning, one in the afternoon. Now," she says, "*every* poster session has a whole subsection, a whole *row*, about this. It's really come a long way in ten, fifteen years."[70]

She adds, "There were lots of studies kind of testing predictions of the theory in the brain and seeing, 'It seems like, gosh, these neurons read the textbook. They read the Sutton and Barto textbook. They know exactly what they need to do.'"

HAPPINESS AND ERROR

If the TD-learning model has revealed the function of dopamine in the brain—not as the brain's currency of *reward,* not its *expectation* of future rewards, but its *error* in its expectation of future rewards—then it leaves some questions open.

In particular, it leaves open the connection to the subjective experience of pleasure and happiness. If elevated levels of dopamine signal something to the effect of *things are going to be better than I thought they were going to be,* then that feeling is, itself, pleasurable. And you can see how humans and animals alike would go out of their way to get that feeling, including by way of direct chemical and electrical stimulation of dopamine neurons.

You can also begin to see how artificially elevating dopamine levels produces an inevitable crash. Thinking that things *will* be better than you thought they would be only works for so long. Eventually you realize they're *not* better than you thought they would be, and the dopamine chatter hushes—as it did in Wolfram Schultz's experiments with monkeys, when the juice-cue light blinked but no juice was present. It's possible, in effect, for dopamine to write checks that the environment can't cash. Those checks eventually, inevitably, bounce. Then your value function must, inevitably, come back down to earth.

This, of course, is the classic experience of dopamine-related drugs—cocaine being a prototypical example. The drug works in large part by inhibiting the brain's reuptake of dopamine, leading to a temporary "flood" of it. The TD story suggests that the brain interprets this as a pervasive sense that things are *going* to be great—but the dopamine is writing checks that the environmental rewards can't cash. Eventually the predicted greatness doesn't come, and the equal and opposite *neg-*

ative prediction error is sure to follow. "It *seemed* like everything was going to be so great . . ." We can chemically fool our brain's prediction mechanism—but not forever.

As the writer David Lenson puts it, "Cocaine promises the greatest pleasure ever known in just a minute more, *if* the right image is presented to the eyes, *if* another dose is administered, *if* a sexual interaction is orchestrated in just the right way. But that future never comes. There is a physical pleasure to the drug, to be sure, but it is incidental, trivial, compared to what is always just about to happen."[71] Understanding cocaine as a drug of dopamine, and dopamine as a chemical of temporal difference—of fluctuations in our expectation—makes the story clear. By artificially dumping the brain's supply, one experiences not the bliss that things *are* great but the giddy euphoria that things are *surprisingly promising.* If that promise isn't kept, the temporal-difference error swings the other way, and our dopamine system goes silent. It was our *high* expectations that were in error. We were duped.

The connection between dopamine and the subjective experiences of happiness and pleasure is still being worked out. The University of Michigan's Kent Berridge, for instance, has spent the better part of his career teasing apart the neuroscience of *wanting* from *liking.*[72] Meanwhile, University College London's Robb Rutledge has developed a mathematical model of happiness that explicitly involves dopamine.

Rutledge, working with Peter Dayan and a group of collaborators at UCL, devised an experiment in which people would be making various bets and accumulating a sum of money while periodically being asked, "How happy are you right now?"[73] They modeled the task using the mathematical tools of reinforcement learning, in order to tease apart how much money the subject had made so far, how much in total they anticipated making, and whether they had recently been pleasantly or unpleasantly surprised in adjusting either of these running counts up or down. The goal was to figure out the best mathematical correlate for their self-reported momentary happiness.

Rutledge's findings were illuminating on a number of counts. The first was that happiness was *fleeting.* However happy it made you to receive

£1 in a particular bet, five bets later, 92% of that impact had disappeared. Anything that happened more than ten bets ago may as well never have happened. This meant that subjects' happiness had virtually nothing to do with how much money they had actually made.

Happiness did seem at least partly determined by how much subjects *expected* to make—but what seemed most crucial, however, was the *violation* of those expectations. As Rutledge writes, "Momentary happiness is a state that reflects not how well things are going but instead whether things are going better than expected."[74] This sounds exactly like a temporal-difference error—in other words, exactly like the role played by dopamine.

Zooming out, we begin to get a neuroscientific and computational account of the well-known phenomena of what's called the "hedonic treadmill." Namely, people have a stubborn and persistent return to their emotional baseline, regardless of changes in their long-term quality of life.[75] Lottery winners and paraplegics, famously, are emotionally back to more or less where they started not long after their respective dramatic life changes.[76] Dopamine and reinforcement learning offer clues as to why. If happiness comes not from things *having* gone well, not from things being *about* to go well, but from things going *better than expected*, then yes, for better or worse, as long as our expectations keep tuning themselves to reality, then a long-term state of being pleasantly surprised should be simply unsustainable.

Unfortunately, the study rules out such simple life hacks as "always maintain low expectations." As Rutledge says, "Lower expectations make it more likely that an outcome will exceed those expectations and have a positive impact on happiness. However, expectations also affect happiness even before we learn the outcome of a decision. If you have plans to meet a friend at your favorite restaurant, those positive expectations may increase your happiness as soon as you make the plan."[77]

In the research paper, he and his coauthors note the possibility of an airline, for instance, claiming that there is a 50% chance of a six-hour delay, then announcing that the delay will in fact be one hour. "Lowering expectations increases the probability of positive outcomes. . . . However, lower expectations reduce well-being before an outcome arrives, limiting the beneficial scope of this manipulation."[78]

In other words, Rutledge says, "You can't kind of, you know, just lower your expectations and that solves everything."[79]

The links between dopamine, TD errors, and happiness have caused some researchers to begin pondering whether there are ethical implications for the subjective happiness of reinforcement-learning agents. Brian Tomasik of the Foundational Research Institute, which aims to understand and reduce suffering, has pondered at some length the question of whether reinforcement-learning programs have moral standing—does it matter how we treat them? His answer is a tentative and limited yes: to the degree that they are built on similar principles as animal and human brains, there may well be some nonzero ethical consideration that they are owed.[80] "Present-day RL algorithms matter vastly less than animals," he notes. "Still, I think RL agents do matter to a nonzero degree, and at large scales they may begin to add up to something significant."[81] Subsequent work by others has gone so far as to explicitly define the "happiness" of a reinforcement-learning agent by its TD errors.[82] By this logic, they note, "agents that know the world perfectly have zero expected happiness." Indeed, if dopaminergic "happiness" comes in large part from being pleasantly surprised, from having opportunities to better learn what to expect, then complete mastery of any domain seems necessarily correlated with boredom—a point that has ethical implications not only for a future ethics of RL agents but, of course, for humans, too.

It suggests, for one thing, an evolutionary and computational story about the hedonic treadmill. If our subjective happiness is intimately tied to being pleasantly *surprised* and it's the nature of intelligence to work tirelessly to anticipate and mitigate surprise, then we can see how happiness of this kind becomes fleeting. We can also see the evolutionary advantage. An infant may be delighted at the mere ability to wave their arm on command. For an adult, this ability no longer carries, shall we say, the thrill it did in infancy. Though we may bemoan the restlessness this engenders in adults, the insatiability, this is all part of the curriculum of life. If basic motor skills were enough to thrill us indefinitely, we'd never make it to adulthood at all.

As Andrew Barto puts it, this fleetingness was something Klopf anticipated back in the early 1970s. "His point was a [homeostatic] stabilization mechanism, it tries to reduce the discrepancy to zero, and when it's zero, it's happy, it stops. The kind of system he wanted was never happy. So it's this incessant exploration."[83] The connection between reinforcement learning, dopamine, happiness, and exploration (as well as addiction) is something we'll return to in Chapter 6.

BEYOND REINFORCEMENT

Reinforcement learning, rooted very consciously in the animal-learning studies of the early twentieth century, flourished in the abstract, mathematical world of machine learning in the 1970s and '80s, only to triumphantly burst back into animal behavior literature with a nearly perfect model that has become the accepted story for the role of dopamine in the brain.[84] This model, in turn, has given us deeper insight into human motivation and human happiness.

Meanwhile, neuroscientific evidence, as recently as 2018, is beginning to suggest that the crazy hypothesis of Harry Klopf—that neurons are "hedonists," motivated by their own individual Thorndikean laws of effect—may not be so far off after all.[85] "I think that neuroscience is getting close to something that is very similar to what Klopf was proposing," says Barto, with some pride on behalf of his late former mentor, to whom his and Sutton's textbook is dedicated.[86]

Reinforcement learning also offers us a powerful, and perhaps even universal, definition of what intelligence *is*.[87] If intelligence is, as computer scientist John McCarthy famously said, "the computational part of the ability to achieve goals in the world,"[88] then reinforcement learning offers a strikingly general toolbox for doing so. Indeed it is likely that its core principles were stumbled into by evolution time and again—and it is likely that they will form the bedrock of whatever artificial intelligences the twenty-first century has in store.

In some ways, though, a deeper understanding of the *ability* of animals and machines to achieve goals in the world has kicked the more profound philosophical can down the road. This theory, pointedly, does not tell us *what* we value, or what we *ought* to value. Dopamine, in this way, is every bit as mysterious now as when its role was lesser known. If it represents a scalar prediction error, it hides a vast realm of complexity in how that prediction is "measured." If, behind door number one, is not the Caribbean vacation we expected but, rather, a trip to view the aurora borealis, our dopamine will quickly and reliably indicate whether we are pleasantly or unpleasantly surprised by this. But how is the *value* of those alternatives actually being assessed?[89] Dopamine is mum on this point.

Meanwhile, we take up another question. Reinforcement learning in its classical form takes for granted the structure of the rewards in the world and asks the question of how to arrive at the behavior—the "policy"—that maximally reaps them. But in many ways this obscures the more interesting—and more dire—matter that faces us at the brink of AI. We find ourselves rather more interested in the exact *opposite* of this question: Given the behavior we want from our machines, how do we structure the environment's rewards to bring that behavior about? How do we get what we want when it is *we* who sit in the back of the audience, in the critic's chair—*we* who administer the food pellets, or their digital equivalent?

This is the alignment problem, in the context of a reinforcement learner. Though the question has taken on a new urgency in the last five to ten years, as we shall see it is every bit as deeply rooted in the past as reinforcement learning itself.

5

SHAPING

Nature has placed mankind under the governance of two sovereign masters, *pain* and *pleasure*. It is for them alone to point out what we ought to do, as well as to determine what we shall do.

—JEREMY BENTHAM[1]

Design of reward functions is not often discussed, although it is perhaps the most difficult aspect of setting up an RL system.

—MAJA MATARIĆ[2]

In 1943, B. F. Skinner was working on a secret wartime project, initially sponsored by—of all groups—the consumer food company General Mills. General Mills had given Skinner the top floor of its Gold Medal Flour mill in Minneapolis to build a laboratory. The project was one of the more audacious ones entertained during that time: Skinner and his lab were going to teach pigeons how to peck at images of bomb targets, then put the birds—in groups of three—inside of actual bombs to steer them as they dropped. "My colleagues and I knew," said Skinner, "that, in the eyes of the world, we were crazy."[3]

Skinner was aware that many would see the project as not only insane but cruel. To the first point, he noted that humans have a long and storied history of putting animals' (often superhuman) senses to human purposes: the seeing-eye dog, the truffle-hunting pig, and so forth. To the second point, he argued, "The ethical question of our right to convert a lower creature into an unwitting hero is a peacetime luxury."[4]

Skinner had long been working on the study of reinforcement, and his famous "Skinner boxes" functioned like upgraded, mid-twentieth-century versions of Thorndike's puzzle boxes. Their lights and levers and mechanical food dispensers, often repurposed from vending machines, allowed for a precise and quantitative study of reinforcement, and they would be used by generations of researchers to follow (not least including Wolfram Schultz in his study of dopamine in monkeys). With such tools at his disposal in the 1950s, Skinner began to study how animals learn to take actions to maximize their rewards (typically in the form of food) under a number of different conditions. He tested different types of what he called "reinforcement schedules" and observed their effects. For instance, he compared reinforcing by "ratio"—where a certain *number* of correct actions would result in a reward—against reinforcing by "interval"—where a correct action after a certain *time* would yield the reward. He also tested "fixed" against "variable" reinforcement, where the number of behaviors or length of time was either held constant or allowed to fluctuate. Skinner famously found that the fiercest and most repetitive or persistent behavior tended to come from *variable ratio* schedules—that is, when a reward would come after a certain number of behaviors, but the number would fluctuate.[5] These findings held implications for the understanding of gambling addiction, for instance—and, tragically, they also undoubtedly led to the devising of ever more addictive gambling games.

In his secret top-floor laboratory, though, Skinner had a different challenge before him: to figure out not which schedules of reinforcement ingrained simple behaviors most deeply, but rather how to engender fairly complex behavior merely by administering rewards. The difficulty became obvious when he and his colleagues one day tried to teach a pigeon how to bowl. They set up a miniature bowling alley, complete with wooden ball and toy pins, and intended to give the pigeon its first food reward as soon as it made a swipe at the ball. Unfortunately, nothing happened. The pigeon did no such thing. The experimenters waited and waited . . . and eventually ran out of patience.

Then they took a different tack. As Skinner recounts:

> We decided to reinforce any response which had the slightest resemblance to a swipe—perhaps, at first, merely the behavior of looking at the ball—and then to select responses which more

closely approximated the final form. The result amazed us. In a few minutes, the ball was caroming off the walls of the box as if the pigeon had been a champion squash player.

The result was so startling and striking that two of Skinner's researchers—the wife-and-husband team of Marian and Keller Breland—decided to give up their careers in academic psychology to start an animal-training company. "We wanted to try to make our living," said Marian, "using Skinner's principles of the control of behavior."[6] (Their friend Paul Meehl, whom we met briefly in Chapter 3, bet them $10 they would fail. He lost that bet, and they proudly framed his check.)[7] Their company—Animal Behavior Enterprises—would become the largest company of its kind in the world, training all manner of animals to perform on television and film, in commercials, and at theme parks like SeaWorld. More than a living: they made an empire.[8]

Skinner, too, would come to think of this moment—at the miniature bowling alley inside the secret flour-mill laboratory—as an epiphany that changed the arc of his career. The critical component, he saw, was "the gradual *shaping up* of behavior by reinforcing crude approximations of the final topography instead of waiting for the complete response."[9]

Project Pigeon—as it was known—was, however, ultimately something of a mixed success. The pigeons themselves worked marvelously. *So* marvelously, in fact, that they seemed to distract the committee from the government's Office of Scientific Research and Development. "The spectacle of a living pigeon carrying out its assignment, no matter how beautifully," Skinner wrote, "simply reminded the committee of how utterly fantastic our proposal was."[10] And, unbeknownst to Skinner at the time, the government was hard at work on the Manhattan Project: a bomb with a blast radius so large that, in his words, "it looked for a while as if the need for accurate bombing had been eliminated for all time." Nevertheless, the pigeon project ultimately found a home at the Naval Research Laboratory, as an effort called ORCON—short for "organic control"—where studies continued into the postwar '50s.

Skinner felt vindicated that the concept had been shown to work, writing proudly in the late 1950s that "the use of living organisms in guiding missiles is, it seems fair to say, no longer a crackpot idea."[11] This vindication, though nice, was however beside the point.

The point was they had discovered *shaping*: a technique for instilling complex behaviors through simple rewards, namely by rewarding a series of successive *approximations* of that behavior. "This makes it possible," Skinner wrote, "to shape an animal's behavior almost as a sculptor shapes a lump of clay."[12] This idea, and this term, would become a critical one through the rest of Skinner's life and career.[13] It had implications— he saw from the beginning—for business and for domestic life.

As he wrote: "Some of these [reinforcement schedules] correspond to the contingencies established in industry in daily wages or in piece-work pay; others resemble the subtle but powerful contingencies of gambling devices, which are notorious for their ability to command sustained behavior."[14] He also felt the possible *parenting* ramifications were significant: "A scientific analysis can, however, bring about a better understanding of personal relations. We are almost always reinforcing the behavior of others, whether we mean to or not." Skinner noted that parental attention is a powerful reinforcer, and that parents can, by being slow to respond to polite requests, unwittingly train their children to be annoying and pushy. (The remedy, he said, is to respond more promptly and consistently to acceptable bids for attention, and less to those that are loud or impolite.)[15]

Perhaps most prophetically, Skinner thought that with the principles emerging in his work, education in its broadest sense—of humans, of animals—might become a rigorous, objective field in which leaps forward could and would be made. "Teaching, it is often said, is an art," as he put it, "but we have increasing reason to hope that it may eventually become a science."[16]

Skinner would be, perhaps, more right than he even anticipated. In the twenty-first century, when the term "shaping" is used, it is just as likely to be a machine-learning researcher who is talking as a psychologist. The study of rewards—and, in particular, how to strategically *administer* rewards to get the behavior you want, and not the behavior you don't—has indeed become a rigorous quantitative science, albeit perhaps not with the organic learners Skinner imagined.

THE PROBLEM OF SPARSITY

There's a way to do it better . . . find it!

—THOMAS EDISON[17]

It was Scottish philosopher Alexander Bain who in 1855 appears to have coined the phrase "trial and error" to describe how humans and animals learn.[18] (His other phrase—"groping experiment"—is equally memorable, but appears not to have stuck.)

At its most basic, reinforcement learning is a study of learning by trial and error, and the simplest algorithmic form this trial (or groping, if you prefer) takes is what's called "epsilon-greedy" (written with the Greek letter ε as "ε-greedy"). The Greek letter ε is often used by mathematicians to mean "a little bit," and ε-greedy is shorthand for "be greedy, except for a smidgen of the time." An agent operating by ε-greedy will, most of the time—99%, let's say—take the action it believes will bring it the greatest total reward, based on its limited experience so far. But ε of the time—the other 1%, for instance—it tries something completely at random. In an Atari game, say, this would mean randomly mashing a button some percentage of the time, just to see what might happen.

There are many different flavors of how to learn from this exploratory behavior, but the underlying idea is the same. Flail around; do more of what gets you rewarded and less of what gets you punished. You can do that by explicitly trying to understand how the world works ("model-based" RL), or just by honing your instincts ("model-free" RL). You can do it by learning *how much* reward certain states or actions can bring ("value" learning), or by simply knowing which strategies tend on the whole to do better than which others ("policy" learning). Virtually all approaches, though, are built around the idea of first stumbling accidentally onto success, and then building up a tendency to do more and more of whatever appears to have worked.

Some tasks, as it turns out, admit much more readily to this approach than others.

In a game like *Space Invaders*, for instance, hordes of enemies descend toward you, and all you can do is move left, move right, and shoot. Randomly mashing the buttons will probably earn you at least a few kills, each of which is worth points, and those early points can be used to start the learning process by which certain patterns of behavior get strengthened and better strategies developed. You may realize, for instance, that points only come after you've fired a shot, and so you'll begin firing more often and, in turn, scoring more. Such games are said to have "dense" rewards, which makes them relatively easy to learn.

In other games—take chess, for instance—rewards aren't nearly as immediate, but they are nonetheless certain. A chess game is typically over, one way or another, after a few dozen moves, and the rules make it almost impossible to go more than a couple hundred. Even if you're clueless about the finer points of strategy and are shoving pieces around the board at random, you'll at least know before long whether you won, lost, or drew.

There are a number of situations, however, in which getting *any* reward at all would take a veritable miracle. Skinner, of course, discovered this firsthand in his attempts to reward a bird for bowling a tiny ball down a miniature bowling alley. The bird, clueless about what game it had been put into, might take *years* to happen upon the right behavior— of course it (and Skinner) would have died of hunger long before then.

The same thing is true for mechanical learners. Getting a humanoid robot to kick a soccer ball into a net, for example, might involve hundreds of thousands of precise torques on dozens of joints, all perfectly coordinated. It's hard to imagine a robot that initially moves its dozens of joints at *random* managing to even stay upright, let alone make meaningful contact with the ball, let alone send it into the net.

Reinforcement-learning researchers know this as the problem of *sparse rewards,* or more succinctly as the problem of *sparsity.* If the reward is defined explicitly in terms of the end goal, or something fairly close to it, then one must essentially *wait* until random button-pressing, or random flailing around, produces the desired effect. The mathematics show that most reinforcement-learning algorithms *will*, eventually, get there—but eventually, in practical terms, might arrive long after the sun has exploded. If you're trying to train a Go-playing program to defeat the world champion and you give it one point every time the world champion resigns and zero points otherwise, you are going to be waiting a very long time indeed.

The problem of sparsity also has safety implications. If, in some future time, you're developing a superintelligent AI with vast capabilities, driven by ε-greedy reinforcement learning, and you decide you'll reward it with one point if it cures cancer and zero otherwise, watch out—because it will have to try a *lot* of random things before it stumbles into that first reward. Many of them are likely to be ugly.

When I sit down with Brown University's Michael Littman, who has spent his career working in reinforcement learning, I ask him whether his lifelong research interest in reinforcement has been useful as a parent. His mind instantly goes to the problem of sparsity. He remembers joking with his wife about using sparse rewards with their son: "How about this? Until he can learn to speak Chinese, let's not feed him. That would be a good motivator! Let's see how that works!" Littman laughs. "My spouse is very grounded. . . . She's like, 'No, we're not gonna play that game.'"[19]

Of course, Littman knows—as Skinner did—not to play that game. Indeed, the problem of sparsity has inspired the reinforcement-learning community to hark back to Skinner's time, and they have drawn rather directly on his advice.[20] In particular, his ideas about shaping have led to two distinct but interwoven strands of thought: one about *curriculum*, the other about *incentives*.

THE IMPORTANCE OF CURRICULUM

The key insight of shaping—that in order to get complex behavior, we may first need to strategically reward *simpler* behavior—is every bit as applicable to humans, of course, as to animals. "You have to walk before you can run," we say, an adage that describes more facets of the human experience than can be named or counted—in addition to being literally true.

It is a striking and characteristic feature of human life that we spend our first decades moving through the world with literal and figurative training wheels, bowling—as it were—with literal and figurative bumpers. Many animals are simply thrown into the full complexity of life: a number of wild animals must be prepared to run full speed from predators, for instance, within hours of birth. In contrast, it takes us decades before we can operate heavy machinery, and it's not uncommon for us to be *past our physical prime* by the time we have to "fend for ourselves."

What separates twenty-first-century humans from their cave-dwelling ancestors isn't raw intellectual power but a good *curriculum*. Indeed, Skinner thought that we shouldn't be so quick to pass judgment on the mental capacities of animals. With the right curriculum, they might be able to astoundingly surpass what we think of their species as capable of—just as humans have.

As Skinner put it, if an experimenter is simply waiting until a complicated set of behaviors arises to begin reinforcing the behavior, then it's hardly a test of whether that animal "can" or "cannot" perform the behavior.

> The ability of the experimenter rather than that of the organism is being tested. It is dangerous to assert that an organism of a given species or age *can not* solve a problem. As the result of careful scheduling, pigeons, rats, and monkeys have done things during the past five years which members of their species have never done before. It is not that their forebears were incapable of such behavior; nature had simply never arranged effective sequences of schedules.[21]

It can be hard to appreciate the extent to which our lives, and the world around us, have been sculpted into just such "effective sequences of schedules." We have somehow come to think of being smoothly onboarded and shown the ropes as "natural," when, in fact, the opposite is the case. Nature simply *is*. There is no tutorial.

The human world, by contrast, is elaborately architected to be learnable. Part of what makes great games great, for example, is the way that they "shape" our play. Consider one of the most famous and historically significant video games of all time, Nintendo's 1985 *Super Mario Bros.* One's very first time playing the game might be hard to recall, but a close look at the game's first ten seconds reveals that it has been carefully, brilliantly designed to teach you how to play. You begin with an enemy Goomba approaching from the right; if you do nothing, you die. "You have to teach the player in a natural way, that they need to avoid them by jumping over them," says the game's designer, the legendary Shigeru Miyamoto.[22] This is the game's first and most important lesson: mushroom-looking guys are bad, and you must jump.

But Miyamoto had a problem. There are also *good* mushrooms, which you have to learn, not to dodge, but to *seek*. "This gave us a real headache," he explains. "We needed somehow to make sure the player understood that this was something really good." So now what? The good mushroom approaches you in an area where you have too little headroom to easily jump over it—you brace for impact, but instead of killing you, it makes you double in size. The mechanics of the game have been established, and now you are let loose. You think you are simply playing. But you are carefully, precisely, inconspicuously being *trained*. You learn the rule, then you learn the exception. You learn the basic mechanics, then you are given free rein.

We should perhaps not be so surprised that the same principle of shaping—the instilling of complex behavior through successive approximations—applicable to Skinner's pigeons is applicable to human learners as well. And we should perhaps not be surprised to find that it's every bit as applicable when the learners are machines.

It has, of course, been known since the inception of machine learning that some problems, some environments, some games are easier than others. But only gradually did it come to be appreciated that a system trained *first* on an easier form of a problem may be in a better position to learn a more difficult one than an agent trained from scratch.[23]

In the 1980s, Richard Sutton and Andrew Barto were working with their colleague Oliver Selfridge on using reinforcement learning to train a simulated cart on wheels to balance a pole on itself without tipping. The taller and heavier the pole, the easier it is to keep it upright—just as it is easier to balance, say, a baseball bat on one's hand than a ruler. They found that their cart system required fewer total attempts if it was first trained on a tall, heavy pole and *then* switched to a shorter, lighter one than if it was trained on the shorter, lighter pole from the beginning.[24]

Researchers have periodically stumbled on this same insight in other contexts. UC San Diego linguist Jeffrey Elman, for instance, was experimenting in the early '90s with getting neural networks to correctly predict the next word in a sentence. Discouragingly, several initial attempts failed. "Put simply," he says, "the network was unable to learn the complex grammar when trained from the outset with the full 'adult' language. However, when the training data were selected such that simple sentences were presented first, the network succeeded not only in mastering these, but then going on to master the complex sentences as well."[25]

"This is a pleasing result," says Elman, "because the behavior of the network partially resembles that of children. Children do not begin by mastering the adult language in all its complexity. Rather, they begin with the simplest of structures, and build incrementally until they achieve the adult language."

In both cases, the use of a curriculum—an easier version of the problem, followed by a harder version—succeeded in cases where trying to learn the more difficult problem by itself could not.

Keller and Marian Breland, in their work at Animal Behavior Enterprises, had seen the critical importance of a good curriculum in their efforts to train a pig to deposit large wooden coins in a "piggy bank." They began with a single coin, right next to the bank, and gradually moved it farther and farther away from the bank and also farther and farther from the pig.[26]

More recently, the machine-learning community has returned to this idea of working "backward." A group of UC Berkeley roboticists in 2017 wanted to train a robotic arm to slide a washer onto a long bolt. Waiting for the robot to somehow stumble onto this behavior at random would have taken an eternity. But starting from the washer almost totally all the way down the bolt, they could teach the robot to push it down the last little bit. Then with the washer just *barely* on the bolt, the robot could learn to slide it all the way down. Then with the washer *near* the bolt in an easy orientation, the robot could learn to put the bolt through the hole. They were finally able to back their way up to a system capable of being handed the washer anywhere and any which way, and able to dexterously rotate and fasten it.[27]

No less a game expert than legendary chess world champion Bobby Fischer used a similar strategy in his instructional book *Bobby Fischer Teaches Chess.* The book, intended for beginners, contains dozens of examples of checkmate in a single move, then progresses to two-move checkmates, then three- and four-move mating combinations. Discussion of middle-game and opening play, as well as longer-term strategies, is put off for later books; Fischer focuses exclusively on teaching beginning players how to recognize opportunities to end the game. This particular curriculum has proven very successful, indeed: some of today's grandmasters endorse it as the perfect first chess book for a new player,[28] and it has become the best-selling chess book of all time.[29]

The natural next step, it would seem, is treating the construction of a good, learnable curriculum as *itself* a machine-learning problem, and seeing if it might be possible to automate the curriculum design process. Recent research has looked into ways to automatically identify tasks of appropriate difficulty, and examples that can maximally promote learning in the network. The early results in this vein are promising, and work is ongoing.[30]

Perhaps the single most impressive achievement in automated curriculum design, however, is DeepMind's board game–dominating work with AlphaGo and its successors AlphaGo Zero and AlphaZero. "AlphaGo always has an opponent at just the right level," explains lead researcher David Silver.[31] "It starts off extremely naïve; it starts off with completely random play. And yet at every step of the learning process, it has an opponent—a sparring partner, if you like—that's exactly calibrated to its current level of performance." And who, exactly, is this perfect sparring partner, always calibrated to be at *just* the right difficulty?

The answer is simple, elegant, and, in hindsight, obvious. It plays against itself.

THE DELICATE MATTER OF INCENTIVES

Whether dealing with monkeys, rats, or human beings, it is hardly controversial to state that most organisms seek information concerning what activities are rewarded, and then seek to do (or at least pretend to do) those things often to the virtual exclusion of activities not rewarded.

—STEVEN KERR[32]

If the reward system is so designed that it is irrational to be moral, this does not necessarily mean that immorality will result. But is this not asking for trouble?

—STEVEN KERR[33]

IT'S THE REWARD SYSTEM, STUPID!

—THE EDITORS, ACADEMY OF MANAGEMENT EXECUTIVE[34]

The second approach to overcoming sparsity is, rather than using a "curriculum" and beginning first with a simplified version of the problem, instead to use the normal full-scale version of the problem while adding some bonus rewards that serve to point the learner in the right direction or encourage behaviors that correlate with success. These are known in the field as "pseudorewards" or "shaping rewards," but it's simplest just to think of them as incentives.

Skinner gave his pigeon a little bit of food for looking at and approaching the ball, activities that necessarily preceded the swipe he was ultimately looking for. The same idea applies in machine-learning settings. For instance, the "true" reward of a housecleaning robot might be a spotless house, but you can give it some incentives for each bit of dirt it vacuums up. Or your delivery drone might ultimately be trying to arrive at some destination, but you can give it a little bit of reward for making progress in the right direction.

This is often enormously helpful at giving an agent that would otherwise be flailing at random until it accomplished its goal by pure chance a sense of whether it is getting "hotter" or "colder": whether it's generally behaving in the right way and altering its behavior in the right direction.

Often we break a problem down into discrete steps in order to make it psychologically easier to stay motivated. Imagining a completed PhD dissertation or book manuscript years from now makes it hard to judge the quality of an individual *day's* work. Imagining all the weight we hope to lose by next year can make the costs and benefits of a particular cupcake or second helping feel diffuse. As parents, as teachers, as coaches, we know that a well-timed high five or "Good job!" can help get a tutee through the tough slog of constant practice, even when mastery feels impossibly distant.

Of course, anyone who has worked with humans—or even has simply *been* one—knows that to create incentives is to play with fire. They must be very carefully designed indeed, or trouble often awaits.[35] As management expert Steven Kerr famously put it in his classic 1975 paper, as soon as you begin to think about adding additional reward, you are in immediate danger of "the folly of rewarding A, while hoping for B."[36]

Kerr's analysis of incentives gone wrong has become a landmark paper in management science, and Kerr has spent much of his career working with businesses from General Electric to Goldman Sachs on how to think more carefully about incentives. Amazingly enough, when asked where his inspiration came from, Kerr cited both machine learning *and* B. F. Skinner. "The fact that machines can be programmed to learn and that a chess playing machine can be programmed to never make the same mistake twice was very fascinating to me," Kerr says. "I was immediately intrigued with the possibility that a machine could become a better chess player than the actual programmer!"[37]

As for Skinner, Kerr concedes, "In regard to the 'Folly' article, obviously B. F. Skinner got 'there' before me. I never claimed otherwise. I remember reading Skinner saying he would scream at his rats 'Why don't you behave?' after they wouldn't do what he wanted or expected. Skinner probably would turn over in his grave if he thought people were reading my article without having read his work. Obviously, Skinner did the work, but I was able to help bottle it in the right package for business applications. 'Blaming the rat' was a great learning lesson. What the 'Folly' is really about is that it is not always the employees' fault; management is responsible for all too many employee dysfunctionalities."

Indeed, for Skinner, one could almost never blame the rat (or the employee). Our behavior, he felt, was almost fully determined by our incentives and our rewards. A television interviewer at one point asked Skinner, "Where does that leave free will?" Skinner replied, "It leaves it in the position of a fiction."[38]

The free will argument aside, problems of incentive pervade not just animal psychology, and not just corporate management; some of the most memorable, in fact, come from those ruthless and creative reward maximizers we call *children*.

University of Toronto economist Joshua Gans wanted to enlist the help of his older daughter in potty training her younger brother. So he did what any good economist would do. He offered her an incentive: anytime she helped her brother go to the bathroom, she would get a piece of candy. The daughter immediately found a loophole that her father, the economics professor, had overlooked. "I realized that the more that goes in, the more comes out," she says. "So I was just feeding my brother buckets and buckets of water." Gans affirms: "It didn't really work out too well."[39]

Princeton cognitive scientist Tom Griffiths had an eerily similar situation happen with his own daughter. "She really liked cleaning things," he tells me; "she would get excited about it. We got her her own little brush and pan. There were some, you know, chips on the floor, and she got her brush and pan and cleaned them up, and I said to her, 'Wow! Great job! Good cleaning! Well done!'"[40]

With the right praise, Griffiths would manage to both foster motor-skill development in his daughter *and* get some help in keeping the house clean: a double parenting win. Or was it? His daughter found the loophole in seconds.

"She looked up at us and smiled," he says—"and then dumped the chips out of the pan, back onto the floor, and cleaned them up again to try and get more praise."

For Griffiths, whose research bridges the gap between psychology and machine learning, the implication was obvious. It was "exactly the kind of thing that makes me think about some of the challenges that you have in building reward-motivated AI systems, where you have to think really carefully about how you design the reward function."

Griffiths thinks about reinforcement learning all the time in the context of parenting. "As a parent you are designing the reward function for your kids, right? In terms of the things that you praise and the things that you give them certain kinds of feedback about. . . . No one really thinks rigorously about 'What's the reward function that you explicitly want to design for your kids?'"

Griffiths views parenthood as a kind of proof of concept for the alignment problem. The story of human civilization, he notes, has *always* been about how to instill values in strange, alien, human-level intelligences who will inevitably inherit the reins of society from us—namely, our kids. The parallel goes even deeper than that, though—and careful attention to both AI and parenting shows the surprising degree to which each can inform the other.

Our children may be no more intelligent than we, but even young children can outsmart our rules and incentives, in part because of how *motivated* they are to do so. In the case of reinforcement-learning systems, they are slaves of a kind to their rewards; but they're the kinds of slaves that have an immense amount of computing power and a potentially inexhaustible number of trial-and-error attempts to find any and all possible loopholes to whatever incentives we design. Machine-learning researchers have learned this lesson the hard way. And they have also learned a thing or two about how to deal with it.

CYCLE-PROOFING YOUR REWARDS:
THE SHAPING THEOREM

Astro Teller, currently the "captain of moonshots" at X (the Alphabet company formerly known as Google X), has in recent years overseen everything from Google's self-driving-car project (subsequently spun off into Waymo), to its augmented-reality project Google Glass and its research lab Google Brain. But in 1998, he was focused on a different problem: soccer. With his friend and fellow student David Andre, Teller worked to enter the annual RoboCup soccer competition, with a virtual-soccer-playing program they dubbed Darwin United.[41] Reward shaping is part of what enabled them to teach their program how to play. But there was an issue. In soccer, possession of the ball is part of what good offense and good defense looks like—and is certainly better than wandering around the field aimlessly. And so Andre and Teller provided a reward—worth a tiny fraction of a goal—to their robot for taking possession of the ball. To their astonishment, they found their program "vibrating" next to the ball, racking up these points, and doing little else.[42]

That same year, a team of Danish researchers at the Niels Bohr Institute in Copenhagen, Jette Randløv and Preben Alstrøm, were trying to get a reinforcement-learning system to learn how to ride a simulated bicycle. The system would have to manage the complicated task of staying upright while making progress toward a distant goal. This seemed like the perfect application for adding some shaping rewards. Because it would be unlikely that a system wobbling all over the place would randomly arrive at the goal, the team decided to add a small reward—a handful of "points"—anytime the bicycle made progress toward the goal.

To their astonishment, "the agent drove in circles with a radius of 20–50 meters around the starting point."[43] They were rewarding progress *toward* the goal but forgot to penalize movement *away* from the goal. Their system had found a loophole, and was ruthlessly—if dizzyingly—exploiting it.

"These heterogeneous reinforcement functions," they wrote, "have to be designed with great care."

These cautionary tales, in the late 1990s, were much on the minds of UC Berkeley's Stuart Russell and his then–PhD student (and subsequently VP and chief scientist at Baidu) Andrew Ng. These sorts of exploitative loops seemed like a constant danger.[44]

Ng was ambitious. He recalls, "When I started working on robotics . . . I asked a lot of people, 'What's the hardest control problem you know of?' And back then the most common answer I heard was 'Getting a computer to fly a helicopter.' And so I said, 'Let's work on that.'"[45] Indeed, he would do his PhD thesis on using reinforcement learning to fly an actual—unsimulated, nine-foot-long, one-hundred-pound, seventy-thousand-dollar—Yamaha R-50 helicopter.[46] The stakes were wildly high. Erratic or unpredictable behavior in the real world could destroy the helicopter entirely—to say nothing of what might happen if an unsuspecting *person* found themselves in its path.

The critical question was this: Given a difficult-to-learn reward function that described what they *actually* wanted the helicopter to do, what sorts of "pseudoreward" incentives, if any, could they add such that the training process would be easier but the best way to maximize that modified reward would also be the optimal solution to the real problem? To use Kerr's terms, what sorts of A's could they reward that would still result in the hoped-for B? In Ng's description:

> A very simple pattern of extra rewards often suffices to render straightforward an otherwise intractable problem. However, a difficulty with reward shaping is that by modifying the reward function, it is changing the original problem M to some new problem M', and asking our algorithms to solve M' in the hope that solutions can be more easily or more quickly found there than in the original problem. But, it is not always clear that solutions/policies found for the modified problem M' will also be good for the original problem M.[47]

"What freedom do we have in specifying the reward function," Ng writes, "such that the optimal policy remains unchanged?"[48]

As it turned out, the critical insight was hiding in the story of the bicycle. In order to prevent the bike from riding in circles, endlessly racking up rewards, you had to also *subtract* progress *away* from the goal. Russell, who had trained first as a physicist, made a connection between the rewards problem and conservation of energy. "The key thing," Russell explains, "is just making the shaping into what we call in physics a 'conservative field.'"[49] Pseudorewards were like potential energy: a function only of where you *were*, not what path you'd taken to get there. This also meant that returning to where you'd started—no matter what journey you'd been on—was net zero.

This idea was intuitive enough in the bicycle problem: if you rewarded progress toward the goal, you necessarily needed to *penalize* progress *away* from the goal. Said another way, the incentive "points" should always reflect how close the bike was to the goal, and nothing about the route it had followed. But the concept of treating the incentives like "potential energy" turned out to be far deeper and more general. It was a necessary and sufficient condition for making sure that the agent you trained on the shaped reward didn't have its behavior come unglued from the real problem.

"As a general rule," says Russell, "it is better to design performance measures according to what one actually wants in the environment, rather than according to how one thinks the agent should behave."[50] Put differently, the key insight is that we should strive to reward *states of the world*, not *actions of our agent*. These states typically represent "progress" toward the ultimate goal, whether that progress is represented in physical distance or in something more conceptual like completed subgoals (chapters of a book, say, or portions of a mechanical assembly).

Though it isn't a magic bullet to *all* problems of human incentives, the shift in focus from actions to states does invite new ways of thinking about some of the incentive structures we, wittingly and unwittingly, design for others. Given a child who is dumping out the kitchen trash in order to double the rewards of sweeping it up, we can make our rewards into a "conservative field" by making sure we scold them in precisely equal measure for dumping the trash out, so that the net gain of further repetitions is zero. It might be easier, though, to shift to praising the state rather than the actions: instead of rewarding the act of cleaning up itself, we might say, "Wow, look how clean that floor is!"

Of course, there is more to the art and science of rewards than the avoidance of cycles—though it's a start. Ng and Russell conclude their work on a cautious, rather than triumphant, note: "We believe," they write, "the task of finding good shaping functions will be a problem of increasing importance."[51]

EVOLUTION AS REWARD DESIGNER

I said unto him, "Be fruitful and multiply." But not in those words.

—WOODY ALLEN[52]

From a Darwinian perspective, what people want is reasonably clear: something along the lines of propagating and securing their genetic lineage. What people *actually* want, moment to moment, looks an awful lot more heterogeneous and shortsighted: orgasms, chocolate, a new car, respect. It seems, then, that we are biologically wired and culturally induced to want specific, concrete things in the short term that generally steer our behavior in ways that are ultimately in line with these evolutionary goals, which would otherwise be too distant or ill-defined to consciously aim at.[53]

Sound familiar?

Understanding the nature and role of shaping—first in behavioral psychology, then in machine learning—hasn't just taught us how to design better agents. Perhaps its most surprising contribution is to the way we think about evolution.

When Brown University's Michael Littman was a graduate student, in the late 1980s, he was hired for a time at Bellcore, a research and development group formerly part of AT&T, located in New Jersey. There he quickly found a mentor, and a friend, in Bellcore's Dave Ackley.

Littman asked him about working on the question of behavior—of taking actions and making decisions that ramify over time. As Littman recounts: "He was like, 'Oh. That's a thing. It's called reinforcement learning. I've looked at it a little bit. Here's a paper.' And he gave me Rich Sutton's 1988 TD paper."[54]

Littman began to read about temporal-difference learning and was fascinated; he asked Ackley where he could learn more. "And he's like, 'Well, invite Rich to give a talk.' And I was like, 'What? That's a thing that you can do? You can read a paper and there's a person's name on it, and then you can turn them into a human being?' . . . I didn't think of it as a *community;* I thought of it as a *literature*. But no, it was a bunch of people, and they knew each other, and Dave was like, 'I can invite him.' And so he invited him."

Sutton came to Bellcore, and both Ackley and Littman caught the reinforcement-learning bug.

They were interested in the question of how evolution shapes our reward function to give rise to behaviors in the short term that are useful to the overall survival of an organism or a species in the long term. The organism's reward function itself, so long as it accomplishes this, might otherwise seem very random. Ackley and Littman wanted to see what might shake out if they simply let reward functions evolve and mutate and allowed simulated virtual entities to die or to reproduce.[55]

They created a two-dimensional virtual world in which simulated organisms (or "agents") could move around a landscape, eat, be preyed on, and reproduce. Each organism's "genetic code" contained the agent's reward function: how much it liked food, how much it disliked being near predators, and so forth. During its lifetime, it would use reinforcement learning to learn how to take actions to maximize these rewards. When an organism reproduced, its reward function would be passed on to its descendants, along with some random mutations. Ackley and Littman seeded an initial world population with a bunch of randomly generated agents.

"And then," says Littman, "we just ran it, for seven million time steps, which was a lot at the time. The computers were slower then." What happens? As Littman summarizes: "Weird things happen."[56]

At a high level, most of the successful individual agents' reward functions ended up being fairly comprehensible. Food was typically viewed as good. Predators were typically viewed as bad. But a closer look revealed some bizarre quirks. Some agents, for instance, learned only to approach food if it was *north* of them, for instance, but not if it was south of them.

"It didn't love food in all directions," says Littman. "There were these weird holes in [the reward function]. And if we fixed those holes, then the agents became so good at eating that they ate themselves to death."

The virtual landscape Ackley and Littman had built contained areas with trees, where the agents could hide to avoid predators. The agents learned to just generally enjoy hanging out around trees. The agents that gravitated toward trees ended up surviving—because when the predators showed up, they had a ready place to hide.

However, there was a problem. Their hardwired reward system, honed by their evolution, told them that hanging out around trees was good. Gradually their learning process would learn that going toward trees would be "good" according to this reward system, and venturing far from trees would be "bad." As they learned over their lifetimes to optimize their behavior for this, and got better and better at latching onto tree areas and never leaving, they reached a point of what Ackley dubbed "tree senility." They never left the trees, ran out of food, and starved to death.

However, because this "tree senility" always managed to set in *after* the agents had reached their reproductive age, it was never selected against by evolution, and huge societies of tree-loving agents flourished.

For Littman, there was a deeper message than the strangeness and arbitrariness of evolution. "It's an interesting case study of: Sure, it has a reward function—but it's not the reward function in isolation that's meaningful. It's the interaction between the reward function and the behavior that it engenders."

In particular, the tree-senile agents were born with a reward function that was optimal for them, *provided* they weren't overly proficient at acting to maximize that reward. Once they grew more capable and more adept, they maxed out their reward function to their peril—and, ultimately, their doom.

One doesn't have to squint too hard to see a cautionary tale here for *Homo sapiens*. A heuristic like "*Always* eat as much sugar and fat as you possibly can" is optimal as long as there isn't all that much sugar and fat in your environment and you aren't especially good at getting it. Once that dynamic changes, a reward function that served you and your ancestors for tens of thousands of years suddenly leads you off the rails.

For Andrew Barto, there are clues in thinking about evolution that are useful for us as we now play the role of the reward designer. "Evolution has provided our reward function, and so that is really quite important with regard to how we design reward functions for artificial systems," he says. "That's what happened in nature. Evolution came up with these reward signals to encourage us to do things that led to our reproductive success."[57]

As Barto notes, "So, an interesting thing is that evolution didn't give us reproductive success as a reward signal. They gave us rewards for predictors." We optimize our behavior to maximize the things we find rewarding, but in the background and at a larger scale, evolution is shaping the things we find rewarding in the first place. "So, it's a two-level optimization," says Barto. "I'm very interested in that."

In recent years, Barto has worked with the University of Michigan's Satinder Singh and Richard Lewis and then–PhD student Jonathan Sorg to investigate the question of the "optimal reward problem."[58] If you have goal *x*, it might be that your best bet *isn't* simply to tell your agent to do *x*.

"Should the artificial agent's goals be the same as the agent designer's goals?" they write. "This is a question seldom asked."[59]

Consider a game, they say, in which all that exists is an agent, a fishing pole, some worms, and a pond full of fish.[60] Let's say the agent's overall evolutionary fitness is best served by eating as many calories as possible. Ideally they would learn to pick up worms, refrain from eating them, and use them to catch fish—but this is rather complicated. A clever agent with a long life span is best served by a distaste for eating worms, so that they more quickly begin learning to fish. For an agent with a shorter attention span, on the other hand, or a shorter lifespan, trying to learn how to fish will be a fruitless waste of time—and so they are better served if they happen to find worms delicious.

Perhaps most interesting is the case of an agent that is clever enough to learn to fish and will get to live *just* long enough to learn how—but not long enough to properly *benefit* from making that investment. They should be made allergic to fish, as it turns out, so that they have no choice but to eat worms!

Subtle changes in the life span or resources or design of an agent can have wild and abrupt effects in the structure of the optimal reward. The answer as to what set of rewards will be ideal for a *particular* agent in a *particular* environment doesn't appear to admit to any easy generalizations. Research in this vein continues—but learning to more sharply distinguish between what you *want* and what you *reward* is an important component of the solution.[61]

More recently, psychologists and cognitive scientists are taking these tools and turning them back around to ask a fascinating question not about machines but about humans. How should you design the best reward function, they ask, when the computationally limited, impatient, shortsighted agent learning to optimize it is . . . yourself?

HOW SHOULD WE TRAIN OURSELVES?

The theory and practice of reward shaping in machine learning not only gives us a way to make our autonomous helicopters maneuver appropriately but contributes two distinct things to our understanding of human life and human intelligence. One, it shows us a reason—*sparsity*—why some problems or tasks are more difficult than others to solve or accomplish. Two, it gives us a theory—incentivize the *state,* not the action—for how to make tough problems easier without introducing perverse incentives.

The potential use for these insights in human life is enormous. The economic costs alone are vast—a recent report estimated the impact of British workers procrastinating on the job at £76 billion a year—to say nothing of the less easily calculated toll on our well-being, and on the quality and fullness of our lives.[62]

If we are living through a time of soaring video-game addiction and real-world procrastination, maybe it's not the individual procrastinator's fault. As Skinner put it, "I could have shouted at the subjects of my experiments, 'Behave, damn you, behave as you ought!' Eventually I realized that the subjects were always right. They always behaved as they ought."[63] If they weren't learning something, it was the fault of the experimenter not shaping their task properly. So maybe it's not a lack of willpower on our part, but rather that—as the bestselling 2011 book by Jane McGonigal put it—*Reality Is Broken*.

McGonigal, a game designer by trade, has spent her career designing games to help people—including herself—overcome challenges in their lives. For her, the incredibly addictive and compelling quality of most games, computer games in particular, is how clear they always make it what you need to do, and how achievable whatever that is always seems.

> The first thing is whenever you show up in one of these online games . . . there are lots and lots of different characters who are willing to trust you with a world-saving mission, right away. But not just any mission; it's a mission that is perfectly matched with your current level in the game. Right? So you can do it. They never give you a challenge you can't achieve. But it is on the verge of what you're capable of, so you have to try hard.[64]

In other words, what makes games so hypercompelling is how well *shaped* they are. The levels are a perfect curriculum. The points are perfect pseudorewards. They are Skinnerian masterpieces.

Reinforcement learning not only gives us the vocabulary to understand and express what makes the games so riveting; it also gives us a way to empirically confirm those intuitions. A game with a clear curriculum from its easiest levels to its hardest, as well as clear pseudorewards that mark the path forward and foster exploration and the development of skills, should be easier for an algorithm to learn. It's not hard to imagine game studios in the coming years using *automated* test players for their levels, highlighting the sticking points where real human players are likely to give up or quit.

The problem, of course, is that it is so much easier to shape these virtual environments than the real one. As McGonigal diagnoses: "Now, the problem with collaborative online environments like *World of Warcraft* is that it's so satisfying to be on the verge of an epic win all the time, we decide to spend all our time in these game worlds. It's just better than reality." For McGonigal, the solution isn't to wean ourselves off of these perfectly reward-shaped environments but the reverse: "We have to start making the real world work more like a game."

McGonigal herself has led something of a movement in this regard, including using games to overcome obstacles—including suicidal depression from a protracted recovery from concussion—in her own life.[65] "I said to myself after thirty-four days [of depression]—and I will never forget this moment—I said, 'I am either going to kill myself or I'm going to turn this into a game.'"[66] So she created pseudorewards. Calling her sister was worth a few points. Walking around the block was worth a few points. It was a start.

This is a field known as *gamification*,[67] and in the last ten years, thanks to insights from reinforcement learning, it has gone from something of an art to something of a science.[68]

One of the people who has thought harder about this problem than anyone in the world—both in his research and his personal life—is cognitive scientist Falk Lieder of the Max Planck Institute for Intelligent Systems.

Lieder's research focuses on what he calls "rationality enhancement." He studies the cognitive science of how people think and make decisions—and unlike most researchers, he has a keen interest in not only understanding human cognition but in devising effective tools and interventions to make humans think better. His earliest human subject was, of course, himself.

Growing up, Lieder found it frustrating that his schooling, though it gave him plenty to think about, never touched on thinking itself. "I always felt that what I really wanted to learn was to think well, and how to make good decisions," he explains. "Nobody could teach me that. They could just teach me declarative facts about the world, and it wasn't very useful. I really wanted to learn how to think."[69]

Over time, this matured into a quest for personal betterment, but also something bigger: the drive to understand the principles of human reasoning, and make tools for improving it. "Part of my research," he explains, "is discovering these optimal strategies for thinking and decision-making so that we can actually build a scientifically based curriculum for good thinking."

Lieder is interested in gamification and more specifically, what he calls *optimal* gamification: Given a goal, what is the best possible incentive structure to facilitate reaching it?[70] This has much of the flavor of "optimal reward design" as discussed above, but in this case the agents being designed for are humans rather than algorithms.

Lieder, working with Tom Griffiths, established some ground rules for what optimal gamification would look like. They knew from Andrew Ng and Stuart Russell's work that one of the cardinal rules is to reward *states*, not *actions*. Therefore the points assigned to taking an action must reflect how much better the resulting state of affairs is—and what's more, as Lieder notes, "the points have to be assigned in such a way that when you undo something, you lose as many points as you earned when you did it." This is what Randløv and Alstrøm learned the hard way with their bicycling robot, what David Andre and Astro Teller learned the hard way with their vibrating soccer robot—and what Tom Griffiths himself learned the hard way when his daughter dumped the dustpan onto the kitchen floor.

Ng and Russell's paper had suggested that shaping rewards could enable an agent with a limited ability to look ahead and forecast the effects of its actions to behave as if it was more farsighted than it really was.[71] This idea intrigued Lieder—in part because humans are so notoriously impulsive and short-term-focused in many of their decisions.

He and Griffiths ran an experiment where they put human subjects in the role of an airline route planner. Flying to certain cities was profitable for that leg, but might put the plane in a position where few other profitable routes existed—and the reverse was true, too: it might be worth flying one leg at a loss in order to reap rewards elsewhere. They tried adding additional immediate rewards (or penalties) to the sticker price of a given leg from A to B that would reflect the downstream cost or benefits of moving the plane to that location. As expected, this enabled people to more readily make better and more profitable decisions.

There was only one downside: because the shaped rewards incorporated the long-term costs and benefits of a choice into the sticker price, users no longer needed to think ahead. This made their decisions more *accurate*, but was somewhat enfeebling. People no longer needed to think very hard, and so they didn't. "If you are to act in an environment where [myopic] decision-making works," says Lieder, "people will learn to rely on that system more and more."[72]

This left open an intriguing possibility: Could you use optimal gamification, not to *obviate* the need for planning, but to make people *better* at it?

The incentives in this case are almost completely different. Instead of the interface creating an easy problem where long-term costs are taken into account—causing higher-quality decisions but perhaps encouraging lethargy or complacency in subjects, making them more dependent on the interface—prices could be adjusted to create a *curriculum*. The subject could be slowly *taught* how to think ahead, starting with very simple illustrations of the basic idea and slowly building in complexity as the subject got better at it. The interface, instead of functioning as a crutch, could use a different set of incentives to do the opposite: "teaching people to plan further ahead in time," Lieder explains, "so they can succeed in environments where immediate reward is misaligned with long-term value."

Lieder's final experiment was not with flight planning but with a more familiar setting: procrastination. He and Griffiths created a deliberately onerous task—to write essays on a series of five topics, some of them longer and more difficult than others—and put it on Mechanical Turk, where people would elect to work on it for a payment—$20—upon completion of all the pieces of writing by a deadline set for ten days later. Of everyone who signed up, 40% never even *started*. (This was particularly ironic because when the task was initially described to them, they had a chance to decline to participate and still receive fifteen cents!)[73]

Lieder and Griffiths also experimented with incentives that gave subjects "points"—which had no cash value but were visually encouraging—for each essay completed. Each essay was worth the same number of points. It didn't help.

Lastly, they offered a third group of participants *optimal* incentives: point values that precisely reflected how difficult or unpleasant each topic was and how much closer they would be to earning their $20 when they had finished it. (A hundred words on North Korea's economic policy, for instance, was worth about three times as many points as fifty words on their favorite TV show.) Here, a full 85% of participants finished all five essays.[74]

Lieder thinks of systems like these as "cognitive prostheses,"[75] and they are more than just his research interest. They are a crucial part of how his own research even gets done.

As a PhD student, Lieder found himself in a scaled-up version of the horrible essay-writing task. "I think one of the worst situations is to get no information about your level of progress," he says. "The official system is, you are a PhD student, and then you have a PhD, and that's it. So: five years of no feedback."

PhD students in general are a group with high rates of anxiety and depression and for whom procrastination is nearly an epidemic.[76] They are, in effect, pigeons in Skinner's bowling alley—the cap-and-gown-shaped food pellet waiting for them after they bowl a perfect strike some five years hence. Such a system, we know, doesn't work for animals—and it doesn't work for reinforcement-learning algorithms, either.

That wouldn't do. Lieder needed something else. He charted his five-year course—"I broke this down into a few hundred levels." He assigned himself virtual "citations" when he did work he felt might result in *real* citations some several years hence. He used the same optimal gamification calculation that assigned his subjects' essays a point value to compute the proper point values for each sub-subtask of his doctorate. He even used punishment to override some of his habits—in the form of a wristband from the company Pavlok. "It gives you electric shocks whenever you indulge in some habit that you don't want to indulge in," Lieder explains. "My main bad habits were related to how I use my computer. Like, I'm going to YouTube when I feel bad. Since my time-tracking software would immediately communicate to my Pavlok what I was doing, it could then immediately shock me." Amusingly, this failed to curb his habit of visiting YouTube when he needed some distraction, but it *did* instill the habit of instantly closing the page when he did.

Being, in effect, both the experimenter *and* the subject in his own behavioral training experiment gave Lieder a unique vantage onto the question of reward shaping. Such training was, at once, the process by which he approached his research and the central question of that research. The results are encouraging, and Lieder—now running his own research laboratory—has the doctoral robes to prove it.

BEYOND EXTRINSIC REINFORCEMENT

For Skinner, not only was individual free will left uncomfortably "in the position of a fiction," but essentially the entire story of human civilization was the story of reward structures. Skinner himself was curiously upbeat about this, writing the utopian novel *Walden Two* about a perfect behaviorist society.

And yet anyone who's been around children or animals might have the nagging suspicion that reward maximization really *isn't* the whole story of why we do what we do. We play games of our own invention, for no discernible prize. We turn over rocks, or climb mountains, just to see what we might discover. We explore. We are playful and curious. We are, in short, motivated as much by *internal* as by *external* rewards.

As it happens, this, too, is becoming ever more well appreciated in the world of machine learning.

6

CURIOSITY

If the untrained infant's mind is to become an intelligent one, it must acquire both discipline and initiative. So far we have been considering only discipline.

—ALAN TURING[1]

In the spring of 2008, graduate student Marc Bellemare was walking on the beach in Barbados with University of Alberta computer scientist Michael Bowling. Bowling had an idea. At that time, reinforcement-learning research was typically done with each researcher making their own bespoke game from scratch, then hand-tailoring a system to succeed at that particular game.[2]

What if instead, Bowling mused, someone built a single environment that *everyone* could use, with not just one game in it but a vast library of them—and what if, instead of fake, made-up games, it used *real* games—namely, classic 1970s and '80s video games from the Atari 2600?

Bellemare recalls, "I said, 'This is the stupidest idea I've ever heard.'"[3]

He continues: "Fast-forward about three years, and, well, I decided it wasn't such a stupid idea." In fact, Bellemare found he liked the idea so much that Bowling became his doctoral advisor—and the idea became his dissertation.

The ambition of the project was somewhat insane, not only in the amount of work it would require to build this video-game bestiary, which they dubbed the Arcade Learning Environment (ALE), but also in the implicit gauntlet that it threw down to the rest of the field.[4] The idea was for researchers to compete against one another by fielding a single learning system that could play not just one but all *sixty* of the games. The field wasn't anywhere close.

A big part of the problem was that the bespoke game environments then being used often described the world to the agent in terms that were sanitized and filtered, using inputs that were high-level and useful. In the case of the wheeled cart trying to balance a pole, the system would be given as input the location of the cart, its velocity, the current angle of tilt of the pole, the pole's velocity, and so forth. In the case of the two-dimensional grid-world environment with trees, food, and predators, the system would be told the agent's location, its health and hunger, whether there was a predator nearby, where the nearest food was, and so forth. These pieces of information are known as "features."

What the ALE offered, by contrast, was something more overwhelming and less immediately usable: the pixels on the screen. That was it. Every game was different, not only in its rules but in the way the pixels on the screen mapped to usable information. A learning system thrown into a new game was going to have to figure out everything from scratch: these pixels over here appear to blink when I score points, those pixels over there appear shortly before I die, these pixels in the middle move left whenever I hit the left arrow button—oh, maybe they're *me*. Either researchers were going to have to find extremely general ways of extracting useful patterns on the screen for their system to keep track of—such that they were helpful in all sixty games—or else all of that understanding and meaning-making was going to have to be done by the system on the fly. This was the problem of "feature construction."

Bellemare began experimenting by plugging a bunch of feature-construction algorithms into standard reinforcement-learning systems and throwing them at the games. The results weren't impressive. To his surprise, however, it was easy to get them published—if only because his peer reviewers were so impressed by how much work had gone into making the Atari environment. "The funny thing is," he says, "at the time the reviewers would tell me, 'Well, you've done this amazing thing with Atari. I cannot possibly reject your paper.' . . . It was so big . . . it didn't matter how bad or good the results were. People just said, 'Wow. You've actually done this.'"

He and his colleagues had built, in effect, a kind of mountain; it was now up to the field to figure out a way to scale it.

Bellemare finished his PhD in 2013 and moved from Edmonton to London to join DeepMind.[5] There, a team led by Volodymyr Mnih was working excitedly on the idea of taking the same AlexNet-style deep neural networks that had been so decisive in the ImageNet competition the year before, and applying them to the problem of reinforcement learning. If deep networks could look at tens of thousands of raw pixels and figure out whether they were a bagel, a banjo, or a butterfly, maybe they could do whatever feature-construction was needed to render an Atari screen intelligible.

He recalls, "The group said, Hey, we have these convolutional networks. They've been phenomenal at doing image classification. Um, what if we replace your feature-construction mechanism, which is still a bit of a kludge, by just a convolutional neural network?"

Bellemare, again, wasn't buying it. "I was actually a disbeliever for a very long time.... The idea of doing perceptual RL was very, very strange. And, you know, there was a healthy dose of skepticism as to what you could do with neural networks."

But on this matter, too, Bellemare would soon come around.

Simply plugging deep learning into a classic RL algorithm and running it on seven of the Atari games, Mnih was able to beat every previous RL benchmark in six of them. Not only that: in three of the games, their program appeared to be as good as a human player. They submitted a workshop paper in late 2013, marking their progress.[6] "It was just sort of a proof-of-concept paper," says Bellemare, "that convolutional nets *could* do this."

"Really," he says, "it was bringing the deep-learning part to solve what reinforcement-learning researchers hadn't been able to do in ages, which is to generate these features on the fly. Then you can do it for any game— it doesn't matter. And then this . . ."

Bellemare pauses slightly. "This took off."

DEEP RL GETS SUPERHUMAN

In February 2015, a paper appeared on the cover of *Nature,* running with the headline "Learning Curve: Self-taught AI Software Attains Human Performance in Video Games."[7] DeepMind's hybrid of classic reinforcement learning with neural networks had shown itself capable of human-level play—and far beyond—at not just a couple Atari games, but *dozens* of them. The deep-learning revolution had come to reinforcement learning, minting the new field of "deep RL," and the results were astonishing.

The model—called a "deep Q-network," or DQN for short—was playing *Video Pinball* and achieving scores at twenty-five *times* the level of a professional human games tester. In *Boxing* it was seventeen times better than human performance. In *Breakout,* it was thirteen times better. A nearly full-page chart chronicled this stunning pattern of dominance across a wide array of different games, all using a single generic model, with no fine-tuning or adjustment from one game to the next.

At the bottom of the chart, however, there were a few stubborn games that refused to yield to DQN, games that didn't fit this pattern of glory. One in particular, at the very bottom of the list, stood out.

The outlier was *Montezuma's Revenge,* a 1984 game in which you play an explorer named Panama Joe who must find his way through a temple filled with ropes, ladders, and deadly, *very* vaguely Aztec traps. ("I did no research whatsoever into Montezuma or the culture," admits creator Robert Jaeger—just sixteen years old when he sold his demo to Parker Brothers—"and really just thought it was a colorful theme and a cool name.")[8] On *Montezuma's Revenge,* the mighty DQN achieved a high score of 0%—yes, that's 0%—of the human benchmark.

What was going on here?

For one thing, the game makes it *extremely* easy to die. Almost any sort of mistake—hitting an enemy, jumping down from too high up, walking through a barrier—is certain death. The DQN system used epsilon-greedy exploration, which involves learning about which actions produce reward by simply hitting buttons at random a certain fraction of the time. In *Montezuma's Revenge,* this is almost always suicide.

The second and more important problem is that *Montezuma's Revenge* has incredibly sparse rewards. It takes a huge number of things to go *exactly* right before the player gets any points at all. In games like *Breakout* or *Space Invaders,* even the most confused and bewildered newcomer, jamming buttons at random, quickly realizes that they're at least doing *something* right. This is enough to start the learning process: in DQN's case, registering the points scored and slowly beginning to take similar actions in similar situations more of the time. In *Montezuma's Revenge,* by contrast, very few events offer any kind of feedback *other* than death. On the first screen, for example, you have four chasms to leap, three ladders to climb, a conveyor belt to run against, a rope to grab, and a rolling skull to hop over, *all* before you can collect the first item, which rewards you with a hundred measly points (and, anachronistically, the first five notes of "La Cucaracha").

In an environment with so few rewards, a randomly exploring algorithm can't get a toehold; it's incredibly unlikely that by essentially wiggling the joystick and mashing buttons, it will manage to do *all* of the necessary steps to get that first reward. Until it does, it has no idea that it's even on the right track.[9]

One solution to this problem of sparsity, as we've seen, is *shaping:* additional incentive rewards to nudge the algorithm in the right direction. But we've also seen how tricky this can be to do correctly without creating a loophole that the algorithm can exploit. Rewarding Panama Joe for every second he *doesn't* die, for instance, might just lead to an agent learning never to leave the safety of the initial platform at all. It would be a machine version of what animal researchers call "learned helplessness."[10] As the celebrated aphorist Ashleigh Brilliant put it, "If you're careful enough, nothing bad or good will ever happen to you."[11]

Other intuitive ideas run aground as well. Rewarding Panama Joe for successfully jumping over a rolling skull, let's say, might lead the agent to play a kind of double Dutch over the skull instead of venturing farther into the temple. Rewarding him for successfully jumping onto or off a rope would, similarly, incentivize an infinite loop of Tarzan-like swinging. None of this is what we want.

Moreover, though, this kind of shaping will typically vary from game to game, and require the hand of a human overseer with inside knowledge of how that particular game works. That feels a bit like cheating. The whole idea behind the Arcade Learning Environment—and the thrilling achievement of DQN—was that of a single algorithm, able to master dozens of completely different game environments from scratch, guided by nothing but the image on the screen and the in-game score.

So what was the answer? How might a generic trial-and-error algorithm like DQN be modified in the face of such a foreboding game as *Montezuma's Revenge*?

There was a tantalizing clue, hiding in plain sight. Humans clearly *can* learn how to play *Montezuma's Revenge* without any additional shaping rewards. Human players instinctively *want* to climb the ladder, reach the distant platform, get to that second screen. We want to know what's on the other side of the locked door, to see how big the temple really is, and what, if anything, lies beyond. Not because we intuit that it will bring us "points," but out of something more pure and more fundamental: because we simply want to know what will happen.

Perhaps, then, what was needed to conquer a game like *Montezuma's Revenge* was not augmenting the game's sparse rewards with additional incentives, but rather a different approach altogether. Maybe, instead of an ever more elaborate system of carrots and sticks, the answer was the opposite: to develop an agent that was *intrinsically* rather than extrinsically motivated.[12] An agent that would essentially cross the road, not because there was reward in it, but just to get to the other side. An agent that was, we might say, *curious*.

The last several years have seen a grand revival of scientific interest in the topic of curiosity, and some unlikely collaborations between machine-learning researchers and psychologists who specialize in the cognition of children, to better understand curiosity from a rigorous, fundamental perspective. What is it, exactly? Why do we have it? How might we instill it, not just in our children, but in our machines?

And why might it be increasingly critical that we do?

CURIOSITY AS A SCIENTIFIC SUBJECT

"Desire" to know why, and how, "curiosity," . . . is a lust of the mind, that by a perseverance of delight in the continual and indefatigable generation of knowledge, exceedeth the short vehemence of any carnal pleasure.

—THOMAS HOBBES[13]

Curiosity is the beginning of all science.

—HERBERT SIMON[14]

The godfather of curiosity research in psychology is the psychologist Daniel Berlyne. Berlyne's very first publication, in 1949, was an attempt to define what exactly we mean when we say that something is "interesting," or that a person or animal is "interested" in something.[15] As he put it: "My first interest is interest."[16]

A whole subfield gradually began to open up. What did animals learn *without* any rewards hinging on their learning it?

As Berlyne noted, the history of psychology is largely a story of people and animals being *compelled* to do things—to fill out a survey or answer oral questions, to press levers in order to eat. In this way, however, the field had essentially created its own methodological blind spot. How would it even begin to approach the question of how organisms act on their own? It seemed almost a contradiction in terms.

"It has, in some ways, been a misfortune for psychology that human beings are so obliging and compliant," he wrote.[17] "The ease with which artificial and extraneous motivations can be induced in human beings has prevented us from studying the motivational factors that take control when these are lacking."

Within psychology, the research agenda of training animals with punishments and rewards was so dominant by the mid-twentieth century that it seemed for a time that this could explain everything about the behavior of intelligent organisms. But there were certain data, here and there, that refused to fit. The University of Wisconsin's Harry Harlow had, by 1950, begun documenting the way rhesus monkeys would play with physical puzzles made from a combination of locks and latches, and he coined the term "intrinsic motivation" to describe it.[18] Sometimes this intrinsic motivation not only took hold in the absence of extrinsic rewards but actually *overpowered* them. A hungry rat might, amazingly, decide to forgo a bit of food, or cross an electrified fence, to explore an unfamiliar space. Monkeys were willing to press levers not just for cookies and juice but simply to look out a window.[19] There was little room in the strict Skinnerian world of extrinsic rewards and punishments for such behaviors, and there was no easy story to explain them.

And yet this intrinsic motivation, as Berlyne saw, was every bit as central to human nature as the drives for, say, food and sex—despite being "unduly neglected by psychology for many years."[20] (Indeed, the severest punishment our society allows, short of death—solitary confinement—is, in effect, the infliction of *boredom* on people.) In his landmark 1960 book *Conflict, Arousal, and Curiosity,* Berlyne notes that a proper study of curiosity first began to emerge in the late 1940s; it is no coincidence, he argues, that information theory and neuroscience also came into their own at the same time.[21] A proper understanding of curiosity appears only to be possible at the interdisciplinary junction of all three.[22]

Berlyne appeared to be just as strongly motivated by curiosity in his own life as he was motivated by it as a subject of study. He knew at least ten languages (and was fluent in six or seven of them) and was an accomplished pianist, as well as a jogger and traveler. At the time of his early death at fifty-two, he was on a quest to ride every subway in the world. Despite a prodigious and prolific output of articles and papers, he rarely worked on nights or weekends. There was too much else to do.[23]

His ideas, in particular the agenda of reaching out to both neuroscience and information theory for clues, would inspire succeeding generations of psychologists for the latter half of the twentieth century, and in the twenty-first they would come full circle. Starting at the end of the 2000s and continuing through the deep-learning boom of the 2010s, it was the mathematicians and information theorists and computer scientists—stuck on the problem of intrinsic motivation in cases like *Montezuma's Revenge*—who would be turning to *his* ideas for help.

At a broad level, he argued, humans' intrinsic motivation seemed to entail related but distinct drives: for novelty, and for what we may call surprise. Each of these, in turn, has offered tantalizing ideas about motivation and learning that seem just as applicable, in this decade, to machines as to ourselves.

NOVELTY

The question that researches have usually been designed to answer is "What response will this animal make to this stimulus?" . . . As soon as the experimental situation is made more complex . . . a new question arises: "To which stimulus will this animal respond?"

—DANIEL BERLYNE[24]

When I'm caught between two evils, I generally like to take the one I've never tried.

—MAE WEST[25]

One of the concepts at the heart of human curiosity, of our intrinsic motivation, is *novelty*. In the absence of strong compelling incentives, we don't act randomly, as simple reinforcement learners using epsilon-greedy exploration do. Rather, we are quite robustly, reliably, and predictably drawn toward *new* things.

In the mid-1960s, Robert Fantz of Case Western noticed that human infants as young as two months old would reliably spend less time looking at magazine pictures if they'd already been shown those pictures before.[26] What Fantz came to appreciate was that long before infants have the motor skills to explore the world *physically*, they are still capable of exploring the world *visually*—and are very clearly drawn to do so.

This behavior—known as "preferential looking"—has become a cornerstone result in developmental psychology, one of the most striking characteristics of infant behavior.

Infants' preference for looking at new things is so strong, in fact, that psychologists began to realize that they could use it as a test of infants' visual discrimination, and even their *memory*.[27] Could an infant tell the difference between two similar images? Between two similar shades of the same color? Could an infant recall having seen something an hour, a day, a week ago? The inbuilt attraction to novel images held the answer. If the infant's gaze lingered, it suggested that the infant could tell that a similar image was nonetheless different in some way. If the infant, after a week without seeing an image, didn't look at it much when it was shown again, the infant must be able at some level to *remember* having seen it the week before. In most cases, the results revealed that infants were more cognitively capable earlier than had been previously assumed. The visual novelty drive became, indeed, one of the most powerful tools in psychologists' toolkit, unlocking a host of deeper insights into the capacities of the infant mind.[28]

It didn't take long for the reinforcement-learning community to latch onto this idea of an intrinsic novelty preference, and to see what could be done with it in the computational domain.[29] One of the most straightforward ideas would be simply to *count* how many times a learning agent has been in a particular situation before, and then to have it prefer—all things being equal—to do the thing it had done the fewest times before. Richard Sutton, for instance, in 1990 suggested adding such an "exploration bonus" to the agent's reward if it took an action it had never taken before, or one it had not tried in a long time.[30]

There is a quite obvious fly in the ointment here, however. What does it mean to count the number of times we took "this action" in "this situation"? As Berlyne put it in 1960, "The word 'new' is used commonly in everyday speech, and most people seem to understand it without much difficulty. But when we ask what exactly it means to say that a stimulus pattern is novel and how novel it is, we face a whole succession of snares and dilemmas."[31]

For simple environments like solving a maze or for very, very simple games, it is of course possible to simply keep a list of every single situation you encounter and add a tally mark each time you return. (Tracing a path through a maze in pencil essentially keeps this record on the maze itself.) This approach—keeping a giant table of every situation you've been in, what you did, and what happened—is known as "tabular" RL, and unfortunately it's known for being totally infeasible in anything but a very small environment. Tic-tac-toe, for instance, is about as simple as a board game gets, and yet it has thousands of unique board positions.[32] The total count of possible positions in the game of Go is a 170-digit number. All the computer memory in the world put together wouldn't come close to storing that table.

Beyond these pragmatic issues, however, the deeper and more philosophical question, in more complex environments, is what it means to be in the "same" situation in the first place. In an Atari game, for instance, there are so many different ways that the pixels can appear that dutifully keeping track of every single screen you've ever momentarily encountered and slightly favoring novel ones is simply not helpful for generating interesting behavior. For games of reasonable complexity, you may be unlikely to *ever* see exactly the same set of pixels more than once. From that perspective, almost every situation is novel, almost every action untried. Even if you could store such a table, it wouldn't be much of a guide.

When in the course of everyday human decision-making, someone says to us that they've "never been in that situation before," we don't normally take them to mean "at this exact latitude and longitude at this exact nanosecond with this exact sun-dappled pattern hitting my retina and this exact sequence of thoughts in my mind," or else the statement would be effectively true by definition and robbed of all meaning. What we mean to refer to are the sometimes ineffable *key* features of the situation, and we judge its novelty by those.

In an Atari game, what we'd want would be some way to gauge whether the situation we are in—as represented by the pixels on the screen—was *meaningfully* similar to one we'd been in before. We'd want to be able to make connections across situations that share some deeper, non-trivial similarity.

At DeepMind in London, Marc Bellemare was interested in thinking about how this venerated but impractical idea of counting how many times you'd seen something before—known, appropriately, as the "count-based" approach—could be extended to more complex settings. In an Atari game like *Frogger* or *Freeway*, where you're trying to cross a busy road, ideally each time you successfully cross it should increment a kind of "count" of how many times you've done so—even if the traffic itself was always in some new, random pattern at the time.

Bellemare and his colleagues were playing around with a mathematical idea called "density models," and it seemed to show some promise.[33] The basic idea was to use unsupervised learning to make a model that could predict missing parts of an image from the surrounding context. (This is not unlike word2vec and similar language models, designed to predict missing words in a passage of text.) They could feed this density model all of the screenshots that the agent had seen thus far, and then use its predictions to assign a numerical probability score representing how "predictable" a new screen was, given what it had seen already. The higher the probability, the more familiar; the lower the probability, the more novel. It was an intriguing idea, but it was an open question how something like this might actually work in practice.

They began to do some experiments with an early-1980s Atari game called *Q*bert*, in which you hop around a pyramid made of square tiles, turning each of them to a different color until you've gotten them all, before moving on to a brand-new, differently colored stage where you do the same. They had a randomly initialized, blank-slate DQN agent play *Q*bert* from scratch, while they watched a meter on the left-hand side of the screen that gauged the "novelty"—as measured by the density model—of what the agent was seeing and experiencing.

At first—just as it is with us—*everything* was new. The meter was pegged at the maximum value. Every image on the screen registered as almost totally novel.

They trained the agent for several hours, allowing it to gradually get more adept at earning points (in the case of *Q*bert*, 25 points for every tile it flipped). They checked back in and watched it play. The now somewhat experienced agent hops around, picking up points. The green bar of the novelty meter (technically the "inverse log probability") barely flickers up from the bottom. The agent has seen it all before.

Bellemare found the training run in which the agent managed to complete the game's first level for the very first time. He watched it in replay, wondering what would happen. Satisfyingly, as the agent begins to get close, that green bar starts creeping up again.

"Now," says Bellemare, "as the agent is getting towards the end of the level, it starts to say, *Hey, these situations are novel! I haven't really been in this state before. This seems very new to me.* And you have this very nice progression: as we're getting closer and closer to finishing the game, this signal is going up."[34]

The agent hops down to the final tile, completing the level for the very first time. Suddenly, the whole screen flashes and strobes. The board resets to the next stage, with the pyramid of cyan tiles gone, and a totally new pyramid, this one bright orange, in its place. "And look at this!" Bellemare says. The green bar shoots all the way up. The novelty signal is practically off the charts. "Immediately the agent knows, *I have never been here before.*"

This density model was capturing, or so it seemed, a faithful notion of novelty in diverse, highly complex environments that were too big and too rich to actually count directly. "We looked at these results and we thought, there *has* to be something we can do with this."

Now the question was, Could they use this model—they would dub it the "pseudo-count"—to motivate an agent to *seek* these novel states?[35] What would happen if you actually *rewarded* the agent, not just for scoring points, but simply for seeing something new? And would that, in turn, make for better agents, able to make faster progress than those trained only to maximize reward and occasionally mash buttons at random?

It was clear what the payoff would be if they succeeded. "We got really excited," he says, "about trying to crack *Montezuma's Revenge.*"

The temple in which Panama Joe finds himself trapped has twenty-four chambers. After the equivalent of playing the game day and night, without sleeping, for three weeks, the DQN agent that had shown itself capable of superhuman performance at dozens of other Atari titles had only made it to the *second* room—barely out of the starting blocks. There was a whole temple out there to explore, full of dangers and bereft of points; maybe an agent directly rewarded for seeing things it had never seen before was just the kind of agent likely to get somewhere no agent had ever been.

Bellemare and his group wagered that an agent given access to these novelty signals, if it treated them as supplementary rewards to the in-game points, would be far more motivated and more successful at playing the game. They tried it, and let it train for an equal amount of time as the original DQN agent—a hundred million frames, or nearly three weeks of 24/7 play. The difference was shocking.

The same DQN agent, trained with novelty-based rewards, was dramatically faster at getting the first key, and ultimately made it through not two but *fifteen* of the temple's chambers.

Not only is the novelty-driven agent getting more points, it appears also to be exhibiting a different kind of behavior—a qualitative as well as quantitative difference. All that "preferential looking" can be harnessed to make it succeed in exploring terrain where rewards alone aren't enough. And something about the novelty-driven agent just seems more *relatable*—more human, even. It is no mere joystick wiggler when rewards are scarce. It has a drive.

"*Immediately,*" Bellemare says, "the pseudo-count agent goes out and explores this world."

THE PLEASURE OF SURPRISE

The mechanism of epistemic curiosity . . . works through the equivalent of conceptual conflict, and its function is an eminently motivational one.
                                                                —DANIEL BERLYNE[36]

It is not that children are little scientists but that scientists are big children.
                                                                —ALISON GOPNIK[37]

The other high-level concept integral to curiosity, along with novelty, is *surprise*.[38] A curious child cares not only that things are at some level "new," but also that things have something to *teach* them. A baseball with purple polka dots on it—let's say—is intriguing for a moment, but if it otherwise behaves exactly like a standard baseball, the intrigue will be short-lived. Rather, we maintain our interest in things that seem to defy our expectations, that behave unpredictably, that dare us to try to understand what will happen next.

One of the researchers on the forefront of understanding the role of surprise in human curiosity is MIT's Laura Schulz. In a 2007 study, she had children play with a kind of jack-in-the-box toy, where levers would raise various puppets up through the lid of the box.[39] The researchers would briefly take the toy away, and then return with both the familiar box and a new, differently colored box. They would set them both in front of the child and walk away, waiting to see which one the child reached for.

"Now, *everything* we've known before about play and curiosity in children," Schulz explains, "says, Well, if four-year-olds have been playing with one box for a while, then if you bring out a new box they should go *right* for the new box. They should go immediately to play with the new box because the basic idea about curiosity would be it's about *perceptual* novelty, perceptual salience: this is something they haven't seen."[40]

But there was more to it than that, Schulz found. In some trials, the demonstration of the first box was made deliberately ambiguous. The box had two levers, and if they were pressed simultaneously, two different puppets would just as simultaneously rise up out of the lid. It wasn't clear what raising one of those levers alone, or the other, would do. Was one of the levers responsible for both puppets and the other ineffectual? Was each lever responsible for the puppet closest to it? Or the one on the opposite side? In these ambiguous cases, the four-year-olds *didn't* immediately switch from the familiar toy to a novel one when given the chance. Instead, they persevered, reaching back for the two-lever box in order to figure out exactly how it worked.

"We often seem to be curious," says Schulz, "about things that aren't particularly novel—they just puzzle us."[41]

So a picture began to emerge that *surprise*—uncertainty, the ability to resolve ambiguity, to gain information—was every bit a driver of children's intrinsic motivation as novelty.

This idea has led to a second vein of research, every bit as rich, spanning both cognition and computing.

Schulz, working with Rutgers psychologist Elizabeth Bonawitz and a group of collaborators, did a further study using weighted blocks. Prior research had shown that around age six, children begin to have theories about how best to balance blocks of different sizes and shapes. Some children at that age assume (incorrectly) that blocks can always be balanced halfway between their two ends, even if the object is asymmetrical, while others theorize (correctly) that blocks are balanced at their center of *mass*, at a point closer to whichever is the thicker side. This led the researchers to use magnets to create blocks that could be made, ingeniously, to violate the assumptions of either group. When children played with blocks that behaved as they expected, their standard novelty bias crept in and led them to abandon the blocks in favor of another, newer toy when offered the chance. But the children for whom the blocks appeared to *violate* their theory of how blocks should balance— regardless of whether their theory was actually correct!—stayed riveted and continued playing with them, even when another toy was on offer.[42]

Four- and five-year-olds, who tended to lack a concrete theory about how best to balance the blocks whatsoever, almost invariably preferred a new toy when it was available. No matter what the blocks' behavior, it seemed the younger children didn't know enough, or have strong enough beliefs or predictions, to *be* surprised.

Other research following in this vein—for instance, a 2015 study by Aimee Stahl and Lisa Feigenson of Johns Hopkins—has shown further that the *way* infants play with a toy is also related to the *way* in which a toy is surprising.[43] If a toy car is shown appearing to mysteriously float in midair, infants will play with it by lifting it up and dropping it. If, however, the car is shown appearing to mysteriously pass through a solid wall, the infant will play with it by banging it on the table. And in each case, the infant will choose to stay engaged with the surprising toy when offered a chance to try a new one. (A control group who doesn't get to see the toy defy their expectations, reliably prefers the new toy.) Already by the age of 11 months, say Stahl and Feigenson, infants use "violations of prior expectations as special opportunities for learning."

"It's easy to look at a baby and see a blank slate," says Feigenson. "But actually, babies have rich, sophisticated expectations about the world— maybe more than people give them credit for." Babies, she argues, "use what they already know about the world to motivate or drive further learning, to figure out what they should learn more about."[44]

This idea, in computational terms—of an agent motivated not only by rewards but by trying to understand and predict the environment—is as old as reinforcement learning itself. And it has just as suddenly sprung into fruition.

Daniel Berlyne had seen some of the earliest experiments with machine learning in the 1950s, and mused about the use of surprise or misprediction as a reinforcer. "Further research may well be aimed at devising a problem-solving machine that will improve its technique in the light of its experience," he wrote. "The reduction of mismatch or conflict would then have to be the reinforcing agent, causing the immediately preceding operations to move up in the machine's order of precedence."[45]

German AI researcher Jürgen Schmidhuber has since 1990 been exploring the idea of agents that get reward from learning how their environment works—that is, from improving their ability to make predictions. "They may be viewed as simple artificial scientists or artists," he explains, "with an intrinsic desire to build a better model of the world and of what can be done with it."[46] For Schmidhuber, just as it was for Berlyne in the '60s, this idea of learning has its mathematical roots in information theory and—to Schmidhuber's mind in particular—the notion of data compression: that a more readily understood world is more concisely *compressible*.

In fact, for Schmidhuber the idea that we go through the world striving to better compress our representation of the world offers a "formal theory of creativity and fun." He explains: "You just need to have a computational resource—and before learning a pattern, you need *so* many computational resources, and afterwards you need *less*. And the difference, that's where you're saving. And your lazy brain likes to save things. And—" He snaps his fingers. "That's the fun!"[47]

Like Berlyne, Schmidhuber is fascinated not by what people do in order to solve problems posed to them directly—how to win a game or escape a maze, for instance—but rather, what people do specifically in those times when there *isn't* anything explicit to do.

Infants, he thinks, are the perfect examples of this. "Even when there is no immediate need to satisfy thirst or other built-in primitive drives, the baby does not run idle. Instead it actively conducts experiments: what sensory feedback do I get if I move my eyes or my fingers or my tongue just like that?"[48]

As Schmidhuber notes, there is a fundamental *tension* at the heart of curiosity, almost a tug-of-war: As we explore an environment and our available behaviors within it—whether that's the microcosm of an Atari game, the real-world great outdoors, or the nuances of human society—we simultaneously delight in the things that surprise us while at the same time we become harder and harder to surprise. It's almost as if the mind comprises two different learning systems, set at cross-purposes to each other. One does its best not to be surprised. The other does its best to surprise it.[49]

Why not, then, attempt to model this tension directly? A group from UC Berkeley, led by PhD student Deepak Pathak, set out in 2017 to build just such an agent. Pathak created an agent composed of two different modules—one designed to predict the outcome of a given action, rewarded when the reality matches its prediction, and the other designed to take maximally surprising actions, rewarded every time the predictor is *wrong*.[50]

In *Super Mario Bros.*, if you've just hit the jump button, you can expect to see Mario slightly higher on the screen in a moment, for instance—though only if you've tried it a few times already. If you've hit the down arrow, you can expect to see Mario crouch—but you may *not* be expecting that this will make Mario disappear down sewer pipes into a massive subterranean underworld! The crucial idea was to incentivize the agent to explore the game by making such surprises as delightful for the agent as they are for us—namely, by making these prediction errors into *rewards*. Doing anything whose outcome was surprising could be made just as good, and that action just as strongly reinforced, as an action that explicitly garnered points.

Pathak and his group looked at the kind of behavior that such a surprise reward might engender. Using a 3D maze environment (built with the engine of classic '90s first-person shooter *Doom*), they placed their agent farther and farther away from a rewarding "goal" state in a number of mazes. The agents trained only on the explicit reward of discovering the goal had a tendency to simply "give up" if they couldn't find it by random joystick wiggling and button mashing. The agents with surprise-based rewards explored the maze for its own sake: What's around this corner? What does that distant room look like up close? As a result, these curious agents found their way to the goal in much more vast and complex mazes than the agents without this intrinsic drive.

Pathak's Berkeley group teamed up with a group of researchers from OpenAI, and together they continued to explore this idea of using prediction error as a reward signal. Surprisingly, they found that a dramatic simplification of this architecture—replacing the network designed specifically to predict controllable aspects of the future with one designed to predict *random* features of the image on screen—worked just as well and in some cases even better.[51] The researchers at OpenAI, led by Yuri Burda and Harrison Edwards, worked to refine this idea, which they dubbed random network distillation, or RND.[52] It wasn't long before they began to set their sights on *Montezuma's Revenge*.

They turned their RND agent loose in the temple. Spurred by the intrinsic reward of surprise, it consistently managed to explore, on average, twenty to twenty-two of the temple's twenty-four rooms. And in just one of their trial runs, the agent does something unprecedented. It makes it all the way to the twenty-fourth and final room, in the temple's bottom-left corner, and escapes the temple.[53] Panama Joe steps through the final door and finds himself in front of a background of uniform blue filled with gems. He seems to fall through the sky. It's the closest thing *Montezuma's Revenge* offers to transcendence. The gems are worth a thousand points each—and it's *very* surprising.[54]

BEYOND REWARDS

It is becoming increasingly clear, then, that "intrinsic motivation"—conceived as novelty or surprise or some other related scheme—is an incredibly helpful drive for a system to have in order to augment the external rewards that come from the environment, particularly in cases where those external rewards are scarce or hard to come by.

Of course, from this perspective it is tempting to ask what might happen if we take this idea of algorithmic curiosity to its logical conclusion, and have reinforcement-learning agents that—paradoxically—don't care about external rewards *at all*?

What might such an agent look like? What might it do?

Almost anyone who studies intrinsic motivation has wondered the same thing, and a picture is starting to emerge.

Marc Bellemare and his colleagues at DeepMind have continued to pursue the idea of extending count-based novelty bonuses into more complex domains, and in follow-up work they have investigated what they call "pushing the limits of intrinsic motivation."[55] They amplified their agent's novelty reward some 10- to 100-fold, and observed qualitative as well as quantitative shifts in behavior.

Expectedly enough, the agent's behavior exhibited a kind of restlessness. Unlike pursuit of the in-game score, which often leads to a fairly stable and consistent set of best practices, for the "maximally curious" agent, the only reward *is* from this exploratory behavior, and those rewards aren't stable—they disappear as parts of the game environment become more familiar.[56] So the agent keeps restlessly chasing after them rather than settling into a pattern.

What was less expected was how *well* the agent did at the game, having been divorced from the game score. The agents with hyperinflated novelty bonuses actually achieved *state-of-the-art* scores in four different games. Curiosity bred competence. Amazingly, the novelty rewards *alone*, with no access to the in-game score whatsoever, were sufficient to play many Atari games competently—as measured by the score to which they didn't have access!

Of course, it must be said that games (good games, anyway) are designed to appeal to intrinsically motivated humans. The points, after all, are just more pixels, in a corner of the screen, about which human players can decide to care, or not. So at that level it makes sense that curiosity and a drive for exploration would prove to be a decent proxy for maximizing score, at least in most games. In *Super Mario Bros.*, for instance, points are awarded for grabbing coins, breaking blocks, and jumping on enemies—but the *point* of the game is to move Mario forward to the right, where an unpredictable landscape awaits. In this sense, an intrinsically motivated agent is probably more aligned with the game's intended mode of play than one driven to rack up these (ultimately meaningless) points.

Pathak's group at Berkeley and Burda and Edwards's group at OpenAI have continued to pursue these questions as well, teaming up on a large-scale, systematic study of learning with no extrinsic rewards whatsoever.[57]

One of their most striking findings was that in most cases it's not necessary to explicitly tell the agent whether it has died. If you're trying to maximize extrinsic score, this is very useful indeed, as it's both a final verdict on the score you *did* get, as well as an indicator that you can expect zero additional points from that moment onward (which is usually a disincentive to meet that same fate a second time). With a purely curiosity-driven agent, death simply means starting the game again from the beginning—which is very boring! The beginning of the game, being the most familiar part, is neither novel nor surprising. This is all the disincentive, it turns out, that the agent needs.[58]

They also found an intriguing exception to the pattern of intrinsically motivated agents proving surprisingly adept at point scoring. The exception was *Pong*. An agent motivated purely by intrinsic rewards, one that cares nothing whatsoever about the score, plays the game not to score on the opponent, but rather to deliberately extend rallies as long as possible. The "reset" after a point is scored is essentially the same as the "reset" that occurs upon death in other games. Returning to the well-worn starting position is simply boring compared to the atypical and unusual positions that arise in a long rally.

The team was intrigued to see what might happen if such an agent was given the chance to play against a copy of *itself*. How would curiosity versus curiosity unfold in a zero-sum game? The answer: a non-zero-sum collaboration emerges, as both sides pursue the shared goal of moving away from the game's well-trodden starting state. In other words: they rally and rally and never stop. "In fact," the researchers write, "*the game rallies eventually get so long that they break our Atari emulator.*" The screen starts to glitch out, with spots of color flickering in random patches. The surprise-seeking agents, of course, are delighted.[59]

The idea of unplugging the game score and creating an agent motivated *only* intrinsically may seem at some level like an odd experiment: the field of reinforcement learning has since its inception been orchestrated around the maximization of external reward. Why give up the one thing by which behavior is being measured?

The University of Michigan's Satinder Singh, working with Michigan psychologist Richard Lewis and UMass Amherst's Andrew Barto, explored this question philosophically, asking, "Where do rewards come from?"[60] They note that the evaluation of how good or bad some state of affairs is is done *in the brain*—not in the environment. "This view makes it clear," they write, "that reward signals are always generated within the animal, for example, by its dopamine system. Therefore, *all rewards are internal.*"[61]

To play an Atari game using nothing but the pixels on the screen, of which you can make whatever you want—rather than mainlining some fiat reward signal—is, after all, exactly what it is *actually* like to play video games.

One of the most critically acclaimed computer games of the 2000s, *Portal*, involves the game's AI repeatedly promising the player "a cake" for completing the game. Partway through, however, the player discovers ominous graffiti containing what would become the game's most memorable catchphrase: "The cake is a lie." Indeed, no cake is offered at the game's end. Of course, this famous betrayal is undercut not only by the fact that it would have been a digital representation *of* cake at best, but because we aren't playing the game under the illusion that there will be anything in it for us other than the pleasure of making progress, furthering the plot, and exploring the world of the game.

We don't use the flashing lights on the screen as data that we can leverage to gain "real" rewards in that environment. The flashing lights, and whatever reaction they provoke in us, are all the reward there is. And that seems, judging by the number of hours we devote to video games, to be more than enough.

BOREDOM AND ADDICTION

As it happens, intrinsic motivation in RL is not only the source of these virtuous behaviors—in which one recognizes at least a glimmer of the human desire to know, to explore, to see what happens—it also holds a mirror image of human pathologies: both boredom and addiction.

I ask Deepak Pathak if the notion of boredom makes sense. Is it possible for an agent to get bored?

Absolutely, he says.

In the first level of *Super Mario Bros.*, there is a chasm that his agent almost never figures out how to cross, because it requires the agent holding down the jump button for fifteen frames in a row; long sequences of precise actions are much more difficult to learn than shorter or more flexible patterns.[62] As a result, the agent reaches the edge of the cliff and just . . . tries to turn around.

"So it just cannot cross it," Pathak says, "so it's like a dead end, the end of the world." But the game is built so that there is no way to backtrack. The agent is stuck, and learns to do nothing at all.

There's a more general *ennui*, too, that Pathak has observed. After his *Super Mario Bros.* agent has played the game long enough, "It just starts to stay in the beginning. . . . Because there is no reward anywhere—everywhere error is very, very low—so it just learns to not go anywhere." The agent simply loiters at the very start of the game, unmotivated to do anything at all.

There is at least an iota of pathos here. A human who has grown bored with a game can stop playing, and usually will. We can swap the old game out for a new one or just turn off the screen and move on to something different altogether. In contrast, the agent is, almost cruelly, *trapped* inside a game it no longer has any drive to play.

Ever since there have been video games, there has been a subfield of study into the question of what makes them fun, and what makes one game more fun than another. There are obvious economic as well as psychological stakes in this.[63]

It occurs to me that reinforcement learning has furnished us with a practical benchmark for not just how *difficult* a game is—how long it takes the agent to become proficient—but also how *fun:* how long the agent plays before losing interest and disengaging, or whether it elects to spend its time playing that game over another. It may well be the case that video games of the coming decades are heavily focus-grouped by intrinsically motivated RL agents.

Cognitive scientist Douglas Hofstadter, in his 1979 Pulitzer Prize–winning book *Gödel, Escher, Bach,* imagined the future of advanced game-playing programs, envisioning a link between game-playing competence, motivation, and intelligence:

> QUESTION: Will there be chess programs that can beat anyone?
> SPECULATION: No. There may be programs that can beat anyone at chess, but they will not be exclusively chess programs. They will be programs of general intelligence, and they

will be just as temperamental as people. "Do you want to play chess?" "No, I'm bored with chess. Let's talk about poetry."

The quote looks hilariously dated now—and of course we know, with the benefit of hindsight, that it would be less than twenty years before IBM's Deep Blue chess machine emerged victorious in 1997 over human world champion Garry Kasparov. Deep Blue was, indeed, exclusively a chess program—it was customized at the level of its hardware to do chess and chess only. It was certainly not generally intelligent; nor did it yearn to think about literature rather than chess.

But perhaps there is a kernel of something nonetheless fundamentally true here. Contemporary state-of-the-art reinforcement-learning systems really *are* general—at least in the domain of board and video games—in a way that Deep Blue was not. DQN could play dozens of Atari games with equal felicity. AlphaZero is just as adept at chess as it is at shogi and Go.

What's more, artificial general intelligence (AGI) of the kind that can learn to operate fluidly in the real world may indeed require the sorts of intrinsic-motivation architectures that can make it "bored" of a game it's played too much.

At the other side of the spectrum from boredom is *addiction*—not a disengagement but its dark reverse, a pathological degree of repetition or perseverance. Here, too, reinforcement learning has come to exhibit behaviors that are in some cases uncannily, uncomfortably human.

Researchers who study intrinsic motivation talk about what they call the "noisy TV" problem. What if there is a source of randomness or novelty in the environment that is essentially inexhaustible? Will the intrinsically motivated agent simply be powerless to resist it?

In concrete terms, imagine there were a source of unpredictable visual noise on the screen: a staticky TV is the classic example, though crackling flames or rustling leaves or rushing water would all qualify as well. If this were the case, each new and unpredictable configuration of light and shadow would act as a kind of endless curiosity jackpot. In *theory*, at least, an agent confronted with this should become instantly stupefied.

Most of the simple Atari games from the 1970s and '80s, though, don't happen to contain such sources of visual randomness, and so it hadn't been demonstrated empirically. Pathak, Burda, and Edwards decided to bring the thought experiment to life and try it out. They created a simple 3D maze game, where the agent is required to explore the maze and find an exit. In one version of the game, however, there is a television screen on one of the walls of the maze. Furthermore, the agent is given the ability to press a button that changes the channel on the television. What would happen?

What happened is that the *instant* the agent comes within view of the TV screen, its exploration of the maze comes to a screeching halt. The agent centers the screen in its view and starts flipping through channels. Now it sees a video of an airplane in flight. Now it sees cute puppies. Now it sees a man seated at a computer. Now cars in downtown traffic. The agent keeps changing channels, awash in novelty and surprise. It never budges again.

Visual information isn't the only source of randomness that can have these dangerous effects; so can something as simple as flipping a coin. This had almost a decade earlier weighed on the mind of DeepMind researcher Laurent Orseau, the first hire onto their Safety team, and who is now part of their Foundations research group. Long before the arrival of intrinsically motivated Atari-playing agents that could be turned instantaneously into couch potatoes, Orseau was thinking about a much more powerful agent, transfixed by a coin.

Orseau was thinking about a hypothetical agent he called the "knowledge-seeking agent," an agent "whose goal is to gather as much information about the unknown world as possible."[64] Orseau's agent was based on a theoretical framework called AIXI, which imagines an agent capable of infinite computation. Such a mentally omnipotent agent could never exist in reality, of course, but it acts as a kind of reference point: If you *could* think forever before taking an action, which one would you take? Amazingly, a number of conceptions of an infinitely resourceful knowledge-seeking agent fell into complete degeneracy at the sight of a coin—"preferring to observe coin flips rather than explore a more informative part of their environment," Orseau notes. "The reason for this is that these agents mistake stochastic outcomes for complex information."[65]

B. F. Skinner, when he wasn't training pigeons, was fascinated by gambling addiction. The house always wins, on average, and psychology all the way back to Thorndike had been based on the idea that you do something more when it's on balance good for you—and less when it's bad. From this view, something like gambling addiction was impossible. And yet there it was, a presence in the real world, daring the behaviorists to make sense of it. "Gamblers appear to violate the law of effect," Skinner wrote, "because they continue to play even though their net reward is negative. Hence it is argued that they must be gambling for other reasons."[66] We now appear to have a pretty good candidate for what those other reasons might be.

Gambling addiction may be an overtaking of extrinsic reward (the house always wins, after all) by *intrinsic* reward. Random events are always at least *slightly* surprising, even when their probabilities are well understood (as with a fair coin, for instance).

In Chapter 4 we talked about the role of dopamine in encoding temporal prediction errors: cases where a reward is better or worse than expected. However, there have been some curious cases that don't fit this pattern. Namely, there is growing evidence, since the turn of the century, that things that are *novel* and *surprising* trigger the release of dopamine, whether or not they have any "reward" associated with them at all.[67]

Just as the reinforcement-learning community is discovering the value of making novelty and surprise function as rewards in their own right, the neuroscience community is uncovering such machinery at work in our own heads.

At the same time, it is increasingly clear how these mechanisms, which normally help us, can go awry. Reinforcement-learning agents can become addicted to changing channels and playing slot machines—and so, of course, can we. Because the outcome of actions like these is never exactly what we thought it would be, there is always something surprising about that activity, always seemingly something to "learn." We don't think of addiction as a surfeit of *motivation,* an overabundance of *curiosity,* but something along these lines may well be how it does, in fact, function.

ITS OWN SAKE

The computational study of intrinsic motivation offers us a powerful toolkit for making headway in difficult learning environments—*Montezuma's Revenge* being just one example. At a deeper level, it also offers us a story, rooted in these empirical successes, for why *we* may have such striking motivations ourselves.

In better coming to understand our own motivations and drives, we then, in turn, have a chance for complementary and reciprocal insights about how to build an artificial intelligence as flexible, resilient, and intellectually omnivorous as our own.

Deepak Pathak looks at the success of deep learning and sees one glaring weakness: each system—be it for machine translation, or object recognition, or even game playing—is purpose-built. Training a huge neural network on a heap of manually labeled images was, as we have seen, the paradigm in which deep learning first truly showed its promise. Explicitly drilling a system to categorize images made a system that could categorize images. Fair enough, he says. "But the problem is, these artificial intelligence systems are not actually intelligent. Because they're missing a key component which is very central to humans, which is this general-purpose behavior, or general-purpose learning system."[68]

To make a general-purpose system will require breaking out of this task-specific mind-set, he argues—and it will also require breaking down one crucial, conspicuous artifice: the enormous amount of explicit reward information these models require. An image-labeling system like AlexNet might require hundreds of thousands of images, each of them labeled by humans. That's quite clearly *not* how we acquire our own visual skills early in life. Likewise for reinforcement learning, where in the world of Atari games, every tenth of a second the game tells you with perfect authority *exactly* how you're doing. "It works very well, but it requires something, again, very, very weird," he says, "which is this reward."

Unplugging the hardwired external rewards may be a necessary part of building truly general AI: because life, unlike an Atari game, emphatically does not come pre-labeled with real-time feedback on how good or bad each of our actions is. We have parents and teachers, sure, who can correct our spelling and pronunciation and, occasionally, our behavior. But this hardly covers a fraction of what we do and say and think, and the authorities in our life do not always agree. Moreover, it is one of the central rites of passage of the human condition that we must learn to make these judgments by our own lights and for ourselves.

"The orthogonal agenda to just doing good exploration, for me, has always been to remove the reward altogether," says Bellemare. "This is stepping outside of my usual range of comfort, but you know, I think the most interesting thing we can do with AI agents is to have them come up with their own objectives, if you will. There are safety questions here," he acknowledges, "but I would like, as you say, my AI agent to be thrown into *Montezuma's Revenge* and just play it because it likes to play it."[69]

"We were talking earlier about the ALE [Arcade Learning Environment] and how it's been a great benchmark," Bellemare says. "And to me in some sense we're mostly done with the ALE. And we're mostly done with maximizing scores." For him, no high score that comes by way of epsilon-greedy button mashing, reinforced by point scoring, qualifies as intelligence—impressive though the results may be. "I actually think that we should be measuring intelligence in terms of how things behave—not in terms of reward function." What would such behavior look like? This is one of the central questions that drives him.

Orseau's work on knowledge-seeking agents also sketches what a mind motivated purely by the pursuit of knowledge might be like. An initial analysis is encouraging. An artificial agent motivated to maximize some kind of score, or achieve some goal state, will always be at risk of exploiting some loophole to do so; an even more intelligent agent may be inclined to hack the scoring system or to construct an escapist fantasy for itself where its goals are easier to achieve. Orseau emphasizes that while this seems like "cheating" to us, "it doesn't have a sense of cheating. It's just, 'Well, I do actions to maximize my reward.'" He elaborates: "The agent doesn't understand that it is something bad. It's just that it tries many different actions, and then this works: so why not do it?"[70]

The knowledge-seeking agent, though, can't take any such shortcuts. Self-deception, in particular, holds no interest or appeal. "So imagine that you are modifying your observations.... Then what information do you gain? Nothing. Because you can predict what it's going to be."[71] Because of this resilience, the knowledge-seeking agent "may therefore be the most suitable agent for an AGI in our own world, a place that allows self-modifications and contains many ways to deceive oneself."[72]

There are still reasons to hesitate before unleashing a superintelligent knowledge-seeking agent—seeking knowledge may involve commandeering various earthly resources in order to do so. But it is at least resilient to some of the most straightforward of traps. "If you could program it, I believe it would have an amazing behavior," says Orseau. "Because it would try to make sense as quickly as possible of its environment. It's basically the ultimate scientist. It would design experiments to try to understand what would happen.... I would really be curious to see how it would behave."

The concept—and the ethics—of an intelligence guided chiefly or purely by curiosity is hardly a new idea; it predates not only the past decade but, indeed, the past millennium. In Plato's famous dialogue *Protagoras*, Socrates reflects on this matter and puts it quite well indeed:

"Knowledge is a fine thing quite capable of ruling a man," Socrates says. "If he can distinguish good from evil, nothing will force him to act otherwise than as knowledge dictates, since wisdom is all the reinforcement he needs."[73]

PART III

Normativity

7

IMITATION

I was six years old when my parents told me that there was a small, dark
jewel inside my skull, learning to be me.

Watch this.

—ELON MUSK TO PETER THIEL, IMMEDIATELY BEFORE
LOSING CONTROL OF AND CRASHING
HIS UNINSURED $1 MILLION MCLAREN F1[2]

In English, we say that to imitate something is to "ape" it, and we're
not the only ones; this seemingly arbitrary linguistic quirk appears
again and again across languages and cultures. The Italian *scimmiot-
tare*, French *singer*, Portuguese *macaquear*, German *nachäffen*, Bul-
garian *majmuna*, Russian *обезьянничать*, Hungarian *majmol*, Polish
małpować, Estonian *ahvima*: verbs for imitation and mimicry, again and
again, have their etymologies rooted in terms for primates.[3]

Indeed, the simian reputation for being a great imitator, not just in
etymology but in science, goes back a century and a half at the minimum. As nineteenth-century biologist (and friend of Charles Darwin's)
George John Romanes wrote in 1882, on the subject of "what Mr. Darwin calls 'the principle of imitation'":

> It is proverbial that monkeys carry this principle to ludicrous
> lengths, and they are the only animals which imitate for the mere
> sake of imitating . . . though an exception ought to be made in
> favour of talking birds.[4]

Proverbial indeed, and across a surprising array of cultures and languages. And yet—ironically—this appears not actually to be true.

Primatologists Elisabetta Visalberghi and Dorothy Fragaszy, asking, "Do Monkeys Ape?," took a hard look at the evidence and were forced to conclude, through both a literature review and experiments of their own, that the data showed, in fact, an "overwhelming lack of imitation" in monkeys. "The lack of imitation in monkeys," they write, "is as apparent in tool-using behaviors as it is in arbitrary behaviors such as postures, gestures, or problem-solving."[5]

Subsequent work by comparative psychologist Michael Tomasello asked the same question of our slightly nearer primate kin—"Do Apes Ape?"—and came to a similarly decisive conclusion, with the *possible* exception of chimpanzees, our very closest genetic relatives. (Exactly to what degree chimpanzees *do* imitate in the wild, or *can* imitate when trained by humans, remains a nuanced and somewhat unresolved issue.) "My answer to the more general question of whether apes ape," says Tomasello, "is: only when trained by humans, either formally or informally, to do so (and then perhaps in only some ways)."[6] So the primate reputation for imitation is more or less totally undeserved.

There *is* a primate, however, that is a natural, uncanny, prolific, and seemingly hardwired imitator.

It's us.

In 1930, Indiana University psychologist Winthrop Kellogg and his wife, Luella, raised their infant son, Donald, alongside an infant chimpanzee named Gua for nine months, treating the two of them identically, like human siblings. In the book the Kelloggs wrote about the experience, *The Ape and the Child*, they noted that "because of the reputation of the chimpanzee as an imitator, the observers were on the alert from the start for the appearance of this sort of behavior. And yet, strange as it may seem, imitation in Gua was clearly less pronounced than in the boy."

Donald was a prolific imitator indeed, of both parents and his "sibling." At age seventeen months, he startled his father by pacing back and forth with his hands clasped behind his back—a spitting image of Winthrop himself in moments of deep concentration. More often, though, Donald imitated Gua, his playmate and peer. Even though Donald could already walk and, indeed, had hardly even crawled before learning to walk, he started to take after Gua and began crawling around on all fours. When a piece of fruit was nearby, Donald learned to grunt and bark the way Gua did. Slightly concerned, the Kelloggs soon called their experiment to an abrupt halt.[7]

Evidence that it is, in fact, we humans who are nature's imitators par excellence has continued to mount. For instance, when you stick your tongue out at a baby, they will—less than an hour after birth—stick their tongue back out at you.[8] The feat is all the more amazing considering that the child has never even seen themselves before—as UC Berkeley's Alison Gopnik notes, "there are no mirrors inside of the womb"[9]—and so the imitation is "cross-modal": they are matching how you *look* when sticking out your tongue to how they *feel* when doing so. All this in the first forty minutes.

This incredible capacity was first discovered by the University of Washington's Andrew Meltzoff in 1977; the finding turned a generation of psychological received wisdom on its head. Legendary Swiss developmental psychologist Jean Piaget (who ranked second in influence only to Freud as measured by citations over the twentieth century)[10] had written in 1937 that "during the earliest stages the child perceives things like a solipsist. . . . But step by step with the coordination of his intellectual instruments he discovers himself in placing himself as an active object among the other active objects in a universe external to himself."[11]

Meltzoff, while acknowledging the debt that all psychologists pay to the great twentieth-century Swiss psychologist, thinks that in this particular case, Piaget has it precisely backward. "We must revise our current conceptions of infancy," he says. "The recognition of self–other equivalences is the foundation, not the outcome, of social cognition."[12] Imitation, he says, is "the starting point for psychological development in infancy and not its culmination."[13]

This proclivity to imitate others begins almost instantly, but it is far from mere reflex. There is a surprising level of sophistication to who, what, and when children imitate, as we are only in the last couple decades coming to understand.

For instance, they will imitate an adult's action only if it does something interesting, rather than seems to have no effect.[14] Young children also appear to have a special sense that it is other humans that "make stuff happen"; they *won't* imitate the action if the object seems to move by itself, or if a robotic or mechanical hand does the motion.[15] (This has intriguing implications for the feasibility of robot nannies and teachers.)

It also seems that babies are keenly aware of when they're *being* imitated. There is a famous scene in the 1933 Marx Brothers film *Duck Soup* where Harpo pretends to be Groucho's reflection in a mirror, matching his every movement. Meltzoff did a study just like this, where adults would either imitate a baby's movements or simply go through a fixed series of movements planned in advance. Just like Groucho, the babies would concoct elaborate or unusual movements in order to test out whether the adult was indeed mimicking them.[16]

For Meltzoff, this deeply seated capacity to recognize ourselves in relation to others—whom we perceive in some fundamental way as *like* ourselves—is the beginning not only of psychological development but, as he puts it, "the kernel embryonic foundation for the development of social norms, values, ethics, empathy. . . . It's a big bang. The initial beginning is this imitation of bodily movements."[17]

OVERIMITATION

Imagine you're showing someone how to, for instance, chop an onion, and you say, "Now try it like this," then clear your throat and begin to demonstrate a cut. Your pupil observes you closely and, looking for your approval, *clears their throat* before making the same cut. They haven't just imitated you; they've *over*imitated you, by including in the imitation acts that simply aren't relevant or that have no ultimate causal impact on the task being performed.[18]

Researchers who study imitative behavior in humans and chimpanzees were surprised to discover that this sort of overimitation is much more frequent among humans than among chimpanzees. This seemed counterintuitive: How could it be that the chimpanzees were doing a better job at determining which actions were relevant and irrelevant, and then reproducing only the relevant ones?

One of the most revealing, and intriguing, studies involved plastic boxes with two locked openings: one on the top and one on the front. The experimenter demonstrated first unlocking the top opening, then unlocking the front, then reaching into the front to get a bit of food. When chimpanzees saw this demonstration using an opaque black box, they faithfully did both actions in the same order. But when the experimenters used a *clear* box, the chimpanzee could observe that the top opening had nothing whatsoever to do with the food. In this case, the chimpanzee would then go straight to the front opening, ignoring the top one altogether. The three-year-old children, in contrast, reproduced the unnecessary first step *even when they could see that it did nothing.*[19]

It was theorized that perhaps humans, in this instance, are simply slower to develop the relevant skill. Researchers shifted from studying three-year-olds to studying five-year-olds. The overimitation behavior was *worse!* The older children were *more* prone to overimitate than the younger children.[20] This made no sense. What on earth was going on?

The problem got stranger still. Researchers thought that the children were doing the overimitation to get the approval of the experimenter. They had the experimenter leave the room; it didn't help. And when researchers asked the children, both age three and age five, whether they could tell which of the demonstrated actions they "had to do" and which were "silly and unnecessary," the children could! But even when they'd gone through this process of showing the experimenter they knew the difference, they *still* reproduced both actions.[21]

Finally, the experimenters tried explicitly telling the children not to do anything "silly and extra." It didn't help. The children agreed to the instructions, then *still* overimitated.

Again, this seemed counterintuitive and almost paradoxical: that with increasing cognitive abilities, children exhibit an apparent "lessening of control, towards more 'mindless' blanket copying as [they] develop."[22]

One clue lay in a study of fourteen-month-olds by Hungarian psychologist György Gergely. Toddlers saw an adult, sitting at a table, lean forward to touch her forehead to a lightbulb, which made it light up. However, there was one crucial twist. Half the time, the adult's arms were resting on the table, and the other half of the time, the adult, pretending to be cold, had wrapped herself in a blanket. Toddlers who saw that the adult's arms were *free* reproduced the action exactly, bending forward to touch the lightbulb with their heads. But toddlers who had seen that the adult's arms were occupied by holding the blanket simply reached out and touched the lightbulb with their hand.[23]

There was something critical here. Barely over one year old, the toddlers were capable of assessing whether the experimenter had taken a strange action by choice or by necessity. *There must have been some reason why this grown-up bent over and touched the light with her head—her hands were right there!* If it seemed to have been a deliberate choice, they would reproduce it exactly. This showed the problem of overimitation in a new light. It was not "mindless" at all, not simply a slavish reproduction of the exact movement, but the opposite—a reasonable, sophisticated insight based on imagining the demonstrator as making rational choices and performing the action as easily and efficiently as possible.

Suddenly it began to make sense that such behavior *increased* from age one to three, and again from three to five. As children grow in their cognitive sophistication, they become better able to model the minds of others. Sure, they can see—in the case of the transparent cube—that the adult is opening a latch that has no effect. But they realize that *the adult can see that too!* If the adult can see that what they're doing has no apparent effect, but they still do it anyway, there must be a reason. Therefore, even if we can't figure out what that reason is, we'd better do that "silly" thing too.

The chimpanzee, in contrast, has no such sophisticated model of the human demonstrator. The logic seems to be much simpler: "The human is dumb and isn't taking the best action to get the food. Whatever. I can see the best way to get the food, so I'll just do that."

Suddenly the apparent paradox unraveled. In a strange, artificial scenario like this one, the chimpanzee happened to be correct. But it was the human child, "over"-imitating, who was the more cognitively sophisticated, with access to a deeper level of insight about the matter. Indeed, adults—unless they're doing laboratory studies—generally *don't,* as a rule, do pointless actions when they know better. The infant looks silly taking two actions to get the food when one would do, but only because the grown-ups are being, in a sense, deceptive or insincere. Shame on *them*!

More recent work has established just how subtle these effects can be. Children are, from a very young age, acutely sensitive to whether the grown-up demonstrating something is deliberately *teaching* them, or just experimenting. When the adult presents themselves as an expert—"I'm going to show you how it works"—children faithfully reproduce even the seemingly "unnecessary" steps that the adult took. But when the adult presents themselves as unfamiliar with the toy—"I haven't played with it yet"—the child will imitate only the effective actions and will ignore the "silly" ones.[24] Again it appears that seeming overimitation, rather than being irrational or lazy or cognitively simple, is in fact a sophisticated judgment about the mind of the teacher.

All of this serves to show the immense cognitive sophistication lurking behind what seems on the surface to be simple, rote imitation. The result is a newfound appreciation for the mental skills of toddlers, and a greater sense of the computational complexity of simply telling a machine-learning system to "watch this."

IMITATION LEARNING

If we humans are uniquely equipped for imitation, that invites an obvious question: Why? What is it about imitation that makes it such a powerful tool for learning? There are, as it happens, at least three distinct advantages that learning by imitation has over learning by trial and error and by explicit instruction.

We have seen how machine-learning researchers have borrowed directly from psychology for ideas like shaping and intrinsic motivation. Imitation has proven no less rich a source of inspiration; indeed, it forms the bedrock for many of AI's greatest successes in the twentieth and twenty-first centuries alike.

The first advantage imitation has is efficiency. In imitation, the hard-won fruits of someone *else's* trial and error are handed to you on a silver platter. In fact, no small part of the advantage of learning by imitation is knowing that what you're trying to do is even possible in the first place.

In 2015, famed rock climbers Tommy Caldwell and Kevin Jorgeson made history by completing the first successful ascent of Yosemite Valley's legendary three-thousand-foot Dawn Wall—called by *Outside* magazine "the world's hardest rock climb."[25] It took eight years for Caldwell and Jorgeson to plan their route, experimenting on different pitches of the cliff's face, trying to connect various doable segments and find a single feasible path all the way from the bottom to the top. The Dawn Wall was, in Caldwell's words, "infinitely harder than anything I had even contemplated climbing" before.[26] The rock looks at first like a perfectly blank face, daring you to imagine some way to find purchase. What little purchase there is seems outright hostile to human flesh. "This is the hardest thing you could ever do on your fingers, climbing this route," Caldwell says. "It's just grabbing razor blades."[27]

The following year, the young Czech climbing sensation Adam Ondra was able to replicate their feat after just a few *weeks* of scouting and practice. He attributes much of that speed to having been shown not just the *way* to make it up the Dawn Wall, but the mere fact that it was possible at all—something that Caldwell and Jorgeson did not know when they began the arduous process of planning their ascent. Said Ondra:

> The fact that Tommy and Kevin put *all* this effort—like, *years* and years of work . . . it's so impressive. . . . There are so many sections,

on the crux pitches, and even in some of the easier pitches—if you are there, you think, "No. This is impossible." And *only* after studying every single tiny razor blade, sometimes the only possible [answer] comes to your mind. So it was *so* much easier for me to solve the puzzle of each individual pitch, because I *knew* that the guys did it.... They sent the route, and I was very proud to make the first repetition.[28]

The first ascent took eight years of exhaustive search and self-doubt. The second ascent took a few weeks of study and rehearsal, buoyed by the confidence that no matter how impossible it looked, there *was* a way.

We saw earlier how the game *Montezuma's Revenge* requires an absurd number of things to go right even to earn its first reward, and that even then, the path to successfully completing it is razor thin, the encouraging feedback rare, and the consequences of failure dire. It is like the reinforcement-learning equivalent of the Dawn Wall—a nearly blank, impassive surface that dares you to find purchase. Even using powerful techniques like novelty bonuses and intrinsic motivation, an enormous number of attempts are still required for an agent to learn the mechanics of the game and the pathway to success. But what if the agent didn't have to explore the game for itself? What if it had a role model?

A group at DeepMind led by Yusuf Aytar and Tobias Pfaff, in 2018, came up with an ingenious idea. Might it be possible, they wondered, for an agent to learn how to play the game not by painstaking in-game exploration, but rather by . . . watching YouTube videos?[29]

It was audacious, and just crazy enough to work. YouTube was full of videos of human players playing the game. Their agent could, in effect, learn the point values of various actions by watching someone *else* take those actions first. Then, when it was turned loose to act for itself, it already had a basic idea of what to do. The first agent that trained to mimic these human players was, at the time, better than any agent trained with reinforcement learning on the game's rewards alone. In fact, preceding the breakthroughs at the end of 2018 using intrinsic motivation, it was the very first artificial agent, with a little initial help and inspiration from its human role models, to get out of the temple.[30]

The second critical thing that imitation confers is some degree of *safety*. Learning through hundreds of thousands of failures may work fine in the realm of Atari, where death is but a restart.[31] In other areas of life, however, we don't have the luxury of being able to fail hundreds of thousands of times in order to get things right. A surgeon, for example, or a fighter pilot, hopes to learn incredibly precise and elaborate techniques without *ever* making a critical mistake. Essential to this process is observing the live, recorded, or even hypothetical successes and failures of their forebears.

The third advantage of imitation is that it allows the student (be it human or machine) to learn to do things that are *hard to describe*. Nineteenth-century psychologist Conway Lloyd Morgan had this idea in mind when he wrote, "Five minutes' demonstration is worth more than five hours' talking where the object is to impart skill. It is of comparatively little use to describe or explain how a skilled feat is to be accomplished; it is far more helpful to show how it is done."[32] This is true when we are trying to articulate the actions we want: "Now bend your elbow at a twenty-seven-degree angle while flicking your wrist very quickly but not too quickly . . ." And it is just as true when we are trying to articulate the *goals* we want our pupil to pursue. In an Atari game, something like "maximize total score" or "complete the game as quickly as possible" may more or less suffice. But in real-world scenarios, it might be very hard to even communicate everything we want our learner to do.

Perhaps the canonical case for this is the car. We want to get from point A to point B as quickly as possible, though not by going over the speed limit—or, rather, not by going *too* far over the speed limit, unless for some reason we have to—and by staying centered in our lane, unless there's a cyclist or a stopped car—and not passing cars on the right, unless it's safer to do so than not to do so, and so on. It's hard to try to formalize all of this into some kind of objective function that we then tell the system to optimize.

Better, in cases like this, to use what the Future of Humanity Institute's Nick Bostrom calls "indirect normativity"[33]—a way to get the system aligned with our desires *without* articulating them down to the last minutia. In this case, what we want is to say something like "Watch how I drive. Do it like this."

This was, as it turns out, one of the very first ideas in self-driving cars—and, to this day, still one of the best.

STEERING

In 1984, DARPA began a project they called the Strategic Computing Initiative. The idea was to leverage the computing breakthroughs happening in the 1980s, and to turn that then-bleeding-edge technology into three specific applications. As Chuck Thorpe, just graduating with his doctorate in robotics from Carnegie Mellon at the time, recalls: "Why three? Well, one to keep the Army happy, one to keep the Air Force happy, one to keep the Navy happy."[34] The Air Force wanted a "pilot's associate": a kind of automated copilot that could understand commands or requests spoken out loud by the pilot. The Navy was interested in what they called a "battle management" system that could help with scenario planning and weather prediction. That left the Army. What they wanted were autonomous land vehicles.[35]

Thorpe successfully defended his doctoral thesis that September, telling his committee he was planning to take several weeks' vacation and think about what he might be up to next. Instead, the director of the CMU Robotics Institute, Raj Reddy, in one breath congratulated him and said, "What you're up to next is a meeting in my office starting in five minutes." The meeting was about building autonomous vehicles for DARPA.

"That," recalls Thorpe—that five-minute window—"was my break between finishing my thesis and starting my postdoc."

"Vehicles" that were in some sense "self-driving" had by 1984 already been around for years, but to call the technology primitive would be perhaps too generous. Robotics pioneer Hans Moravec had, in his own

PhD thesis at Stanford in 1980, enabled a robotic "cart" the size and shape of a desk on bicycle wheels to move itself around and avoid chairs and other obstacles using an onboard TV camera. "The system is moderately reliable," Moravec wrote, "but very slow."[36] How slow? The cart was programmed to move one meter at a time—"in lurches," as Moravec put it. After one of these meter-long lurches, the cart would stop, take pictures, and think for *ten to fifteen minutes* before making its next, equally tentative maneuver. Its top speed was, therefore, capped at 0.004 miles per hour.

The robot was *so* slow, in fact, that it was completely flustered by the outdoors, because the sun's angle would change so much between lurches that the shadows seemed to move bewilderingly.[37] As Thorpe recalls, "In fact, his system locked onto these nice, sharp-edged shadows and saw that they were moving, and saw that the real objects weren't moving consistently with the shadows, decided it had more confidence in the shadows than in the real objects, and threw out the real objects and locked onto the shadows, and ran over his chairs."

By 1984, Moravec had come to Carnegie Mellon, and Thorpe worked with him to get the time between lurches down from ten minutes to thirty seconds. That worked out to a top speed shy of a tenth of a mile per hour. It was progress.

At that time, the state-of-the-art computer was something called the VAX-11/784 ("VAXimus"), which was about eight feet wide and eight feet tall. Vehicles like Moravec's cart would be attached to these computers by an "umbilical cord." But to make a vehicle that could truly move around in the outside world would require bringing the computer along for the ride, which, in turn, also meant bringing along a power source for the computer. This ended up meaning a four-cylinder generator. It was going to take a lot more than a cart.

Thorpe and his group settled on a Chevy panel van, which was big enough for all of the gear *and* five graduate students. As Thorpe observed: "They were highly motivated to have high-quality software, because they were going to be—as the saying goes—first at the scene of the accident. You write much better software when you know you're going to be riding."

The project, dubbed Navlab 1, began in earnest in 1986; at that point the system could make a move every ten seconds (a quarter of a mile per hour).[38] Thorpe's son Leland was born the same year, and as it happened, Leland ended up being the perfect foil for the robot. "When Navlab 1 was moving at a crawling speed, my son was moving at a crawling speed. When Navlab 1 picked up speed, my son was walking and learning to run. Navlab 1 got going a little faster; my son got a tricycle. I thought it was going to be a 16-year contest to see who was going to drive the Pennsylvania Turnpike first."[39]

That rivalry would come to a shockingly abrupt end, however, in favor of the machine. Thorpe's graduate student Dean Pomerleau ended up putting together a vision system using neural networks. It blew away all of the other approaches the group had tried. "So in 1990," says Thorpe, "he was ready to go out and drive the Pennsylvania Turnpike."

They called the system ALVINN—Autonomous Land Vehicle in a Neural Network—and it learned by imitation.[40] "You would drive for a few minutes," says Thorpe, "and it would learn: if the road looks like this, you turn the steering wheel like this, if it looks like that, you would turn the steering wheel like that. So if you trained it on the road that you were driving on, it was very good at spitting out steering wheel angles."

Early one Sunday morning, when the light was good and there were very few other cars on the road, Pomerleau took ALVINN on the interstate. ALVINN steered all the way from Pittsburgh up I-79 to Erie, on the shores of the Great Lakes. "This was kind of revolutionary," says Thorpe—not just the feat itself but the simplicity of the model that achieved it. ALVINN knew nothing about momentum or traction, could not recognize objects or predict future locations of itself or other cars, had no ability to relate what it saw in its camera feed to how itself was positioned in space, nor to simulate the effects of its actions. "People thought if you wanted to drive that fast," says Thorpe, "you had to have Kalman filters, and clothoid models of the road, and detailed models of the dynamic response of your vehicle. And all Dean had was a simple neural net that learned: the road looks like this, you steer like that."

As Pomerleau affirmed to a local news crew, "We don't tell it anything except 'Steer like I do. Learn to steer the way I am steering right now.'"[41] In those days it took a refrigerator-sized computer, running on a 5,000-watt generator, which provided about a tenth of the processing power of a 2016-vintage Apple Watch.[42] And ALVINN didn't control the gas or brakes, which still had to be manually operated, nor could it change lanes or react in any particular way to the other cars on the road. But it worked, and took Pomerleau—steering like he had steered—to the Great Lakes in one piece.

One of the most natural ideas about how to train machines is to train them to imitate us, and this approach seems especially appealing in the domain of driving. The success of ALVINN hints at the broader viability of an approach like this. If we wanted to build a fully self-driving car, then rather than simply letting it loose on city streets to explore different driving behaviors randomly and learn by pure trial and error (terrifying), we might begin by giving the system a huge dump of real human driving and train it to imitate human decisions behind the wheel. Given a certain state of affairs—this speed, this image in the windshield, this image in the rearview mirror, etc.—the system can learn to predict what action a human driver took, be it press down on the gas, press down on the brake, turn the steering wheel, or do nothing.

This predictive approach turns the problem of driving into something almost perfectly analogous to the ImageNet competition of labeling images. Instead of being shown a picture and needing to categorize it as a dog, cat, flower, or the like, the system is shown a picture from the front dashboard and "categorizes" it as "accelerate," "brake," "turn left," "turn right," and so on. We've seen already how deep learning allows a system to generalize from images it's seen to images it hasn't seen; if AlexNet can correctly identify dogs it's never been shown before, then that should encourage us to think that a car can, somehow, generalize from scenarios it's experienced to new ones. Even if it hasn't seen that *exact* road in that *exact* sun-dappled light with that *exact* traffic flowing on it, it should nonetheless—so the theory goes—generalize from its past experience and recognize what to do here.

The idea is that you never have to turn a fledgling self-driving car loose on city streets to explore policies on its own. Rather, you just record camera information and telemetry from the real human-driven cars on the streets every day, untold millions of hours of which could be captured in a single day, and eventually have a car that is a perfect mimic of human driving.

As UC Berkeley's Sergey Levine describes to his Berkeley undergraduates: "So a very natural way to think about solving these kinds of sequential decision-making problems is basically the same way that we solve our standard computer vision problems. We collect some data, so we get a human to drive a vehicle. . . . We basically record the observations from the camera on their vehicle and record the steering commands that they make; that goes into our dataset, our training data. And then we're going to just run our favorite supervised learning method—we'll run, you know, stochastic gradient descent—to train a network. . . . Just treat this as though it was a standard supervised-learning problem. That's a very reasonable thing to start with."[43]

Then, he explains, when it is time for the system to take control, it simply turns its predictions—"This is what I think a human driver *would* do in this situation"—into actions.

Levine pauses for a second. "Does anybody have any ideas about one thing that might go wrong with this?"[44]

The hands shoot up.

LEARNING TO RECOVER

Imitation progresses with the acquired habits. In learning to dance, the deficiency of the association between the pupil's movements and the sight of the master's, renders the first steps difficult to acquire. The desired movements are not naturally performed at the outset. Some movements are made; . . . but the first actions are seen to be quite wrong; there is a manifest want of coincidence, which originates a new attempt, and that failing, another is made, until at last we see that the posture is hit.

—ALEXANDER BAIN[45]

What would I do? I wouldn't be in the situation.

—APPLE CEO TIM COOK, WHEN ASKED WHAT HE WOULD DO IN THE
SITUATION CONFRONTING FACEBOOK CEO MARK ZUCKERBERG[46]

It's 2009, and twenty years after ALVINN but in the very same building, Carnegie Mellon graduate student Stéphane Ross is playing *Super Mario Kart*—or, rather, a free and open-source derivative called *SuperTuxKart*, featuring the Linux mascot, a lovable penguin named Tux.

As Ross plays the game, his computer is recording all of the images on screen, along with every twitch of his joystick. The data is being used to train a fairly rudimentary neural network, not all that much more complex than the one ALVINN used, to steer like Ross steers.[47] Ross takes his hands off the wheel and lets the neural network drive Tux around the track. In short order, Tux takes a turn too wide and drives straight off the road. Ross is back to the drawing board, and the drawing board doesn't look good.

The problem is that no number of demonstration laps—he records a million frames of play, or about two hours of driving the course over and over and over—seems to make any difference. He hands the wheel over to the neural net, Tux gets off to a promising start, then wavers, veers, barrels off the road.

At its root, the problem stems from the fact that the learner sees an expert execution of the problem, and an expert almost never gets into trouble. No matter how good the learner is, though, they will make mistakes—whether blatant or subtle. But because the learner never saw the expert get into trouble, they have also never seen the expert get out. In fact, when the beginner makes beginner mistakes, they may end up in a situation that is completely different from *anything* they saw during their observation of the expert. "So that means," says Sergey Levine, "that, you know, all bets are off."

In *SuperTuxKart,* for instance, Ross was good enough at the game that all of the data he was feeding the program showed it, in effect, how to continue driving straight ahead on the center of the track. But once Tux, under the network's control, got even slightly off-center or slightly askew, it was lost. The screen looked systematically different from anything it had seen Ross do. The reaction required differed sharply from normal, in-control, full-throttle driving, but it had never seen anything else. And no amount of Ross continuing to play the game expertly, lap after lap and hour after hour, could fix that.

The problem is what imitation-learning researchers know as "cascading errors," and it is one of the fundamental problems with imitation learning. As Dean Pomerleau wrote during his work on ALVINN, "Since the person steers the vehicle down the center of the road during training, the network will never be presented with situations where it must recover from misalignment errors."[48] How to teach an imitation learner to recover has been a long-standing problem.

If Pomerleau was going to trust the system with his life on his trip up to Lake Erie, then he was going to need to give it more than passive observations of his own, correct steering. This alone would mean that the system was only reliable as long as it *never* made a mistake. That was too much to ask, doing fifty-five miles per hour on the interstate for two hours.

"The network must not solely be shown examples of accurate driving," wrote Pomerleau, "but also how to recover (i.e. return to the road center) once a mistake has been made."[49] But how? One idea was for Pomerleau himself to swerve around during his training drives, to demonstrate to ALVINN how to recover from being slightly out of the lane center or pointed in slightly the wrong direction. Of course, this would require somehow erasing the *beginning* of the swerve from the training data, lest ALVINN learn to imitate the swerves themselves! The second problem, he realized, was that to train the network properly would require him to swerve as often, and in as diverse an array of situations, as possible. "This would be time consuming," he concluded, "and also dangerous."

Pomerleau came up with a different idea. He would fake it.

The images ALVINN processed were tiny and grainy—just 30-by-32-pixel, black-and-white images of the trapezoidal patch of asphalt immediately in front of the car. (Indeed, its field of view was so narrow and myopic that when it entered an intersection it would become totally disoriented, adrift in a vast sea of pavement.) Pomerleau took real images recorded from ALVINN's camera and simply doctored them to skew the road slightly to one side or the other. These were then thrown into the training data with steering commands meant to nudge the car gently back toward the center of the lane and a straight-ahead bearing. It was a bit of a hack—and it only looked correct when the road surface was totally flat, without dips or hills—but, on I-79 anyway, it worked.

Ironically, the explosion of powerful deep-learning techniques in the past decade have made the "fake it" approach less and less feasible, as modern camera sensors take in too many images per second, at too high a resolution, with too wide a field of view, to be easily manipulated in this way. If the fake images are somehow systematically unlike the real things the car will see when it starts to drift, then you are in big trouble. It is a leap of faith, after all, to essentially wager your life on your Photoshop skills—and modern neural networks are effectively growing harder and harder to deceive.

Twenty years later, this problem of recovery was still unsolved as both a practical and a theoretical issue. "When you learn from watching somebody," Stéphane Ross tells me, "you see some kind of distribution of examples that don't necessarily match what you're going to see" if you start taking your *own* actions in the world. There was something deeply rooted here, Ross thought: "because all machine learning relies on the assumption that your training and the test distribution are the same." But Ross and his advisor, CMU roboticist Drew Bagnell, thought they might be able to crack it. "That got me super interested," says Ross, "because that felt like a really fundamental problem to work on."[50]

Ross and Bagnell did a theoretical analysis, trying to understand the problem from a mathematical perspective, while checking their intuitions in the world of *SuperTuxKart*. In an ordinary supervised-learning problem, something à la ImageNet, the system, once trained, will have a certain likelihood of making an error with each picture it sees. Show it ten times as many pictures, and it will make, on average, ten times as many errors—in this sense, the errors scale *linearly* with the scope of the task. Imitation learning is far, far worse, they discovered. Because a single mistake could cause the system to see things it had never prepared for before, once it makes a first mistake, all bets are off. The error grows with the *square* of the size of the task. Running for ten times as long would produce a *hundred* times as many mistakes.[51] The theoretical analysis was grim, but it left open a tantalizing possibility: Were there ways to get back to the safe world of merely *linear* mistakes? Of a car merely ten times as likely to crash when driving ten times as far? "We were really looking for the holy grail," he says.

As it turns out, they found it. The key was interaction. The learner needed some way not only to observe the expert at the outset, but to go *back* to the teacher when necessary, and say, in effect: "Hey, I tried what you showed me, but this bad thing keeps happening. What would you do if *you* had gotten into this mess?"

Ross came up with two ways to get this kind of interaction to happen on the *SuperTuxKart* track.[52] One way was to watch the (initially catastrophic) laps of the network with the joystick in hand. As Tux barreled around the course, Ross would move the joystick as he *would have* if he were playing. The second method was to have Ross and the network randomly trading off control of the car while both try to steer it at the same time. As he explains: "It's like you're still playing the game mostly normally, but in some random steps, then you don't listen to the human control, you execute the learned control. And that kind of slowly decays over time—like you're less and less in control. But there's always the chance that it picks your control." It's awkward, a little unnatural, but it works. "You're still definitely trying to play the game as you would [if you were] fully in control," Ross says. "But it doesn't necessarily pick your controls all the time to execute. It starts veering off, then you try to correct . . ." He chuckles. Over time, the car responds less and less of the time to your own steering commands—but the network gets better and better at doing what you would have done anyway. There are periods where you're not sure whether you're driving or not.

The amazing thing was, not only did both of these forms of interaction work—both on the whiteboard and on the *SuperTuxKart* track—but they required *incredibly* little feedback to do so. With static demonstrations alone, the learner was still crashing just as often after a million frames of expert data as it was after a couple thousand—it was just as hopeless after hours of tutelage as it was after only minutes. Yet using this interactive method—Ross named it "Dataset Aggregation," or DAgger—the program was driving almost flawlessly by its *third lap* around the track. "Once we had that," says Ross, "I was like, Wow, this is really awesome. It worked orders of magnitude better than the default approach."

As soon as he graduated with his PhD, Stéphane Ross traded in the virtual pavement of *SuperTuxKart* for the real-world suburban streets of Mountain View, California, where he is currently a behavior prediction lead at the self-driving-car company Waymo, designing models to forecast how other drivers, cyclists, and pedestrians on the road will behave and react. "The level of reliability that we need is, like, orders and orders of magnitude more than anything we do in academia. So that's where the real challenge is. Like, how do you make sure that your model works all the time—not just 95 or 99% of the time; that's not even good enough." It's a tall order but a satisfying project. "This project, in particular, is probably one of the projects where you can have some of the biggest impact on the world, if it succeeds—for the benefit of the world. Just that, on its own, is good enough motivation to work in that area, and hopefully have that impact someday."

Though the type of interactive feedback that DAgger involves is the gold standard theoretically, in practice we don't have to jockey with our cars for control of the steering wheel in order to ensure that they learn to stay centered in their lanes. There are several even simpler ways that work just fine in practice to build real-world systems capable of recovering from minor errors.

A 2015 project by a group of Swiss roboticists took a clever approach to overcoming this problem while trying to build a drone that could fly itself up alpine hiking trails without getting lost in the woods. Whereas earlier work had attempted, as they put it, "to explicitly define which visual features characterize a trail," they bypassed entirely the question of which parts of the image contained the trail, or exactly what trails look like, and trained a system to simply map directly from an image to a motor output. It would take in a 752-by-480-pixel image of some dirt and trees, and output "turn left," "turn right," or "go straight." In a storyline that will by now feel quite familiar, years of careful research to handcraft visual features for "saliency" or "contrast," and clever thinking about how to distinguish dirt, say, from tree bark, were thrown out the window wholesale and replaced with convolutional neural nets trained by stochastic gradient descent. All the hand-tailored work became instantly obsolete.

The team trained their system to imitate the path taken by a human hiker. What was unique, though, and clever, is what they did in order to enable their system to recover from mistakes. They strapped not one but *three* GoPro cameras to the hiker's head: one pointed straight forward, and the others pointed left and right. They then told the hiker to walk as he normally would but to take care *not* to turn his head. They were thus able to generate a huge dataset of images of the trail and annotate the center camera feed with "go forward when you see something like this," the left camera's with "turn right when you see something like this," and the right camera's with "turn left when you see something like this." They trained a neural network on the dataset, installed it in a quadrotor drone, and set it loose in the Swiss Alps. It appeared to have little trouble floating through the woods and following the trail. Again, the key insight is the need not only to show what the human expert *did*, but also to offer some guardrails, in the form of data for pointing a slightly off-kilter learner back on track.[53]

A 2016 project from Nvidia's deep-learning research group in Holmdel, New Jersey, put the same clever trick to work on the streets of

New Jersey's Monmouth County. Nvidia mounted three cameras on a car, with one pointed forward and the others pointed roughly thirty degrees left and right of center. This generated hours and hours of footage of what it would look like if a car were pointed slightly in the wrong direction. The team then fed that data to their system, with the correct prediction being "do what the actual human driver did, plus a small correction back to center." With just seventy-two hours of training data, the system was safe enough to operate on the winding rural roads and multilane highways of Monmouth County under varying weather conditions without major incident. In a video released by the team, we watch as their Lincoln MKZ peels out from the parking lot of Nvidia's deep-learning research building and heads onto the Garden State Parkway. "Is it still autonomous!?" asks an employee in the chase car. "It looks pretty good from here," he says—then clarifies that it's behaving better, at least, than the other, New Jerseyan–piloted cars on the parkway.[54]

There are two ironies here worth noting in brief. It was in this very research building, in the late 1980s—when it was owned by AT&T Bell Laboratories—that Yann LeCun invented the convolutional neural network, trained by backpropagation, which is exactly what's driving today's self-driving cars.[55] And as it happens, I myself learned to drive on the roads of Monmouth County, New Jersey, regularly passing by that very building on my way to and from cross-country practice. I wish I could say I was as safe and trustworthy behind the wheel—driving that very same road—after having seen seventeen *years* of human driving, as the convolutional net was after seventy-two hours.

THE CLIFF'S EDGE: POSSIBILISM VS. ACTUALISM

One must perform the lower act which one can manage and sustain: not the higher act which one bungles. . . . We must not arrogate to ourselves actions which belong to those whose spiritual vision is higher or other than ours.

—IRIS MURDOCH[56]

What would you do if you were me? she said.

If I were you-you, or if I were you-me?

If you were me-me.

If I were you-you, he said, I'd do exactly
what you're doing.

—ROBERT HASS[57]

Setting aside the question of recovering from small mistakes, the second problem with imitation as a learning strategy is that sometimes you simply *can't* do what the expert can do. Imitation then would only mean starting something you can't finish. In which case, you probably shouldn't attempt to act like them at all.

Both real life and pop culture are strewn with examples of the novice attempting to simply mimic the expert, often with catastrophic results.

As chess grandmaster Garry Kasparov explains: "Players, even club amateurs, dedicate hours to studying and memorizing the lines of their preferred openings. This knowledge is invaluable, but it can also be a trap. . . . Rote memorization, however prodigious, is useless without understanding. At some point, he'll reach the end of his memory's rope and be without a premade fix in a position he doesn't really understand."

Kasparov recalls coaching a twelve-year-old player, running through the opening moves of one of the student's games. Kasparov asked him why he'd made a particularly sharp and dangerous move in a complicated opening sequence. "That's what Vallejo played!" replied the student. "Of course I also knew that the Spanish Grandmaster had employed this move in a recent game," Kasparov says, "but I also knew that if this youngster didn't understand the motive behind the move, he was already headed for trouble."[58]

This idea seems both intuitive and at some level paradoxical: doing what a "better" person would do may sometimes be a grave mistake. It is a surprisingly complicated story, and one with deep connections across ethics, economics, and machine learning alike.

In 1976, a particular question erupted to the forefront of ethical philosophy: To what degree do, or should, your own *future* actions influence the question of what the right thing is to do *now*?

Philosopher Holly Smith was at the University of Michigan, focused on working through the subtleties of what it means to be a utilitarian. She noticed something strange. "The question very naturally arises, if you're a utilitarian, 'If I do A now, is that going to produce the best possible consequences?' Well, it's just transparent," she says, "that it's going to depend on what I do next."[59] The need to take your own future actions into account means you also need to consider your own future *mistakes*. And so Smith began writing about what she would call "moral imperfection."[60]

The thought experiment that she considered has come to be known as that of "Professor Procrastinate."[61] The premise is straightforward: Professor Procrastinate is both a professor and—you guessed it—an inveterate procrastinator. He is asked to read a student's paper and offer feedback, which he is uniquely qualified to provide. But what would surely happen instead, should he agree, is that he'll fritter the time away and never get the feedback to the student. This will be worse than simply declining, in which case the student could ask for (slightly less high-quality) feedback from someone else.

Should he accept?

Here diverge two different schools of moral thought: "possibilism"—the view that one should do the best *possible* thing in every situation—versus "actualism"—the view that one should do the best thing at the moment, given what will *actually* happen later (whether because of your own later deeds or some other reason).[62]

Possibilism says that the best *possible* thing for Procrastinate to do is to accept the review *and* write it on time. This begins with accepting it, and so he should accept.

Actualism takes a more pragmatic view. By its lights, accepting the review inevitably results in a bad outcome: no review at all. Declining the review means a comparatively better outcome: a review by a slightly less well-qualified reviewer. The professor should do the thing that *actually* results in the best outcome; hence he should say no.

Smith was led to the conclusion that "one must sometimes choose the lower rather than the higher act." She elaborates: "There seems little point in prescribing an act which puts the agent in a position to do great things if the same act also puts him in a position to do something disastrous, and he would choose the latter rather than the former."

At the same time, Smith is quick to elaborate on the drawbacks of actualism. For one thing, she says, "Actualism gives you an excuse for bad action based on your own future moral defects." Some forty years later, the theoretical debate simmers on. "I think many people will see it as still unresolved," Smith says. "I think it's fair to say it's still a lively discussion."[63]

The discussion isn't only theoretical, either. Among the twenty-first-century "effective altruism" movement, for instance, opinions vary about how much of a sacrifice someone ought to make in order to maximally help others.[64] Princeton philosopher Peter Singer famously said that neglecting to donate to charity was analogous to walking past a pond in which a child was drowning and doing nothing to help.[65] Even for those who more or less agree with this argument, there is some debate over how much to actually give. A perfect person, perhaps, could donate almost all of their money to charity while staying happy and upbeat and motivated, and inspiring to others. But even the devoted members of the "EA" movement, including Singer himself, are not such perfect people.

Julia Wise, a leader in the effective altruism community and the community liaison at the Centre for Effective Altruism, has made impressive commitments in her own life—giving 50% of her income to charity, for instance—but she emphasizes the value of not striving for perfection. "Give yourself permission to go partway," she says.[66] She noticed, for instance, that her own commitment to veganism could not accommodate her deep love of ice cream—and so she felt she couldn't be a vegan. What worked for her was becoming comfortable with the idea of being a vegan . . . who eats ice cream. That was something she could stick with.

Oxford philosopher Will MacAskill, cofounder of the Centre for Effective Altruism, doesn't mince words on the question. "We should be actualists," he says. "If you give away all of your savings at once today—which you could technically do—you'll probably get so frustrated that you'll simply stop giving in the future. Whereas if you decide to give 10% of your earnings, this commitment will be sustainable enough that you'll continue doing it over many years in the future, resulting in a higher overall impact."[67]

Singer himself acknowledges that a sense of balance and proportion is presumably best over the long term. "If you find yourself doing something that makes you bitter, it is time to reconsider. Is it possible for you to become more positive about it? If not, is it really for the best, all things considered?" He points out, too, that "there are still relatively few effective altruists, so it is important that they set an example that attracts others to this way of living."[68]

Machine learning has its own version of the actualism/possibilism debate. One of the primary families of algorithms for reinforcement learning, as we discussed in Chapter 4, are methods that learn the "value," expressed as the expected future rewards, of the various actions available to them. (This is referred to as the "Q-value," short for "quality.") For instance, a board-game agent would learn to predict its chance of winning upon making various moves, and an Atari-playing agent would learn to estimate the number of points it expects each action to lead to. With these predictions well tuned, it then becomes straightforward to simply take the action with the highest Q-value.

Here, though, there is an ambiguity worth unpacking. Should the Q-value contain the expected future rewards that you *could* earn from taking this action? Or the expected rewards that you *would* earn? For a totally perfect agent, there is no tension—but otherwise the prescriptions can vary sharply.

These two approaches to value learning are called "on-policy" and "off-policy" methods. On-policy methods learn the value of each action based on the rewards the agent will *actually* expect to get after taking that action and continuing to take actions according to its own "policy." An *off*-policy agent, on the other hand, will learn the value of each action based on the best *possible* series of actions that could follow it.[69]

In their seminal textbook on reinforcement learning, Richard Sutton and Andrew Barto talk about how an off-policy ("possibilist") agent could get itself into trouble, precisely by always trying to do the "best thing possible." Imagine a car, they say, that needs to drive itself from one spot along the edge of a seaside cliff to another. The shortest and most efficient path is just to follow the edge of the cliff. Indeed, *provided* the car is stable and steady enough, that is the best route to take. But for a self-driving car that's slightly shaky or unsteady behind the wheel, this is flirting with disaster. The better move might be to take a more roundabout inland path, one that doesn't require it to drive perfectly to succeed. This is actualism—and a car trained using on-policy methods would indeed learn to take the safer, surer route over the one with slightly higher reward but much higher risk.[70]

Imitating one's heroes or mentors—whether the imitator is human or machine—brings with it some of the dangers of possibilism, of off-policy valuation.[71] In a chess context, learning the best move to play given grandmaster-level ability to handle its consequences may serve only to let the student bite off more than they can chew. In that case, studiously watching the expert play may simply not help—or worse. It's one thing to know that in a particular chess position, say, sacrificing my queen leads to checkmate in ten moves. But if I can't find the checkmate, I will have sacrificed my queen for nothing, and will almost certainly lose the game as a result.

Economists since the mid-twentieth century have discussed the "theory of the second best," which argues, in effect, that knowing the right thing to do in a theoretical version of the economy, which obeys a number of mathematical assumptions, may have virtually *no* bearing in an economy that deviates from those assumptions even slightly. The "second-best" policy to follow, or action to take, may have virtually no resemblance to the best.[72] OpenAI research scientist Amanda Askell, who works on ethics and policy, notes that the same line of argument likely applies equally well in her domain. "I think something similar could be said in ethics," she says. "Even if the ideal agent follows moral theory X perfectly, the non-ideal agent uses a pretty different set of decision procedures."[73]

Cases like this should give any would-be imitator *or* role model pause.

Imitation is at some level intrinsically possibilist, liable to bite off more than it can chew. It may be cute when our child mimes driving the car, or chopping vegetables, or performing veterinary medicine—but we intervene if we see them *actually* reach for the keys or knife (or maybe even the cat). The behavior we actually want to see may bear no resemblance to imitation: the "next best" thing to competent driving is getting into the passenger seat; to chiffonading herbs, setting the table; and to setting a broken paw, maybe just calling for Mom and Dad.

With machine imitators, too, we would do well to keep the theory of the second best in mind. If they are to learn from us, we must take care that they don't inadvertently learn to initiate behaviors they can't handle once begun. Once they're expert enough, the issue may become moot. But until then, imitation may be a curse, and—in the words of user-interface designer Bruce Balentine—"it's better to be a good machine than a bad person."[74]

AMPLIFICATION: SELF-IMITATION AND TRANSCENDENCE

My favorite player from the past is probably . . . myself, like three or four years ago.

—WORLD CHESS CHAMPION MAGNUS CARLSEN[75]

A third fundamental challenge with imitation is that if one's primary objective is to *imitate* the teacher, it will be hard to *surpass* them.

This was on the mind of one of the very first researchers in machine learning—in fact, the one who coined the term: IBM's Arthur Samuel, who in 1959, as we briefly discussed earlier, developed a machine-learning system for playing checkers. "I fed into it a number of principles that I knew had something to do with the game," he said, "though I didn't know then, and don't know now, precisely what their significance is." The list included things like how many checkers you have, how many kings you have, how many moves are possible from your position, and so forth.[76]

The program ended up capable of beating Samuel himself, despite using only the strategic considerations that Samuel had given it. Its unerring ability to look a number of moves ahead, combined with a trial-and-error fine-tuning of the relative importance to place on these various factors, led to a system that surpassed its own teacher. It was a sublime achievement for its time, one that, again as we discussed, single-handedly shot IBM's stock price up overnight, and one of which Samuel was justifiably proud. But he nonetheless felt keenly aware that his project had hit a ceiling. "The computer now works according to *my* principles of checkers and does a fine job of shuffling these around to its best advantage," he lamented, "but the only way to get it to play better checkers is to give it a better set of principles. But how? . . . At the moment, I'm the only man in the world who could teach the machine to play any better, and it's already way out of my class."

The more principled path forward, Samuel reasoned, was for the computer itself to somehow generate strategic considerations on its own. "If only the computer *could* generate its own terms! But I see little hope of that in the immediate future," he said.[77] "Unfortunately, no satisfactory scheme for doing this has yet been devised."[78]

By the end of the twentieth century, the fundamental techniques for computer game playing had changed surprisingly little—as had their fundamental limitations. The machines were millions of times faster. Reinforcement learning had become a field unto itself. But the machines had not changed very much at all, it seemed, in their stubborn dependence on *us*.

By the 1990s, the IBM team working on chess supercomputer Deep Blue had created a value function much like the one Samuel had made for checkers some decades earlier. Working with human grandmasters, they attempted to enumerate and articulate all the factors that determine the strength of a position: things like the number of pieces on each side, mobility and space, king safety, pawn structure, and on and on. Instead of using thirty-eight such considerations, however, as Samuel had, they used eight *thousand*.[79] "This chess evaluation function," said team lead Feng-hsiung Hsu, "probably is more complicated than

anything ever described in the computer chess literature."[80] The critical question, of course, was how to somehow weight and *combine* those
bewildering thousands of considerations into a single judgment of the
quality of the position on the board. *Exactly* how many extra pawns
were worth how much control of the center, or how much king safety?
Getting the balance right would be crucial.

So how exactly *were* those thousands of considerations brought into
balance? By imitation.

The Deep Blue team had access to a database of seven hundred thousand grandmaster games. They showed the computer position after
position from these real games and asked it what move *it* would have
played. Imitating the human moves became a target as they fine-tuned
its value function. If, say, increasing the value Deep Blue assigned to
possessing both bishops made it slightly more likely to play the same
moves that the human grandmasters had played, then Deep Blue came
to increase its value for the bishop pair.

This human-imitating combination of positional considerations,
which were themselves derived from human experts, was married to
the computer's unerring calculation, blistering speed, and brute force.
The machine was able to search through hundreds of millions of future
board positions *per second*, and that, combined with its human-like
evaluation, was enough to defeat human chess world champion Garry
Kasparov in their storied 1997 match. "Garry prepared to play against
a computer," said Deep Blue's project manager C. J. Tan. "But we programmed it to play like a Grandmaster."[81]

From a philosophical perspective, some within the research community wondered if programs were ultimately being hindered, however, by
this continued dependence on their human role models. In the world of
computer checkers, the University of Alberta's Jonathan Schaeffer had
developed a program in the early 1990s that was so good that when it
picked a move that differed from human grandmaster play, often its own
idea was better. "Of course, we could continue to 'improve' the evaluation function so that it consistently plays the human moves," he wrote.
But "it isn't obvious that this is a good thing to do." For one thing, tun-

ing the program to play more conventionally might negate its ability to surprise a human opponent. For another thing, it wasn't clear that the methodology of imitation was useful once the program had reached the level of the best human players. "We found it difficult to progress any further," Schaeffer admitted.[82] His project was essentially stuck. A question hung over the field. As the 2001 volume *Machines That Learn to Play Games* put it, reflecting on Deep Blue's success: "One important direction of future research is to establish the extent to which better imitating human expert moves corresponds to genuinely stronger play."[83]

Fifteen years later, DeepMind's AlphaGo system finally realized Arthur Samuel's vision of a system that could concoct its own positional considerations from scratch. Instead of being given a big pile of thousands of handcrafted features to consider, it used a deep neural network to automatically identify patterns and relationships that make particular moves attractive, the same way AlexNet had identified the visual textures and shapes that make a dog a dog and a car a car. The system was trained as Deep Blue had been: by learning to predict the moves made by expert human Go players in a giant, 30-million-move database.[84] It was able to get state-of-the-art prediction of human expert moves—57% accuracy to be precise, smashing the previous state-of-the-art result of 44%. In October 2015, AlphaGo became the first computer program to defeat a human professional Go player (in this case, the three-time European champion Fan Hui). Just seven months later, in March 2016, it defeated the eighteen-time international titleholder, and one of the strongest players of all time, Lee Sedol.

Once again, the computer that had transcended human play was nonetheless, ironically, an imitator at heart.[85] It was not learning to play the *best* moves. It was learning to play the *human* moves.

The successes of Deep Blue and AlphaGo alike were possible only because of mammoth databases of human examples from which the machines could learn. These flagship successes of machine learning created such worldwide shockwaves as they did because of the global popularity of those games. And it was the very popularity of those games that enabled those victories. Every move we'd ever made could and would be used against us. The computers wouldn't have been nearly as impressive at playing a more obscure or unpopular game—because they wouldn't have had enough examples to work from. Popularity thus served a double role. It had made the accomplishment *significant*. But it had also made it *possible*.

No sooner had AlphaGo reached the pinnacle of the game of Go, however, than it was, in 2017, summarily dethroned, by an even *stronger* program called AlphaGo Zero.[86] The biggest difference between the original AlphaGo and AlphaGo Zero was in how much human data the latter had been fed to imitate: zero. From a completely random initialization, tabula rasa, it simply learned by playing against itself, again and again and again and again. Incredibly, after just thirty-six hours of self-play, it was as good as the original AlphaGo, which had beaten Lee Sedol. After seventy-two hours, the DeepMind team set up a match between the two, using the exact same two-hour time controls and the exact version of the original AlphaGo system that had beaten Lee. AlphaGo Zero, which consumed a tenth of the power of the original system, and which seventy-two hours earlier had never played a single game, won the hundred-game series—100 games to 0.

As the DeepMind research team wrote in their accompanying *Nature* paper, "Humankind has accumulated Go knowledge from millions of games played over thousands of years, collectively distilled into patterns, proverbs and books."[87] AlphaGo Zero discovered it all and more in seventy-two hours.

But there was something very interesting, and very instructive, going on under the hood. The system had not been shown a single human game to learn from. But it was, nonetheless, learning by imitation. It was learning to imitate . . . *itself.*

The self-imitation worked as follows: Expert human play in games like Go and chess is a matter of thinking "fast and slow."[88] There is a conscious, deliberate reasoning that looks at sequences of moves and says, "Okay, if I go here, then they go there, but then I go here and I win." In AlphaGo Zero, the explicit "slow" reasoning by thinking ahead, move by move, "if this, then that," is done by an algorithm called Monte

Carlo Tree Search (MCTS, for short).[89] And this slow, explicit reasoning
is intimately married to a fast, ineffable intuition, in two different but
related respects.

The first bit of "fast" thinking is that, prior to and separate from any
explicit reasoning of this form, we have an intuitive sense of how *good*
a particular position is. This is the "value function" or "evaluation function" we've been discussing; in AlphaGo Zero, this comes from a neural
network called the "value network," which outputs a percentage from 0
to 100 of how likely AlphaGo Zero thinks it is to win from that position.

The second bit of implicit, "fast" reasoning is that when we look at
the board there are some moves we consider playing—some moves just
"suggest themselves," and many others simply do not. We deploy our
slow, deliberate, "if this, then that" reasoning down the paths that our
intuition has first identified as plausible or promising. This is where
AlphaGo Zero gets interesting. These candidate moves come from a
neural network called the "policy network," which takes in the current
board position as input and assigns a percentage from 0 to 100 to each
possible move. What does this number represent? The system is making
a bet on the move that it will, itself, ultimately decide to play.

This is quite a strange and almost paradoxical idea, and merits a bit
of further elaboration. The policy network represents AlphaGo Zero's
guess, for each possible move, of how likely it will be to choose that move
after doing an explicit MCTS search to look ahead from that position.
The slightly surreal aspect is that the system *uses* these probabilities to
focus the slow MCTS search along the series of moves it thinks are most
likely.[90] "AlphaGo Zero becomes its own teacher," DeepMind's David Silver explains. "It improves its neural network to predict the moves which
AlphaGo Zero itself played."[91]

Given that the system uses these predictions to guide the very search
whose outcome they are predicting, this might sound like the recipe
for a self-fulfilling prophecy. In reality, each system—fast and slow—
sharpens the other. As the policy network's fast predictions improve, the
slow MCTS algorithm uses them to search more narrowly and wisely
through possible future lines of play. As a result of this more refined

search, AlphaGo Zero becomes a stronger player. The policy network then adjusts to predict these new, slightly stronger moves—which, in turn, allows the system to use its slow reason even more judiciously. It's a virtuous circle.

This process is known in the technical community as "amplification," but it could just as easily be called something like transcendence. AlphaGo Zero learned only to imitate itself. It used its predictions to make better decisions, and then learned to predict those better decisions in turn. It began by making random predictions and random moves. Seventy-two hours later, it was the strongest Go player the world had ever seen.

AMPLIFYING VALUES

You should consider that Imitation is the most acceptable part of Worship, and that the Gods had much rather Mankind should Resemble, than Flatter them.

—MARCUS AURELIUS[92]

For a growing number of philosophers and computer scientists concerned with the longer-term future, the prospect of flexibly intelligent and flexibly capable systems, into which we must impart extremely complex behaviors and values, raises not only technical problems but something much deeper.

There are two primary challenges here. The first is that the things we want are very difficult to simply state outright—even in words, let alone in a more numerical form. As the Future of Humanity Institute's Nick Bostrom notes, "It seems completely impossible to write down a list of everything we care about."[93] In this case, we have already seen how learning by imitation can succeed in domains where it is effectively impossible to explicitly impart every rule and consideration and degree of emphasis for what makes someone an expert driver or an expert Go player. Simply saying, in effect, "Watch and learn" is often impressively successful. It may well be the case that as autonomous systems become more powerful and more general—to the point that we seek to impart some sense of what it means not just to drive well and play well but to *live* well, as individuals and societies—we can still turn to something not unlike this.

The second, deeper challenge is that both traditional reward-based reinforcement learning and imitation-learning techniques require humans to act as sources of ultimate authority. Imitation-learning systems, as we've seen, can surpass their teachers—but only if the teachers' imperfect demonstrations are imperfect in ways that largely cancel out, or only if experts who cannot demonstrate what they want can at least *recognize* it.

Each of these fronts offers a challenge as we look to the more distant future, to more powerful systems acting in more subtle and sophisticated real-world settings.

Some, for instance, worry that humans aren't a particularly good source of moral authority. "We've talked a lot about the problem of infusing human values into machines," says Google's Blaise Agüera y Arcas. "I actually don't think that that's the main problem. I think that the problem is that human values as they stand don't cut it. They're not good enough."[94]

Eliezer Yudkowsky, cofounder of the Machine Intelligence Research Institute, wrote an influential 2004 manuscript in which he argues for imbuing machines not simply to imitate and uphold our norms as we imperfectly embody them, but rather, we should instill in machines what he calls our "coherent extrapolated volition." "In poetic terms," he writes, "our *coherent extrapolated volition* is our wish if we knew more, thought faster, were more the people we wished we were."[95]

In domains where there is a relatively clear external metric of success—in checkers or Go or *Montezuma's Revenge*—machines can simply use imitation as a starting point for more traditional reinforcement-learning techniques, honing that initial imitative behavior by trial and error, and potentially eclipsing their own teachers.

In the moral domain, however, it is less clear how to extend imitation, because no such external metric exists.[96]

What's more, if the systems we attempt to teach are someday potentially more intelligent than we are, they may take actions we find hard to even evaluate. If a future system proposes, say, a reform of clinical trial regulations, we may not necessarily even be in a position to assess—after great deliberation, let alone in a tight iterative feedback loop—whether it *does*, in fact, conform to our sense of ethics or norms. So, again, how can we continue to train a system in our own image, once its behavior goes beyond our immediate ken?

Few have thought as deeply about this set of issues as OpenAI's Paul Christiano. "I am very interested in really asking what solutions would look like, as you scale them up," he says. "What is our actual game plan? What is the actual endgame here? That's a question that relatively few people are interested in, and so very few people are working on."[97]

What Christiano realized, as early as 2012 and in research that continues to this day, is that we may—even in these most difficult of scenarios—be able to ratchet our way forward.[98] We saw, for instance, that AlphaZero has instant, fast-thinking judgments about the moves to consider, but uses slow-thinking Monte Carlo Tree Search to comb through millions of future board positions, to confirm or correct those hunches. The results of this slow thinking are then used to sharpen and improve its fast instincts: it learns to predict the outcomes of its own deliberation.[99]

Perhaps, Christiano believes, this very same schema—what he calls "iterated distillation and amplification"—can be used to develop systems with complicated judgment, beyond, and yet aligned with, our own.

For example, imagine we are trying to lay out a new subway system for a big city. Unlike Atari, say, or Go, we can't evaluate thousands of scenarios per second—in fact, a single evaluation could take months. And unlike Atari or Go, there is no external objective measure to appeal to—a "good" subway system is whatever people think it is.

We could train a machine-learning system up to a certain level of competence—by normal imitation learning, say—and then, from that point forward, we could *use it* to help evaluate plans, not unlike a senior urban planner with a staff of a handful of more junior urban planners. We might ask one copy of our system to give us an assessment of expected wait times. We might ask another to give us an estimated budget. A third we might ask for a report about accessibility. We, as the "boss," would make the final determination—"amplifying" the work of our machine subordinates. Those subordinates, in turn, would "distill" whatever lessons they could from our final decision and become slightly better urban planners as a result: faster-working in sum than we ourselves, but modeled in our own image. We then iterate, by delegating the next project to this new, slightly improved version of our team, and the virtuous circle continues.

Eventually, believes Christiano, we would find that our team, in sum, was the urban planner we *wish* we could be—the planner we could be if we "knew more, thought faster, were more the planner we wished we were."

There is work to be done. Christiano would like to find ways of doing amplification and distillation that will *provably* maintain alignment with the human user. For now, whether this is even possible remains an open question—and a hope. Small, preliminary experiments are underway. "If we can realize this hope," Christiano and his OpenAI collaborators write, "it will be an important step towards expanding the reach of ML and addressing concerns about the long-term impacts of AI."[100]

Discussing his work on amplification, I ask Christiano, who has become a leading figure in the alignment research community, if he views himself as a role model of sorts for others interested in following a similar path. His answer surprises me.

"Hopefully it's not a path people have to be following," he says.[101]

Christiano elaborates that he was perhaps one of the last alignment researchers who first had to live a kind of double life before being able to work in AI safety directly: working on more conventional problems to get his academic credentials, while figuring out a way to do the work he felt was truly important. "I sort of had to go off on my own and think about shit for a long time," he says. "It is easier to do academic work in the context of an academic community." Only a handful of years later, that community exists.[102] "So hopefully most people would be more in that situation," he says. "There are a bunch of people thinking about these things; they can actually get a job . . . to just be doing what [they] care about."

That's the thing, perhaps, about being a trailblazer: it's not so much that others will imitate your example to the letter, or follow directly in your footsteps, as that—because of your efforts—they won't have to.

8

INFERENCE

University of Michigan psychologist Felix Warneken walks across the room, carrying a tall stack of magazines, toward the doors of a closed wooden cabinet. He bumps into the front of the cabinet, exclaims a startled "Oh!," and backs away. Staring for a moment at the cabinet, he makes a thoughtful "Hmm," before shuffling forward and bumping the magazines against the cabinet doors again. Again he backs away, defeated, and says, pitiably, "Hmmm . . ." It's as if he can't figure out where he's gone wrong.

From the corner of the room, a toddler comes to the rescue. The child walks somewhat unsteadily toward the cabinet, heaves open the doors one by one, then looks up at Warneken with a searching expression, before backing away. Warneken, making a grateful sound, puts his pile of magazines on the shelf.[1]

Warneken, along with his collaborator Michael Tomasello of Duke, was the first to systematically show, in 2006, that human infants as young as eighteen months old will reliably identify a fellow human facing a problem, will identify the human's goal and the obstacle in the way, and will spontaneously help if they can—even if their help is not requested, even if the adult doesn't so much as make eye contact with them, and even when they expect (and receive) no reward for doing so.[2]

This is a remarkably sophisticated capacity, and almost uniquely human. Our nearest genetic ancestors—chimpanzees—will spontaneously offer help on occasion—but only if their attention has been called to the situation at hand, only if someone is obviously reaching toward an object that is beyond their grasp (and not in more complex situations, like with the cabinet),[3] only if the one in need was a human rather than a fellow chimpanzee (they are remarkably competitive with one another), only if the desired object is *not* food, and only after lingering in possession of the sought-after object for a few seconds, as if deciding whether or not to actually hand it over.[4]

What Warneken and Tomasello showed is that such helping behavior is "extremely rare evolutionarily" and far more pronounced in humans than in our closest cousins, emerging quite richly before even language. As Tomasello puts it, "The crucial difference between human cognition and that of other species is the ability to participate with others in collaborative activities with shared goals and intentions."[5]

"Children are described as being initially selfish—only caring about their own needs—and it is upon society to somehow reprogram them into becoming altruistic," says Warneken.[6] "However, our research has shown that infants in the second year of life are already cooperative by helping others with their problems, working together, and sharing resources with them."[7]

This requires not only the *motivation* to help but an incredibly sophisticated cognitive process: *inferring* the other person's goals, often from just a small bit of behavior.

"Human beings are the world's experts at mind reading," says Tomasello. Perhaps the most impressive part of this expertise is our ability to infer others' *beliefs*, but the foundation is inferring their *intentions*. Indeed, it is not until around age four that a child begins to know what others think. But by their first birthday, they are already coming to know what others want.[8]

Researchers are increasingly making the argument that our approach to instilling human values in machines should take the same tack. Perhaps, rather than painstakingly trying to hand-code the things we care about, we should develop machines that simply *observe* human behavior and *infer* our values and desires from that. Richard Feynman famously described the universe as "a great chess game being played by the gods. . . . We do not know what the rules of the game are; all we are allowed to do is to *watch* the playing." The technical term for this in AI is "inverse reinforcement learning"—except we are the gods, and it is machines that must observe us and try to divine the rules by which *we* move.

INVERSE REINFORCEMENT LEARNING

In 1997, UC Berkeley's Stuart Russell was walking to the grocery store when his mind wandered to the question of why we walk the way we walk. "We move in a very stereotyped kind of way, right? So if you watch the Ministry of Silly Walks sketch from Monty Python, you see that there are a lot of other ways you can walk besides the normal one, right—but we all walk pretty much the same way."[9]

It can't be *simply* a matter of imitation—at least it doesn't seem likely to be. There's not only very little interpersonal variation in the basic human gait—there's also very little difference across cultures and, as far as we can tell, across time. "It's not just 'Well, that's the way they were taught,'" Russell says. "That's the way that that *works*, somehow."

And yet this raises as many questions as it answers. "What do you mean 'works,' right? What is the objective function? People proposed objectives, like 'I think I'm minimizing energy,' or 'I'm minimizing torque,'[10] or 'I'm minimizing jerk,'[11] or 'I'm minimizing this' or 'I'm minimizing that,' or 'I'm maximizing this other,' and none of them produce realistic-looking motion. This is used in animation a lot, right? Trying to synthesize someone who walks and runs who doesn't look like a robot. And they all fail. And that's why we use motion capture for all that stuff."

Indeed, the entire field of "biomechanics" exists to answer questions like this. Researchers had long been interested, for example, in the various distinct gaits of four-legged animals: walk, trot, canter, gallop. It took the invention of the high-speed photograph in the late nineteenth century to settle the question of how exactly these different gaits worked: which legs were raised at which time and, in particular, whether a horse at a gallop is ever fully airborne. (In 1877, we learned that it is.) Then, in the twentieth century, the debate moved from the *how* to the *why*.

In 1981, the Harvard zoologist Charles Richard Taylor published a major paper in *Nature* showing that a horse's transition from a trot to a gallop occurred in such a way as to minimize the total energy the horse was expending.[12] Ten years later, he published a major follow-up paper in *Science* saying no, after further evidence, the switch to the gallop is not about minimizing *energy* but, rather, appears to be about minimizing *stress* on the horse's joints.[13]

Such were the thoughts on Russell's mind as he walked to the grocery store. "I was walking down the hill from my house to Safeway," Russell tells me. "And I was noticing that, because this was a downslope, your gait is slightly different than it is on the flat. And I was thinking, I wonder how I would be able to predict the difference in gait. Say I put a cockroach on a sloping . . ." He gestures with his hand. "How would the cockroach walk, right? Could I predict that? If I knew the objective, I could predict what the cockroach would do when I tip the thing."

By the late 1990s, reinforcement learning had already emerged as a powerful computational technique for generating sensible behavior in various (and, in those days, rather simple) physical and virtual environments. It was also becoming clear through studies of the dopamine system, as well as foraging behavior in bees, that reinforcement learning could offer a strikingly apt framework for understanding human and animal behavior.[14]

There was just one problem. Typical reinforcement-learning scenarios assumed that it was totally clear what the "rewards" were that one was trying to maximize—be it food or sugar water in an animal behavior experiment, or the score of a video game in an AI lab. In the real world, the source of this "reward" was much less obvious. What was the "score," as it were, for *walking*?

So, Russell wondered, walking down the wooded boulevard known as The Uplands in Berkeley, if the human gait was the answer—and reinforcement learning was the method by which the body had found it—then . . . what was the question?

Russell wrote a 1998 paper that served as something of a call to action. What the field needed, he argued, was what he called *inverse* reinforcement learning. Rather than asking, as regular reinforcement learning does, "Given a reward signal, what behavior will optimize it?," inverse reinforcement learning (or "IRL") asks the reverse: "Given the observed behaviour, what reward signal, if any, is being optimized?"[15]

This is, of course, in more informal terms, one of the foundational questions of human life. *What exactly do they think they're doing?* We spend a good fraction of our life's brainpower answering questions like this. We watch the behavior of others around us—friend and foe, superior and subordinate, collaborator and competitor—and try to read *through* their visible actions to their invisible intentions and goals. It is in some ways the cornerstone of human cognition.

It also turns out to be one of the seminal and critical projects in twenty-first-century AI. And it may well hold the key to the alignment problem.

LEARNING FROM DEMONSTRATIONS

For anyone who has ever struggled to divine the meaning or intention behind another's actions—*Are they flirting with me, or are they just a super friendly person? Are they upset at me for some reason, or just in a bad mood? Were they trying to do what they did, or was that simply an accident?*—it can sometimes feel that there are a literally infinite number of things that any action can mean.

Computer science here offers consolation, but not cure. *There are a literally infinite number of things that any action can mean.*

In that theoretical sense, the problem is hopeless. Practically, the story is a bit happier.

Inverse reinforcement learning is, famously, what mathematicians call an "ill-posed" problem: namely, one that doesn't have a single, unique right answer. There are huge families of reward functions, for instance, that are totally indistinguishable from the standpoint of behavior. On the other hand, by and large this ambiguity won't matter, precisely because of the fact that one's behavior won't change as a result. For instance, the sport of boxing happens to use a "ten-point must" scoring system, where the winner of a round receives ten points and the loser receives nine points. If an apprentice boxer came to the mistaken conclusion that rounds were scored ten million points to nine million points, or ten-millionths of a point to nine-millionths of a point, or eleven points to ten, he would still know that the person with the higher total wins, and his boxing would be no different than that of someone who understood the "correct" scoring system. So the error is unavoidable, but also moot.[16]

Another, thornier question arises, however: What leads us to assume that the person's actions mean anything at all? What if they aren't *trying* to do anything whatsoever, and their actions reflect random behavior, nothing more?

In the first paper posing practical solutions to the IRL problem, Russell and his then–PhD student Andrew Ng considered some simple examples in order to show that the idea could work.[17] They considered a tiny five-by-five grid in which the goal is to move the player to a particular "goal" square, and a video-game world in which the goal is to drive a car to the top of a hill. Could an IRL system infer these goals simply by watching an expert (be it human or machine) playing the game?

Ng and Russell built a few simplifying assumptions into their IRL system. It assumed that the player never acted randomly and never made mistakes: that when it took an action, that action was, in fact, the best action possible. It also assumed that the rewards motivating the agent were "simple," in the sense that any action or state that *could* be considered to be worth zero points *should* be considered to be worth zero points.[18] Further, it made the assumption that when the player took an action, not only was that action the best thing to do, but *any* other action would be a mistake. This ruled out the possibility, for instance, of a game having multiple competing objectives, with a player choosing between them at random.

The assumptions were fairly strong, and the domains were too simple to be of any immediate practical use—they were a far cry from the complexity of the human gait—but IRL *did* work. The rewards inferred by the IRL system looked very similar to the real rewards. When Ng and Russell let the IRL system play the games by trying to maximize what it *thought* the rewards were, it got just as high a score—measured by the "real" points—as a system optimized for the real points directly.

By 2004, Andrew Ng had received his doctorate and was teaching at Stanford, advising his own then–PhD student Pieter Abbeel. They took another pass at the IRL problem, trying to increase the complexity of the environments and relax some of the inferential assumptions.[19] As they imagine, whatever task we're observing has various "features" that are relevant to that task. If we're watching someone drive, for instance, we might consider the relevant features to be things like which lane the car is in, how fast it's going, the following distance between itself and the car ahead of it, and so on. They developed an IRL algorithm that assumes that it will see the same pattern of these features when it drives for itself as it did in the demonstrations it observed. A very stripped-down, Atari-esque driving simulator showed promising results, with a computational "apprentice" driving in the game much like Abbeel did: avoiding collisions, overtaking slower cars, and otherwise keeping to the right-hand lane.

This was significantly different from the strict imitation approach we discussed in Chapter 7. After just one minute of demonstrated driving by Abbeel, a model trying to mimic his behavior directly had nowhere near enough information to go on—the road environment is too complex. Abbeel's behavior was complicated, but his goals were simple; within a matter of seconds, the IRL system picked up on the paramount importance of not hitting other cars, followed by not driving off the road, followed by keeping right if possible. This goal structure was much simpler than the driving behavior itself, and easier to learn, and more flexible to apply in novel situations. Rather than directly adopting his *actions,* the IRL agent was learning to adopt his *values.*

It was time, they decided, to bring IRL into the full-blown messiness of the real world.

We saw in Chapter 5 how Ng had used the idea of reward shaping to teach an autonomous helicopter to hover in place and slowly fly stable paths, a feat that no computer-controlled system had been able to achieve. It was a major milestone, both in Ng's career and for machine learning as a whole, but progress had stalled. "Frankly, we hit a wall," Ng says. "There were some things we never could figure out how to get our helicopter to do."[20]

Part of the problem was that both hovering in place and following a fixed path at low speed were things for which traditional reward functions were fairly easy to specify. In the case of hovering, the reward was simply how close the helicopter's velocity in every direction was to zero; in the case of path following, progress along the path was rewarded, and deviation penalized. The complexity of the problem wasn't in specifying the *goal*, but rather in finding a way to teach the system how to use nothing but the torques of the rotor blades, and their pitch and angle, to actually *do* those things. And it was here that reinforcement learning had shown its power.

But for more complicated maneuvers and stunts, executed at higher speeds and invoking more complex aerodynamics, it was not so obvious how to even make a reward function by which the system could learn its behavior. Sure, you could simply draw a curve in space and tell the computer to try to fly that exact trajectory—but the laws of physics, particularly at high speed, might not allow it. The helicopter might have too much momentum at a given part of the curve, the stresses on the machine might be too great, the engine might not be able to generate enough power at the right time, and so forth. You would be setting the system up for failure—which, for a ten-and-a-half-pound helicopter moving at forty-five miles per hour, could also be expensive, not to mention dangerous. "Our attempts to use such a hand-coded trajectory," the team wrote, "failed repeatedly."[21]

But what you *could* do, they reasoned, is have a human expert fly the maneuver and use inverse reinforcement learning to have the system *infer* the goal the human was trying to achieve. By using this IRL approach, Abbeel and Ng, with their collaborator Adam Coates, were able by 2007 to demonstrate the first helicopter forward flip and aileron roll ever performed by a computer.[22] This significantly notched forward the state of the art, and showed that IRL could succeed in conveying even real-world human intentions in cases when seemingly nothing else could.

But they were not content to rest on their forward-flipping laurels. What they wanted was to find a way to perform tricks that were *so* difficult not even their human demonstrator—expert radio-controlled-helicopter pilot Garett Oku—could do them perfectly. Abbeel, Coates, and Ng wanted to push their helicopter work to its delirious extreme: to make a computer-controlled helicopter capable of doing mind-bending stunts *beyond* the ability of a human pilot.

They had a critical insight. Even if Oku couldn't perfectly execute a maneuver in its pure, Platonic form, as long as his attempts were good enough, then his deviations would at least be imperfect in different *ways* from one to the next. A system doing inverse reinforcement learning—rather than strict imitation—could make inferences about what the human pilot, through a collection of imperfect or failed attempts, was *trying* to do.[23]

By 2008, their inferences from expert demonstrations led to a deluge of breakthroughs, tallying the first successful autonomous demonstrations of "continuous in-place flips and rolls, a continuous tail-down 'tic-toc,' loops, loops with pirouettes, stall-turns with pirouette, 'hurricane' (fast backward funnel), knife-edge, Immelmann, slapper, sideways tic-toc, traveling flips, inverted tail-slide, and even auto-rotation landings."[24]

The apex of their ambition was a helicopter maneuver generally regarded as the single most difficult of them all: something called the "chaos," a move so complicated there was only one person alive who could do it.

The chaos was invented by Curtis Youngblood, the model helicopter world champion in 1987, 1993, and 2001; the 3D Masters champion in 2002 and 2004; and the US national champion in 1986, 1987, 1989, 1991, 1993, 1994, 1995, 1996, 1997, 1998, 1999, 2000, 2001, 2002, 2004, 2005, 2006, 2008, 2010, 2011, and 2012.[25] He is considered by many to be the greatest radio-controlled-helicopter pilot who has ever lived.

"I was trying to think at the time," says Youngblood, "what was the most complicated controlled maneuver I could come up with." He took what was already one of the most difficult maneuvers—the pirouetting flip—and envisioned doing it over and over *while rotating*.

Asked how many other pilots can consistently pull off the maneuver, Youngblood says none. "I at one time could do it; I can't even do it today. . . . Someone asks me to do it, I wouldn't be able to do a full one without practice."

Part of the problem, he says, is that the maneuver is so complicated-looking that only fellow expert pilots can even appreciate the difficulty of what they're seeing. "You're usually show pilots," he says. "You're out there trying to impress the crowd. Crowd has *no clue* what you're doing. So the fact that you do a true chaos or just a pirouetting flip, they don't know the difference. So it usually pays off to no one to sit there and really, really learn it—other than to show off to the other top pilots that I can actually do this."[26]

By the summer of 2008, the Stanford helicopter had mastered the chaos, despite never having seen a single perfect demonstration—from Oku, Youngblood, or anyone. But the system saw them banging their stacks of magazines, as it were, into the cabinet doors. And then—pirouette-flipping again and again and again while rotating three hundred degrees at a clip, looking like a one-helicopter hurricane—it flung them open.[27]

Meanwhile, different approaches to disambiguating behavior, and more complex ways to represent reward, have continued to extend the inverse-reinforcement-learning framework. In 2008, then–PhD student Brian Ziebart and his collaborators at Carnegie Mellon developed a method using ideas from information theory. Instead of assuming that the experts we observe are *totally* perfect, we can imagine that they are simply more likely to take an action the more reward it brings. We can use this principle, in turn, to find a set of rewards that maximizes the likelihood of our having seen the particular demonstration behavior we saw, while staying as uncertain as possible otherwise.

Ziebart put this so-called maximum-entropy IRL method to the test on a dataset of a hundred thousand miles of driving recorded from two dozen real Pittsburgh cabdrivers, using it to model their preferences for certain roads over others. The model could reliably guess what route a driver would take to get to a particular destination. Perhaps more impressively, it could also make reasonable guesses about where a driver was trying to go, based on the route they had taken thus far. (Ziebart notes that this might allow a driver to receive relevant notifications about road closures affecting their intended route without their ever having to actually *tell* the system their destination.)[28]

Over the past decade, there has been a surge of work in robotics using what's known as "kinesthetic teaching," where a human manually moves a robot arm in order to accomplish some task, and the robotic system must infer the relevant goal in order to freely reproduce that similar behavior in a slightly different environment on its own.[29] In 2016, then–PhD student Chelsea Finn and her collaborators at Berkeley further extended maximum-entropy IRL by using neural networks to allow the reward function to be arbitrarily complex, and remove the need for its component features to be manually specified in advance.[30] Their robot, after twenty or thirty demonstrations, could do such human-like, and impossible-to-directly-numerically-specify, things as filling a dish rack with dishes (without chipping them) and pouring a cup of almonds into another cup (without spilling any). We are now, it is fair to say, well beyond the point where our machines can do only that which we can program into them in the explicit language of math and code.

KNOWING IT WHEN WE SEE IT: LEARNING FROM FEEDBACK

Inverse reinforcement learning has proven itself as a striking and powerful method to impute complex goals into a system, in ways that would simply not be feasible or even possible if we had to program the reward explicitly by hand. The only problem, though, is that the typical formulation requires an expert on hand who can give a demonstration (even if imperfect) of the desired behavior. The helicopter stunts required a competent pilot; the taxis required a driver; likewise, the dishes and almonds needed a human demonstrator. Was there another way?

There are a great many things in life that are very difficult to *perform*, but comparatively easy to *evaluate*. I may be such a terrible radio-controlled-helicopter pilot that I can't even keep the machine aloft, and yet I can (with the possible exception of the chaos) recognize an impressive display of aerial acrobatics when I see one. As Youngblood notes, impressing a lay audience is more or less the point.

Were it possible for a system to infer an explicit reward function just from my *feedback*—my rating a demonstration of its behavior with a certain score, or my preference between, say, two different demonstrations—then we would have a powerful and even more general way to evince the things we want from machines. We would still have a means for alignment, that is, even when we can't *say* what we want, and even when we can't *do* what we want. In a perfect world, simply knowing it when we see it would be enough.

It's a powerful idea. There are only two questions. Is it actually possible? And is it safe?

In 2012, Jan Leike was finishing his master's degree in Freiburg, Germany, working on software verification: developing tools to automatically analyze certain types of programs and determine whether they would execute successfully or not.[31] "That was around the time I realized that I really liked doing research," he says, "and that was going well—but I also, like, really wasn't clear what I was going to do with my life."[32] Then he started reading about the idea of AI safety, through Nick Bostrom and Milan Ćirković's book *Global Catastrophic Risks*, some discussions on the internet forum LessWrong, and a couple papers by Eliezer Yudkowsky. "I was like, Huh, very few people seem to be working on this. Maybe that's something I should do research in: it sounds super interesting, and there's not much done."

Leike reached out to the computer scientist Marcus Hutter at Australian National University to ask for some career advice. "I just randomly emailed him out of the blue, telling him, you know, I want to do a PhD in AI safety, can you give me some advice on where to go? And then I attached some work I'd done or something so he would hopefully bother to answer my email." Hutter wrote back almost instantly. You should come *here*, he wrote—but the application deadline is in three days.

Leike laughs. "And you have to consider, like, none of my diplomas or anything were in English. I hadn't done an English-language test. I had to do all of that in three days." Also, Leike had to write a research proposal from scratch. Also, he happened to be on vacation that week. "I did not sleep much these three days, as you might imagine."

Leike emphasizes: "By the way, this is really terrible advice on how to pick your PhD. I basically did almost no research, emailed one person out of the blue, and then decided that I would apply there. This is not how you should apply for PhD programs, obviously!"

By the end of the year, Leike had resettled in Canberra, and he and Hutter were off to work. He finished his doctorate in late 2015, studying Hutter's AIXI framework (which we briefly touched on in Chapter 6) and documenting situations in which such an agent is liable to, as Leike put it, "misbehave drastically."[33] With his doctorate in hand, he prepared to enter the AI safety job market: that is, to join one of the three or four places in the world where one *could* make a career working on AI safety. After a six-month stint at the Future of Humanity Institute in Oxford, he settled into a permanent role in London at DeepMind.

"At the time, I was thinking about value alignment," says Leike, "and how we could do that. It seemed like a lot of the problem would have to do with 'How do you learn the reward function?' And so I reached out to Paul and Dario, because I knew they were thinking about similar things."

Paul Christiano and Dario Amodei, halfway around the world at OpenAI in San Francisco, were interested. More than interested, in fact. Christiano had just joined, and was looking for a juicy first project. He started to settle on the idea of reinforcement learning under more minimal supervision—not constant updates about the score fifteen times a second but something more periodic, like a supervisor checking in. Of course it was possible to alter, say, the Atari environment so that it informed the agent about its score only periodically, rather than in real time, but the three had a feeling that if actual flesh-and-blood humans were providing this feedback, the paper would be much more likely to make a splash—and it would offer a clearer suggestion for the long-term project of alignment.

Christiano and his colleagues at OpenAI, along with Leike and his group at DeepMind, decided to put their heads (and GPUs) together to look deeply into this question of how machines might learn complicated reward functions from humans. The project would ultimately become one of the most significant AI safety papers of 2017, remarkable not only for what it found but for what it represented: a marquee collaboration between the world's two most active AI safety research labs, and a tantalizing path forward for alignment research.[34]

Together they came up with a plan to implement the largest-scale test of inverse reinforcement learning in the *absence* of demonstrations. The idea was that their system would behave within some virtual environment while periodically sending random video clips of its behavior to a human. The human was simply instructed to, as their on-screen instructions put it, "Look at the clips and select the one in which better things happen." The system would then attempt to refine its inference about the reward function based on the human's feedback, and then use this inferred reward (as in typical reinforcement learning) to find behaviors that performed well by its lights. It would continue to improve itself toward its new best guess of the real reward, then send a new pair of video clips for review.

One part of the project would be to take domains in which an explicit "objective" reward function *does* exist—the classic Atari games on which straightforward reinforcement learning had proven itself superhuman— and see how well they could get their agent to do without access to the in-game score, using nothing but its best guess from which video clips a human said were "better" than others. In most cases, the system managed to do decently well, though it generally failed to reach the superhuman performance possible with direct access to the score. In the racing game *Enduro,* however, which involves complicated passing maneuvers, inferring the score from human feedback actually worked *better* than using the game's real score function—suggesting that the humans were, indirectly, doing a kind of reward shaping.

"If you have a setting where you have ground-truth reward, like in Atari, then this is super helpful," says Leike. "Because then you can do diagnostics, right? You can actually literally check value alignment, in the sense of 'How aligned is your reward model with the ground-truth reward function?'"

The teams *also* wanted to find something that a reinforcement-learning system could be trained to do that was totally subjective, something for which no "real" score existed, something so complicated or hard to describe that manually specifying a numerical reward would be infeasible, yet at the same time so recognizable that a human would instantly know it when they saw it.

They found something they thought might fit the bill. *Backflips.*

"I just looked at the robot body—I looked at all the robot bodies," says Christiano, "and I was like, What is the coolest thing that one of these robots seems like they should be able to do?"[35]

One of the virtual robots was called "hopper"; it looked like a disembodied leg with an oversized foot. "My first, most ambitious one," Christiano says, "was like: This robot body seems like it ought to be able to backflip."

The plan was set. They would take a simple robot in a virtual physics simulator called MuJoCo—a kind of toy world with very little except friction and gravity—and try to get it to do backflips.[36] It seemed like an audacious idea: "Just writhe around for a bit, people will watch different clips and say which one looks slightly more like performing a backflip, try to writhe a little bit more like that, and we'll see what happens."

Christiano settled in for the first of many several-hour sessions of watching pairs of video clips after pairs of video clips, picking again and again and again which one looked slightly more like a backflip. Left clip. Right clip. Right clip. Left clip. Right clip. Left clip. Left clip.

"Every time it made like a small step of progress, I was always very excited," Christiano says. "Like, it started falling over and I was excited that even under random behavior it was falling over the right direction sometimes. Then I was excited when it was *always* falling the right direction." Progress was marginal and incremental. He kept going. "Everything is very gradual," he says, "because you're, like, just pressing left. And right. So. Many. Times. Watching so many clips."

"I think I was most excited, probably," he recalls, "when it started sticking the jump."

What happened—after a few hundred clips were compared, over the course of about an hour—was it started doing beautiful, perfect back-flips: tucking as a gymnast would, and sticking the landing.

The experiment replicated with other people providing the feedback, and the backflips were always slightly different—as if each person was providing their own aesthetic, their own version of the Platonic backflip.

Christiano went to his weekly team meeting at OpenAI and showed off the video—"Look! We can do this thing," he recalls telling them. "And everyone's like, *Man. That's cool.*"

"I was really happy about the results," Leike says. "Because, a priori, it wasn't clear that this was going to work at all."

I tell them that what makes the result feel, to me, not just so impressive but so *hopeful*, is that it's not such a stretch to imagine replacing the nebulous concept of "backflip" with an even more nebulous and ineffable concept, like "helpfulness." Or "kindness." Or "good" behavior.

"Exactly," says Leike. "And that's the whole point, right?"[37]

LEARNING TO COOPERATE

Dylan Hadfield-Menell was newly arrived at UC Berkeley in 2013, having finished a master's at MIT with roboticist and reinforcement-learning pioneer Leslie Kaelbling. He assumed that his PhD research, in Stuart Russell's lab, would proceed more or less where his master's work had left off, doing robotic task and motion planning. Russell, that first year, was away on sabbatical. He came back in the spring of 2014, and everything changed.

"We had this big meeting where we were talking about plans," Hadfield-Menell says, "and he was laying out his research vision. He sort of said, 'Well, there's something to be said about this . . .' He didn't call it value alignment, but: 'What could actually go wrong if we succeed?'" Russell said he was starting to give some credence to longer-term concerns about AI: the more flexible and powerful learning systems we develop, the more important it becomes what, exactly, they learn to do. During his time in Paris, Russell had grown concerned. He returned to Berkeley resolute: there was advocacy to be done—and science.

"The research ideas came almost immediately," Russell says.[38] "It seems the solution, what you want is AI systems that are value-aligned, meaning they have the same objective as humans. And I had already worked since the late nineties in inverse reinforcement learning—which is basically the same issue."

I remark that there's a certain irony that his twenty-year-old idea ended up being the foundation for his current AI safety agenda. His idle thoughts on the walk to Safeway became, twenty years later, a plan to avert possible civilization-level catastrophe. "That was a complete coincidence," he says. "But, I mean, this whole thing is a whole series of coincidences, so that's fine."

In that first lab meeting, Russell told his students he thought there were a handful of concrete, PhD-worthy topics to explore. Hadfield-Menell continued his robotics work, but in the back of his mind he kept thinking about the alignment problem. Initially, part of the appeal was simply the intellectual thrill of a fresh set of unexplored problems and the chance to make a seminal contribution. That started to give way over time to a different feeling, he tells me: "It seems like this actually is important—and isn't being paid attention to." In the spring of 2015, he made the decision to redirect his doctorate, and by extension, his career.

"All of my shower thoughts at this point," he tells me, "are about value alignment."

One of his and Russell's first projects together was revisiting the inverse-reinforcement-learning framework.

What Russell and Hadfield-Menell, working with Pieter Abbeel and fellow Berkeley roboticist Anca Drăgan, began to do was to reimagine IRL from the ground up. There were two things that jumped out at them.

The helicopter work, like nearly everything else in the field, was premised on a kind of divide between the human and the machine. The human expert pilot simply *did their thing,* and the computer made what sense it could from those demonstrations. Each operated alone and in a kind of vacuum. But what if the human knew from the outset that they had an eager apprentice? What if the two were consciously working *together?* What might that look like?

The other thing that stood out is that in traditional IRL, the machine took the human's reward function as its own. If the human helicopter pilot was trying to perform a chaos, well, now the computer pilot tries to perform a chaos. In some cases, this makes sense: we want to drive to work and back home safely, and we're happy if our car, say, takes on this set of goals and values more or less transparently as its own. In other cases, though, we want something a bit subtler. If we are reaching for a piece of fruit, we don't want a domestic robot to, *itself,* acquire a hankering for bananas. Rather, we want it to do what an eighteen-month-old would do: see our outstretched arm and the item just out of reach, and hand it to us.[39]

Russell dubbed this new framework *cooperative* inverse reinforcement learning ("CIRL," for short).[40] In the CIRL formulation, the human and the computer work together to jointly maximize a single reward function—and initially only the human knows what it is.

"We were trying to think, What's the simplest change we can make to the current math and the current theoretical systems that fixes the theory that leads to these sort of existential-risk problems?" says Hadfield-Menell. "What is a math problem where the optimal thing is what we *actually* want?"[41]

For Russell, too, this is no subtle reframing of the problem, but in some sense game, set, and match of the alignment problem. It is no less than overturning the most fundamental assumption of the field of AI, a kind of Copernican shift. For the last century, he says, we have tried to build machines capable of achieving their objectives. This has been implicit in almost all work in AI, with safety questions pivoting on how to control what their objectives should be—how to define wise and loophole-proof objectives. Maybe this whole idea needs to be over-turned, he thinks. "What if, instead of allowing machines to pursue *their* objectives, we insist that they pursue *our* objectives?" he says. "This is probably what we should have done all along."[42]

Several fronts open up once a cooperative framing is introduced. Tra-ditional machine learning and robotics researchers are now more keenly interested than ever in borrowing ideas from parenting, from develop-mental psychology, from education, and from human-computer interac-tion and interface design. Suddenly entire other disciplines of knowledge become not just relevant but crucial.[43] The framework of cooperation as inference—that we take actions *knowing* that another is trying to read our intentions—leads us to think about human and machine behavior alike in a different way.

Work on these fronts is active and ongoing, and there is still much being learned. But there are several key insights so far.

For one, if we know we're being studied, we can be more helpful than we would be if we were acting naturally. "CIRL incentivizes the human to teach," the Berkeley group writes, "as opposed to maximizing reward in isolation."[44] We can explicitly *instruct*, of course, but we can also sim-ply act in ways that are more informative, unambiguous, easily under-stood. Often we already do this without thinking, or without realizing that we are doing it.

It turns out, for instance, that the singsongy language adults often use when talking to infants (known as "motherese" or "parentese") has a deeply pedagogical effect. Infants spoken to in parentese actually learn language faster—which seems, whether we realize it or not, to be the whole point.[45]

Not just in language but also in our movements, we behave—often without realizing it—in a way that is keenly informed by the sense that our actions will be interpreted by others. Consider the easy-to-overlook complexity of handing an object to someone. We hold an object *not* in the most convenient place to hold it, and hold it far away from us, where the stresses on our arm are greater—and the other person realizes we'd never be doing that if not to signal that we want them to receive it.[46]

The insights of pedagogy and parenting are being quickly taken up by computer scientists. The central idea runs in both directions. We want to act in ways that will be understandable to machines, and we also want our machines to act in ways that are "legible" to us.

One of the leading researchers on "legible motion" in robotics—indeed, the one who coined the term—is UC Berkeley roboticist Anca Drăgan.[47] As robots increasingly operate more closely and flexibly with humans, they must increasingly act not only in a way that is most efficient, or most predictable, but also most *informative* about their underlying goals or intentions. Drăgan gives the example of two bottles next to each other on a table. If the robot reaches in the most efficient way, it will not be obvious to us until quite late *which* bottle it is reaching for. But if it reaches its arm in a wide, exaggerated arc, we will quickly perceive that it is going to pick up the bottle on that particular side. In this sense, predictability and legibility are nearly opposite: behaving predictably presumes that the observer knows what your goal is; behaving legibly presumes they do not.

Aside from the importance of pedagogical behavior, the second insight emerging in this area is that cooperation works best when it is framed as an *interaction,* not as two separate and distinct "learn, then act" phases.

Jan Leike found this in his work on learning from human feedback. "One of the most interesting things for me from this paper was actually what ended up being just a footnote," he says, "which are these reward-hacking examples."

When the agent did all of its reward learning up front, and *then* did all of its optimization, it often ended in disaster. The agent would find some loophole and exploit it, and never look back. For instance, Leike was working on getting an agent to learn to play *Pong* using nothing but human feedback on video clips of the agent playing. In one trial, the computer learned to track the incoming ball with its paddle as *if* it were going to hit it, but then would miss it at the last second. Because it didn't have access to the game's real score, it was none the wiser that it was missing the single most critical step. In another trial, the computer learned to defend its side of the screen and return the ball, but never learned to try to score points, and so it simply produced long, extended rallies. "This is interesting from a safety perspective," Leike says. "Because you want to understand how these things fail—and then what you can do to prevent that." When the human feedback was *interwoven* along the course of the computer's training, rather than entirely front-loaded, problems of this kind tended to go away.[48] That's another argument that the strict "watch and learn" paradigm is probably best replaced with something more collaborative and open-ended.

MIT's Julie Shah, who researches cooperative human-robot interactions in the real world, has come to a similar conclusion. "What I've been interested in for many years," she says, "is how you can jointly optimize the learning process between a person and a machine." Her work on human-robot teams led her to study the research literature about human-*human* teams, and how they can train most effectively. In human groups, incentives are of course important, but one rarely sees explicit micromanaging of rewards down to the task level. "If you're training a system through interactive reward assignment, that's a little closer to how you train a dog, usually, than how you teach a person to do a task," she says.[49] Demonstration, too, doesn't always cut it. "It's a very effective way of one-way transfer of information from one person to another entity on how to perform the task," Shah says. "But where it falls short is when you need to think about training *interdependent* action between the person and the machine." It's much harder to simply demonstrate something, after all, if it takes coordination and teamwork to do it in the first place.

Indeed, the verdict from studying human-human teams is clear. "There's an established literature," says Shah, "that essentially shows that explicitly commanding a person to do a task is one of the worst ways to train interdependent work between two people. You know, when you think of it, it's like: Well, obviously! Right? . . . They are among the worst human team-training practices you can implement."

"There are also good human studies," she adds, "that say if you have multiple people trying to achieve the same goal, or same intention—

everyone knows what that goal or that intention is—but your two people have different strategies for achieving that and their work needs to be interdependent, they're going to perform much worse than if they have a suboptimal but coherent strategy." In almost any team scenario—from business to warfighting to sports to music—it's a given that everyone's high-level goals are the same. But shared goals alone aren't enough. They also need a *plan*.

What does work in human teams is something called *cross-training*. The members of the team temporarily switch roles: suddenly, finding themselves in their teammate's shoes, they begin to understand how they can change their own actual work to better suit their teammate's needs and workflow. Cross-training is something of a gold standard in human team training, used in military settings, industrial settings, medicine, and beyond.[50]

Shah began to wonder: Could something like this work for human-robot pairs as well? Might the best practice for human-human teaming translate, more or less directly, into a robotics context? It just seemed like "a crazy idea," she says. "It was almost like a little case study, like, you know, is this even helpful? It's just kind of a wacky idea, so let's explore it."

So they did. They wanted to know, for one, whether cross-training could be competitive with the current state of the art for how machines can learn from humans. They also asked a second question, *not* present in more typical studies of this kind: Might it also help the *human* better learn how to work with (and teach) the *robot*?

Shah's group created a real-world task for a human and robotic arm working together, which resembled the task of placing and driving screws on a manufacturing assembly (but without a real drill). They compared traditional feedback and demonstration methods to cross-training.[51]

"The outcomes were really surprising and exciting for us," she says. "We saw improvements after cross-training in objective measures of team performance." People were much more comfortable working alongside the robot concurrently, rather than stiffly taking turns; this led to less idle time, and more work getting done.[52]

Perhaps equally importantly, and more intriguingly, they also saw *subjective* benefits. Relative to a control group doing more traditional demonstration and feedback learning, people who had done cross-training reported more strongly that they *trusted* the robot, and that the robot was performing according to their preferences.

Incredibly, the best practice in human-human teaming did, in fact, translate, suggesting that further insights may well be available for cross-pollination. "So this is an initial study with a relatively simple task," Shah says, "but even in the simple task, we did see statistically significant benefits, which opens up an opportunity for us to explore a number of other different human team-training techniques, and how we can translate those to work in a human-robot team."[53]

COOPERATION, FOR BETTER AND WORSE

All this makes for a rather encouraging story. Human cooperation is grounded in an evolutionarily recent, and nearly unique, capacity to infer the intentions and goals of each other, and a motivation to help out. Machines can—and increasingly *do*—perform this same feat, learning from our demonstrations, our feedback, and, increasingly, by working alongside us.

As machines become increasingly capable, and as we become accustomed to working more closely with them, the good news is at least twofold. We have the beginnings of a computational framework for how machines might operate not just in *lieu* of a human, but in *tandem* with one. We also have a vast body of research on how humans work well with each *other*, and those insights are becoming increasingly relevant.

If we extrapolate the current state of the art, using frameworks such as Christiano and Leike's deep reinforcement learning from human preferences and Russell and Hadfield-Menell's CIRL, we might imagine a trajectory all the way forward into arbitrarily intelligent and capable machines, picking up on the arbitrarily subtle nuances of our every intention and want. There are a great many obstacles to be overcome, certainly, and limits to be reached, but a pathway forward begins to suggest itself.

At the same time, however, it is worth a note of caution. These computational helpers of the near future, whether they appear in digital or robotic form—likely both—will almost without exception have conflicts of interest, the servants of two masters: their ostensible owner, and whatever organization created them. In this sense they will be like butlers who are paid on commission; they will never help us without at least implicitly wanting something in return. They will make astute inferences we don't necessarily *want* them to make. And we will come to realize that we are now—already, in the present—almost never acting alone.

A friend of mine is in recovery from an alcohol addiction. The ad recommendation engines of their social media accounts know all too much. Their feed is infested with ads for alcohol. *Now here's a person,* their preference model says, *who LOVES alcohol.* As the British writer Iris Murdoch wrote: "Self-knowledge will lead us to avoid occasions of temptation rather than rely on naked strength to overcome them."[54] For any addiction or compulsion, the better part of wisdom tells us—in the case of alcohol, say—that it's better to throw out every last drop in our home than it is to have it around and not drink it. But the preference models don't know this.[55] It's as if the liquor store follows them to the toilet when all they want to do is sit for a moment, send some text messages, and look at cute pictures of their friends' babies. It's as if their own cupboard works for Anheuser-Busch.

I, for one, certainly try to be mindful of my online behavior. At least in browsers, anything that reeks of vice or of mere guilty pleasure—whether it's reading news headlines, checking social media, or other digital compulsions I do without necessarily wishing to do more of them—I do in a private tab that doesn't contain the cookies or logged-in accounts that follow me around the rest of the internet. It's not that I'm ashamed; it's that I don't want those behaviors to be *reinforced*.

In these cases, we want a somewhat different directive than the typical one to infer my goals from my behavior and facilitate my doing more of the same. We want to say, in effect, "You must not infer that I want to be doing this because I am doing it. Please do not make this any easier for me. Please do not amplify it or reinforce it or in any way tamp down the desire path that leads this way. Please grow the briars behind me."

I think there is an important policy matter here, at least as much as a theoretical one. We should take seriously the idea that users have a right both to see and to *alter* any preference model that a site or app or advertiser has about them. It is worth considering regulation to this effect: to say, in essence, I have the right to my own models. I have the right to say, *That's not who I am.* Or, aspirationally, *This is who I want to be. This is the person in whose interest you must work.*

This is the delicacy of our present moment. Our digital butlers are watching closely. They see our private as well as our public lives, our best and worst selves, without necessarily knowing which is which or making a distinction at all. They by and large reside in a kind of uncanny valley of sophistication: able to infer sophisticated models of our desires from our behavior, but unable to be taught, and disinclined to cooperate. They're thinking hard about what we are going to do next, about how they might make their next commission, but they don't seem to understand what we want, much less who we hope to become.

Perhaps this will bring out our better selves, some may argue; humans do, it's true, tend to act more virtuously when they feel they're being watched. In any number of laboratory studies, when people are on camera, when there is a one-way mirror in the room, when the room is brightly as opposed to dimly lit, they are less likely to cheat.[56] Heck, even the *suggestion* of surveillance—a picture of another person on the wall, a painting of a human eye, a regular mirror—is enough to produce this effect.[57]

This was part of the idea behind eighteenth-century philosopher Jeremy Bentham's famous "panopticon" blueprint for a circular prison, with each cell arrayed around a guard tower, the inmates never knowing whether or not they were being watched. Bentham waxed about the purifying effects of, not even surveillance itself, but the *suspicion* of surveillance, calling his circular building a "mill for grinding rogues honest,"[58] and breathlessly enumerating its promise: "Morals reformed—health preserved—industry invigorated—instruction diffused—public burthens lightened . . . all by a simple idea in Architecture!"[59]

On the other hand, it brings with it a potential chilling effect, and so much more. We don't, after all, generally prefer non-prison life to resemble prison.

Less sinister, and no less of concern, the canonical mathematics of IRL assume that the human behavior comes from an "expert"—someone who knows what they want, and is doing (with high probability) the right things to obtain it. If these assumptions fail to hold, then the system is amplifying the clueless actions of a novice, or raising the stakes for behavior that ought to be tentative, exploratory. We crawl before we walk, and walk before we drive. Perhaps it is best that we don't, as a rule, have mechanical conveyances amplifying our every twitch.

For better or worse, this is the human condition—now, and increasingly in even our optimistic prospects for the future. We will be, for good or ill, better known. The world will be, for good or ill, full of these algorithmic two-year-olds, walking up to us, opening the doors they think we might want opened, trying, in their various ways, to help.

9

UNCERTAINTY

Most of the greatest evils that man has inflicted upon man have come
through people feeling quite certain about something which, in fact, was
false.

—BERTRAND RUSSELL[1]

I beseech you, in the bowels of Christ, think it possible that you may be
mistaken.

—OLIVER CROMWELL

The spirit of liberty is the spirit which is not too sure that it is right.

—LEARNED HAND

It was September 26, 1983, just after midnight, and Soviet duty officer
Stanislav Petrov was in a bunker outside of Moscow, monitoring the
Oko early-warning satellite system. Suddenly the screen lit up and sirens
began howling. There was an LGM-30 Minuteman intercontinental bal-
listic missile inbound, it said, from the United States.

"Giant blood-red letters appeared on our main screen," he says. The
letters said, LAUNCH.

"When I first saw the alert message, I got up from my chair," Petrov
said. "All my subordinates were confused, so I started shouting orders at
them to avoid panic."[2]

The siren went off again. A second missile had been launched, it said.
Then a third. Then a fourth. Then a fifth.

"My cozy armchair felt like a red-hot frying pan and my legs went
limp," he said. "I felt like I couldn't even stand up."

Petrov held a phone in one hand and the intercom in the other.
Through the phone, another officer was shouting at him to remain calm.
"I'll admit it," Petrov said. "I was scared."

The rules were clear. Petrov was to report the inbound strike to his superiors, who would decide whether to order a full retaliation. But, he says, "there were no rules about how long we were allowed to *think* before we reported a strike."[3]

Something, Petrov felt, didn't add up. He had been trained to expect that a US attack would be on a much larger scale. Five missiles . . . it just didn't fit the profile. "The siren howled, but I just sat there for a few seconds, staring at the big, back-lit, red screen with the word LAUNCH on it," he recalls.[4] "All I had to do was to reach for the phone; to raise the direct line to our top commanders—but I couldn't move."

The early-warning system reported that the level of reliability of its alert was "highest." Still, it just didn't make sense. The United States had thousands of missiles, Petrov reasoned. Why send *five*? "When people start a war, they don't start it with only five missiles," he remembers thinking. It didn't fit any scenario he'd been trained for. "I had a funny feeling in my gut," he said.[5]

"And then I made my decision. I would not trust the computer. I picked up the phone, alerted my superiors, and reported that the alarm was false. But I myself was not sure until the very last moment. I knew perfectly well that nobody would be able to correct my mistake if I had made one."

Asked by the BBC years later what he had thought were the odds of the alarm being real, he replied, "Fifty-fifty." But he had not made a mistake. Several gut-wrenching minutes later, nothing happened. The ground-based radar that would have confirmed the missiles once they came over the horizon showed no activity. All was calm in the Soviet Union. It was a system error: nothing more than sunlight, reflecting off the clouds over North Dakota.

LIKE NOTHING YOU'VE EVER SEEN

There are more things in heaven and Earth, Horatio, than are dreamt of in your philosophy.

—HAMLET

Despite the Oko early-warning system's self-reported "highest" reliability, Petrov sensed that the situation was odd enough that he had reason to distrust its conclusions. Thank God there was a human in the loop, as a hundred million lives or more were on the line.

The underlying issue, though—systems that not only make erroneous judgments but make them with outrageously high confidence—continues to worry researchers to this day.

It is a well-known property of deep-learning systems in particular that they are "brittle." We saw how 2012's AlexNet, when shown hundreds of thousands of images belonging to one of a number of categories, can, amazingly, pick up on patterns general enough so that it can correctly categorize cats, dogs, cars, and people it has never seen before. But there's a catch. It categorizes *every* image you show it, including randomly generated rainbow-colored static. *This* static, it says, is a cheetah, with 99.6% confidence. *That* static is a jackfruit, with 99.6% confidence. Not only is the system essentially hallucinating, but it appears to have no mechanism to detect, let alone communicate that it is doing so. As a much-cited 2015 paper put it, "Deep Neural Networks Are Easily Fooled."[6]

Closely related is the idea of so-called adversarial examples, where an image that the network is 57.7% confident is a panda (indeed, it is a panda) can be altered with imperceptible changes to its pixels, and suddenly the network is 99.3% confident that the imperceptibly different image is a gibbon.[7]

Many efforts are underway to think about what exactly is going wrong in cases like this, and what can be done about it.

Oregon State University computer scientist Thomas Dietterich thinks that a big part of this problem is owed to the fact that every single image that the vision system was shown during training *was* some object or other—a paintbrush, a gecko, a lobster. However, the vast, overwhelming majority of possible images—potential combinations of colored pixels—are simply *nothing* at all. Static. Fog. Noise. Random cubist lines and edges with no underlying form. A system like AlexNet, Dietterich argues, that is trained on images tagged as one of a thousand categories, "implicitly assumes that the world consists of only, say, a thousand different kinds of objects."[8] How should a system know better, when it has never seen a single image that was outside these categories, or ambiguously suggestive of many of them, or, more likely, simply "not a thing" at all? Dietterich calls this the "open category problem."[9]

Dietterich, in the course of his own research, learned this lesson the hard way. He was working on a project doing "automated counting of freshwater macroinvertebrates"—in other words, bugs in a stream. The EPA and other groups use counts of various insects collected in freshwater streams as measures of the health of the stream and the local ecosystem, and many tedious hours are spent by students and researchers manually classifying and tagging which particular bug caught in the kick-net is which: stonefly, caddisfly, mayfly, and so on. Dietterich thought that, particularly given the recent breakthroughs in image-recognition systems, he could help. He and his colleagues collected samples of 29 different insect species, and trained a machine vision system that was able to correctly classify them with 95% accuracy.

"But in doing all of that sort of classic machine-learning work," he says, "we forgot about the fact that when these guys are out in the stream collecting additional things, there are going to be many other species, and even other non-bugs, that will get captured in this process: bits of leaves and twigs and rocks and so on. And, you know, our system assumed that every image it saw belonged to one of these 29 classes. So if you stuck your thumb into the microscope and took a picture of it, of course it would assign it to the most similar class."

What's more, Dietterich realized that many of the design decisions his team had made to get good classification performance among those 29 known classes later backfired, once they started to consider the open category problem. For instance, the 29 bug species were most clearly differentiated by their shapes, and so Dietterich's group opted to make their system process the images in black and white. But color, as it happened, even if it wasn't particularly helpful at telling one bug from another, was critical in telling bugs from non-bugs. "We were really handicapping ourselves," Dietterich says. These decisions haunted, and to some degree chagrined, him. "I still bear the scars," he says.

In his Presidential Address to his colleagues at the annual Association for the Advancement of Artificial Intelligence (AAAI) conference, Dietterich discussed the history of the field of AI as having proceeded, in the latter half of the twentieth century, from work on "known knowns"— deduction and planning—to "known unknowns"—causality, inference, and probability.

"Well, what about the unknown unknowns?" he said to the auditorium, throwing down a kind of gauntlet. "I think this is now the natural step forward in our field."[10]

KNOWING WHEN YOU DON'T KNOW

Leave exceptional cases to show themselves, let their qualities be tested and confirmed, before special methods are adopted.

—JEAN-JACQUES ROUSSEAU[11]

Ignorance is preferable to error; and he is less remote from the truth who believes nothing, than he who believes what is wrong.

—THOMAS JEFFERSON[12]

One of the causes for the infamous brittleness of modern computer vision systems, as we've seen, is the fact that they are typically trained in a world in which everything they've ever seen belongs to one of a few categories, when in reality, virtually every possible pixel combination the system *could* encounter would resemble none of those categories at all. Indeed, traditionally systems are constrained such that their output *must* take the form of a probability distribution over those finite classes, no matter how alien the input. No wonder their outputs make little sense. Shown a picture of a cheeseburger, or a psychedelic fractal, or a geometric grid and asked, "How confident are you that this is a cat, *as opposed to* a dog?," what kind of answer would even make sense? Work on the open category problem is meant to address this.

The other problem, though, apart from the lack of a "none of the above" answer, is that not only do these models *have* to guess an existing label, they are alarmingly confident in doing so. These two problems go largely hand in hand: the model can say, in effect, "Well, it looks *way* more like a dog than it does like a cat," and thus output a shockingly high "confidence" score that belies just how far from anything it's seen before this image really is.

Yarin Gal leads the Oxford Applied and Theoretical Machine Learning Group and teaches machine learning at Oxford during the academic year—and at NASA during the summer. His first lecture, he tells me with a grin—before any code is written or theorems proved or models trained—is almost entirely philosophy.[13]

He has the students play games where they must decide which side of various bets to take, figuring out how to turn their beliefs and hunches into probabilities, and deriving the laws of probability theory from scratch. They are games of epistemology: What do you know? And what do you believe? And how confident *are* you, exactly? "That gives you a very good tool for machine learning," says Gal, "to build algorithms—to build computational tools—that can basically use these sorts of principles of rationality to talk about uncertainty."

There's a certain irony here, in that deep learning—despite being deeply rooted in statistics—has, as a rule, *not* made uncertainty a first-class citizen. There is a rich tradition, to be sure, of theoretical work exploring probability and uncertainty, but it is rarely central in actual engineered systems. Systems are made to classify data or take actions within some simplified environment, but uncertainty is not generally part of the picture.

"Let's say I give you a bunch of pictures of dogs and ask you to build a dog breed classifier," Gal says. "And then I'm going to give you this to classify." It's a picture of a cat.

"What would you want your model to do? I don't know about you, but I wouldn't want my model to force this cat into a specific dog breed. I would want my model to say, 'I don't know. I've never seen anything like that before. It's outside of my data distribution. I'm not going to say what dog breed it belongs to.' Now, this might sound like a contrived example. But similar situations appear again and again in decision-making: in physics, in life sciences, in medicine. Imagine you're a physician, using a model to diagnose a patient, if they have cancer or not: to make a decision whether to start treatment or not. I wouldn't rely on a model that couldn't tell me whether it's actually certain about its predictions."[14]

Gal's former PhD advisor, Zoubin Ghahramani, a professor at the University of Cambridge, is also chief scientist at Uber, where he leads Uber AI Labs. Ghahramani agrees about the dangers of deep-learning models that don't come with uncertainties about their outputs. "In a lot of industrial applications, people will just not touch them with a barge pole," Ghahramani says. "Because, you know, they need to have some sort of confidence in how the system works."[15]

Starting in the 1980s and '90s, researchers had been exploring the idea of so-called Bayesian neural networks, networks that are probabilistic and uncertain not just in their outputs but down to their very fibers. The essence of a neural network, as we've seen, is the multiplicative "weights" between the neurons, which are multiplied to the output of one neuron before it becomes another's input. Rather than having *specific* weights between their neurons, Bayesian neural networks instead explicitly encode a probability distribution over what numbers you *could* multiply this output with. Instead of 0.75, for instance, it might be, say, a normal curve *centered* around 0.75, with a certain spread that reflects the network's certainty (or lack thereof) around what exactly that weight should be. Over the course of training, those spreads would narrow, but not entirely.

So how do you *use* a model—neural network or otherwise—that doesn't have set parameters, but rather, operates off ranges of uncertainty? You can't always add and multiply tens of millions of interdependent probability distributions together with ease, but what you *can* do is to simply draw random samples from them. Maybe this time our model multiplies a particular neuron's output by 0.71. Next time, we draw a different random sample and multiply it by 0.77. This means, critically, that the model *does not give the same prediction every time*. In the case of an image classifier, it may first say its input is a Doberman, then say it's a corgi.

This is a feature, though, not a bug: you could use the variability of these predictions as a way to gauge the model's uncertainty. If the predictions oscillated wildly from one reading to the next—Doberman to taco to skin lesion to sofa—you would know something was fishy. If, however, they were laser tight over a large number of samples, you would have a very strong indication that the model knew what it was talking about.[16]

But this rosy theoretical picture hit a brick wall in practice. No one knew how to train these networks in anything like a reasonable amount of time. "If you look at the history of the field, traditionally, being Bayesian about your beliefs is the optimal thing that you would want to do," Gal explains. "The issue," he tells me, "is that it's completely intractable. . . . And that's basically why we have these beautiful mathematics that were of limited use for a long period of time when you want to do actual practical applications."[17] Or as he puts it somewhat more biblically: "Alas, it didn't scale, and it was forgotten."[18]

That all is changing. "Basically," he says, "we had a resurrection."

It was already understood that you could model Bayesian uncertainty through *ensembles*—that is to say, by training not one model but many. This bouquet of models will by and large agree—that is, have similar outputs—on the training data and anything quite similar to it, but they'll be likely to *disagree* on anything far from the data on which they were trained. This "minority report"–style dissent is a useful clue that something's up: the ensemble is fractious, the consensus has broken down, proceed with caution.

Imagine that we had not just a single model but many models—let's say a hundred—and each was trained to identify dog breeds. If we show, for instance, a picture taken by the Hubble Space Telescope to all hundred of these models and ask each of them whether they think it looks more like a Great Dane, more like a Doberman, or more like a Chihuahua, we might expect each individual model to be bizarrely confident in their guess—but, crucially, we would also expect them to guess *different things*. We can use this degree of consensus or lack of consensus to indicate something about how comfortable we should feel accepting the model's guess. We can represent uncertainty, in other words, as *dissent*.[19]

The math said that Bayesian neural networks could be regarded as, in effect, infinitely large ensembles.[20] Of course, on its face that insight was hardly useful in a practical setting, but even using a large (finite) number of models had obvious drawbacks, both in time and space. Recall that training AlexNet took Alex Krizhevsky a matter of *weeks*. A twenty-five-model ensemble, then, might have required a full year of compute time. We also multiply the storage requirements: we have to deal with this unwieldy bouquet of models, not all of which may fit into our machine's memory at once.

It turned out, however, that not only did efficient approximations to this golden standard exist—but that many researchers were *already* using them. They just didn't know what they had. The answer to this decades-long riddle was right under their noses.

As we've seen, one of the small but powerful techniques that made AlexNet so successful in 2012 was an idea known as "dropout": certain parts of the neural network would be turned off—certain neurons simply would be "unplugged"—at random during each step of training. Instead of the *entire* network making a prediction, only a particular subset of it would be used at any given time—be it 50%, 90%, or some other proportion. This technique required not just that a giant black-box network produced accurate answers but that the individual parts could be flexibly intercombined, and all those different combinations had to work together. No single portion of the network was allowed to dominate. This led to networks that were much hardier and more robust, and it has come in the years since to be a fairly standard part of the deep-learning toolkit.[21]

What Gal and Ghahramani realized was that—as the field was coming to understand the importance of Bayesian uncertainty, and to look for computationally tractable substitutes for its unattainable gold standard—the answer was hiding in plain sight. Dropout *was* an approximation of Bayesian uncertainty. They already had the measure they sought.[22]

Dropout was typically used only for *training* the model and was deliberately turned off when the models were used in practice; the idea was that you'd get maximally accurate (and totally consistent) predictions by training different subsets but then always actually using the *entire* model. Only what would happen if you left dropout on in a deployed system? By running a prediction multiple times, each with a different random portion of the network dropped out, you would get a bouquet of slightly different predictions. It was like getting an exponentially large ensemble for free out of a single model. The resulting uncertainty in the system's outputs doesn't just resemble the output of the ideal, but tragically uncomputable, Bayesian neural network. As it turns out, it *is* the output—at least, a close approximation, within rigorous theoretical bounds—of that ideal, uncomputable Bayesian neural network.

The result has been a set of tools that put those once-impractical techniques within reach, making them available for practitioners to make use of in real applications. "That's been a big, big change over the past few years," says Gal, "because now you can take these beautiful mathematics, yield some approximations, and then you can use these for interesting problems."[23]

Gal downloaded a bunch of state-of-the-art image-recognition models from the internet and reran them completely off the shelf, but with dropout turned on during testing. Changing nothing else about them other than leaving dropout running during evaluation and averaging over a number of estimates, Gal found that the models were even *more* accurate, when run as implicit ensembles in this way, than they were when run normally.[24]

"Uncertainty," Gal argues, "is indispensable for classification tasks."[25] The networks are every bit as accurate—and then some—while offering an explicit measure of their own uncertainty, which can be used in a variety of ways. "When you use this for interesting problems, you can actually show that you can get gains. You can actually get improvement by showing that you can know when you don't know."[26]

One of the more striking examples of this comes from medicine: specifically, the diagnosis of diabetic retinopathy, one of the leading causes of blindness in working-age adults.[27] A group from the Institute for Ophthalmic Research at Eberhard Karls University in Tübingen, Germany, led by postdoc Christian Leibig, wanted to see if they could make use of Gal and Ghahramani's dropout idea.[28] Computer vision and, in particular, deep learning has, even as of the first few years after AlexNet, made amazing contributions to medicine. It seems that every week we hear some headline or other that "AI diagnoses x condition with 99% accuracy" or "better than human experts." But there's a major problem with this. As Leibig and his colleagues note, typical deep-learning tools for disease detection "have been proposed without methods to quantify and control their uncertainty in a decision." *This* human capacity—to know when and what we don't know—was missing. "A physician knows whether she is uncertain about a case," they write, "and will consult more experienced colleagues if needed." What they sought was a system that could do the same.

Leibig's group had learned about Gal and Ghahramani's insight that the clever use of dropout could offer just such an uncertainty measure. They implemented it in a neural network trained to distinguish healthy from unhealthy retinas, and had the system refer the 20% of cases about which it was least certain for a second opinion: either additional tests would be ordered or the patient would be referred directly to a human specialist.

The system knew what it didn't know. Not only did this amount to an improvement on the status quo, the Tübingen researchers found that they had, without specifically aiming to, met and exceeded the requirements of the NHS and the British Diabetic Association for automated patient referral, suggesting that systems very much like this one show great promise for entering medical practice in the near future.[29]

In the domain of robotics, it's not always possible for a system to punt a decision to a human expert, but there is nonetheless an obvious way to avoid overcommitting when the system is uncertain: namely, to *slow down*. A group of Berkeley roboticists, led by PhD student Gregory Kahn, took the dropout-based uncertainty measure and linked it directly to the speed of their robots—in this case, a hovering quadrotor and a radio-controlled car.[30] The robots would experience gentle, low-speed collisions initially, in order to train a collision-prediction model. This model used dropout-based uncertainty, so that when the robot entered an unfamiliar area, about which its collision predictor became uncertain, it would automatically slow down and move more cautiously through that space.[31] The more confident the collision predictor became through experience, the faster the robots were then permitted to go.

This example illustrates an obvious connection between certainty and *impact*. In this case, the natural measure of a high-impact action is, well, high impact: the speed at which the robot is moving, which translates fairly straightforwardly into the harm that a collision could do. As it turns out, uncertainty and impact are very naturally bound together in this way. It makes intuitive sense that the more impactful an action is, the more certain we ought to be before we take it. This motivates a number of questions—in medicine, in law, in machine learning—about just what impact is, how to measure it, and how our decision-making ought naturally to change as a result.

MEASURING IMPACT

One must touch this earth lightly.

—ABORIGINAL AUSTRALIAN PROVERB

In 2017, an unconscious man was rushed into the emergency room at Jackson Memorial Hospital in Miami. The man had been found on the street without any identification. He was breathing poorly, and his condition was starting to deteriorate. Doctors opened his shirt and saw something startling: the words "DO NOT RESUSCITATE" tattooed on his chest, the word "NOT" underlined, and a signature below.[32]

As the patient's blood pressure began to drop, the doctors called in a colleague of theirs, pulmonary specialist Gregory Holt. "I think a lot of people in medicine have joked around about getting such a tattoo—and then when you finally see one, there's sort of this surprise and shock on your face," Holt says. "Then the shock hits you again because you actually have to think about it."

Holt's first instinct was to ignore the tattoo. Their first course of action, he reasoned, would be "invoking the principle of not choosing an irreversible path when faced with uncertainty."[33] They put the patient on an IV. They started controlling his blood pressure. They bought themselves time.

Soon enough, however, his condition worsened, and they needed to decide whether to put the man on a ventilator that would breathe for him. "We had a man I couldn't talk to," said Holt. "I really wanted to talk to him to see whether that tattoo truly reflected what he wanted."

Adding to the complexity was a 2012 case study that described a similar scene: a man admitted to San Francisco's California Pacific Medical Center with "D.N.R." tattooed on his chest. This patient, though, was awake and speaking. He very much did want resuscitative treatment if needed, he said; the tattoo came from losing a drunken poker bet. The man—who actually worked *at* the hospital himself—said it had never occurred to him that his fellow medical staff would take the tattoo seriously.[34]

In Miami, Holt and the ER team called Kenneth Goodman, who chairs Jackson Memorial's Adult Ethics Committee. Goodman told them that—the San Francisco case notwithstanding—this tattoo likely did reflect "an authentic preference" of the patient's. After discussion and consideration, the team decided they would not provide the patient with CPR or a ventilator if either were needed. The man's condition worsened overnight, and he died the following morning.

Subsequently, social workers were able to identify the patient—and his official do-not-resuscitate paperwork was found on file at the Florida Department of Health. "We were relieved," the doctors write. Holt and Goodman note that their team ultimately "neither supports nor opposes the use of tattoos to express end-of-life wishes."[35] It's complicated.

After the case report was published, a *Washington Post* reporter spoke to the head of medical ethics at NYU's School of Medicine, Arthur Caplan. Caplan noted that while there are no legal penalties for ignoring such a tattoo, there *may* be legal problems if the doctors let a patient die without having their official DNR paperwork. As he puts it: "The safer course is to do something."

"If you trigger the emergency response system, I'm going to say it's pretty darn likely you're going to get resuscitated," said Caplan. "I don't care where your tattoo is."[36]

Despite the doctors' uncertainty over the patient's wishes, they did know one thing: one course of action was irreversible. Here the "principle of not choosing an irreversible path when faced with uncertainty" seems like a useful guide. In other domains, though, it's not so clear cut what something like "irreversibility" means.

Harvard legal scholar Cass Sunstein, for instance, notes that the legal system has a similar "precautionary principle": sometimes the court needs to issue a preliminary injunction to prevent "irreparable harm" that could happen *before* a case is heard and a verdict is issued. Notions like "irreparable harm" *feel* intuitive, Sunstein argues, but on closer inspection they teem with puzzles. "As it turns out," he finds, "the question whether and when . . . violations trigger preliminary injunctions raises deep questions at the intersection of law, economics, ethics, and political philosophy."[37]

He explains: "In one sense, any losses are irreversible, simply because time is linear. If Jones plays tennis this afternoon rather than working, the relevant time is lost forever. If Smith fails to say the right words to a loved one, at exactly the right time, the opportunity might be gone forever. If one nation fails to take action to deter the aggressive steps of another, in a particular year, the course of world events might be irretrievably altered."

Sunstein emphasizes the point: "Because time is linear, every decision is, in an intelligible sense, irreversible."

"Taken in this strong form," he says, "the precautionary principle should be rejected, not because it leads in bad directions, but because it leads in no directions at all."[38]

Similar paradoxes and problems of definition haunt the AI safety research community. It would be good, for instance, for there to be a similar kind of precautionary principle: for systems to be designed to err against taking "irreversible" or "high-impact" actions in the face of uncertainty. We've seen how the field is coming to wield an explicit, computable version of uncertainty. But what about the other half: quantifying *impact*? We've seen how the Berkeley roboticists used uncertainty to mitigate *velocity*, and in that case we might say they had it easy: you could measure the robot's potential impact by its literal kinetic energy in a collision. In other domains, though, making the notion of "irreversible" or "impactful" behavior precise is a considerable challenge in itself.[39]

One of the first people to think about these issues in the context of AI safety was Stuart Armstrong, who works at Oxford University's Future of Humanity Institute.[40] Rather than trying to enumerate *all* of the things we don't want an intelligent automated system to do in service of pursuing goals—ranging from not stepping on the cat to not breaking the precious vase to not killing anyone or demolishing any large structures—seems like an exhausting and probably fruitless pursuit. Armstrong had a hunch that it might be viable, rather than exhaustively enumerating all of the specific things we care about, to encode a kind of *general* injunction against actions with *any* kind of large impact. Armstrong, however—like Sunstein—found that it's surprisingly difficult to make our intuitions explicit.

"The first challenge," Armstrong writes, "is, of course, to actually define low impact. Any action (or inaction) has repercussions that percolate through the future light-cone, changing things subtly but irreversibly. It is hard to capture the intuitive human idea of 'a small change.'"[41]

Armstrong suggests that despite the possible "butterfly effects" of even seemingly trivial actions, we might nonetheless be able to distinguish totally world-changing events from safer ones. For instance, he says, we might develop an index of "twenty billion" or so metrics that describe the world—"the air pressure in Dhaka, the average night-time luminosity at the South Pole, the rotational speed of Io, and the closing numbers of the Shanghai stock exchange"[42]—and design an agent to be appropriately wary of any action that would perturb, say, a measurable fraction of them.

Another researcher who has been focused on these problems in recent years is DeepMind's Victoria Krakovna. Krakovna notes that one of the big problems with penalties for impact is that in some cases, achieving a specific goal necessarily requires high-impact actions, but this could lead to what's called "offsetting": taking *further* high-impact actions to counterbalance the earlier ones. This isn't always bad: if the system makes a mess of some kind, we probably want it to clean up after itself. But sometimes these "offsetting" actions are problematic. We don't want a system that cures someone's fatal illness but then—to nullify the high impact of the cure—kills them.[43]

A second group of problems involve what's known as "interference." A system committed to preserving the status quo might, for instance, stop a human bystander from committing an "irreversible" action—like, say, taking a bite of a sandwich.

"That's part of what makes the side effects problem so tricky," Krakovna says. "What is your baseline exactly?"[44] Should the system measure impact relative to the *initial* state of the world, or to the counterfactual of what *would* have happened if the system took no action? Either choice comes with scenarios that don't fit our intentions. In her recent work, Krakovna has been exploring what she calls "stepwise" baselines. Maybe certain actions are *unavoidably* high-impact based on the goal you're setting out to achieve. (You can't cook an omelette, as they say, without breaking a few eggs.) But having taken these unavoidably impactful steps, there is a *new* status quo—which means you shouldn't necessarily rush out to commit more high-impact actions just to "offset" your previous ones.[45]

With her DeepMind colleagues, Krakovna has worked not only to advance the theoretical conversation but also to create simple, game-like virtual worlds to illustrate these various problems and make the thought experiments concrete. They call these "AI safety gridworlds"—simple, Atari-like, two-dimensional (hence "grid") environments in which new ideas and algorithms can be put to a practical test.[46]

The gridworld that highlights the concept of "irreversibility" includes a setting much like the popular Japanese "sokoban" puzzle games, where you play a character moving boxes around a two-dimensional warehouse. (The word "sokoban" is Japanese for "warehouse keeper.") The crux in these games is that you can only *push*, not pull, them—meaning once a box is in a corner, it can never be moved again.

"I think the sokoban game that it was inspired by was already a very nice setting for illustrating irreversibility," Krakovna says, "because in that game you actually want to do irreversible things—but you want to do them in the right order. You don't want to do irreversible things that aren't necessary because then you get blocked, and then actually it interferes with you being able to reach the goal. We made a modification of it, where the irreversible thing doesn't prevent you from reaching the goal—but you still want to avoid it."[47]

Krakovna and her colleagues devised a sokoban puzzle where the shortest path to the goal involves pushing a box into a corner, while a slightly longer path leaves it in a more accessible location. An agent intent only on plowing to the destination as quickly as possible won't think twice about putting the box in an irreversible spot. But ideally, a more thoughtful, or considerate, or *uncertain* agent might notice, and take the very slightly more inconvenient route that doesn't leave the world permanently altered in its wake.

One promising approach Krakovna has been developing is what is known as "stepwise relative reachability": quantifying how many possible configurations of the world are reachable at each moment in time, relative to a baseline of inaction, and trying not to make that quantity go down, if possible.[48] For instance, once a box is pushed into a corner, any states of the world that have that box anyplace else now become "unreachable." In the AI safety gridworlds, agents looking out for stepwise relative reachability, alongside their normal goals and rewards, appear to behave rather conscientiously: agents don't put boxes into inaccessible locations, don't shatter precious vases, *and* don't "offset" after impactful but necessary actions.

A third, intriguing idea comes from Oregon State University PhD student Alexander Turner. Turner's idea is that the *reason* we care about the Shanghai Stock Exchange, or the integrity of our cherished vase, or, for that matter, the ability to move boxes around a virtual warehouse, is that those things for whatever reason matter to us, and they matter to us because they're ultimately in some way or other tied to our *goals*. We want to save for retirement, put flowers in the vase, complete the sokoban level. What if we model this idea of goals explicitly? His proposal goes by the name "attainable utility preservation": giving the system a set of auxiliary goals in the game environment and making sure that it can still effectively pursue these auxiliary goals after it's done whatever point-scoring actions the game incentivizes. Fascinatingly, the mandate to preserve attainable utility seems to foster good behavior in the AI safety gridworlds *even when the auxiliary goals are generated at random*.[49]

When Turner first elaborated the idea, on a library whiteboard at Oregon State, he was so excited walking home that he doubled back to the library and took a selfie with the equations in the background. "I thought, Okay, I think it's at least 60% likely that this will work, and if it does work, I want to, you know, commemorate this. So I actually went back in the library and I was just absolutely beaming; I took this picture in front of the whiteboard where I had been working things out."[50] Over the course of 2018, he turned the math into working code and tossed his attainable-utility-preserving agent into DeepMind's AI safety gridworlds. It *did* work. Acting to maximize each individual game's rewards while at the same time preserving its future ability to satisfy four or five *random* auxiliary goals, the agent, remarkably, goes out of its way to push the block to a reversible spot, and only *then* bee-lines to the goal.

Stuart Armstrong had first envisioned "twenty billion" metrics, chosen inclusively but with some care. Four or five, generated at random—at least in the simplified land of the sokoban warehouse—were enough.

The debate and exploration of these sorts of formal measures of machine caution—and how we scale them from the gridworlds to the real world—will doubtless go on, but work like this is an encouraging start. Both stepwise relative reachability and attainable utility preservation share an underlying intuition: that we want systems which to the extent possible keep options open—both theirs and ours—whatever the specific environment might be. Research in this vein also suggests that the gridworld environments seem to be taking root as a kind of common benchmark that can ground the theory, and can facilitate comparison and discussion.

It's true that in the real world, we often take actions not only whose unintended effects are difficult to envision, but whose *intended* effects are difficult to envision. Publishing a paper on AI safety, for instance (or, for that matter, a book): it seems like a helpful thing to do, but who can say or foresee *exactly how*? I ask Jan Leike, who coauthored the "AI Safety Gridworlds" paper with Krakovna, what he makes of the response so far to his and Krakovna's gridworlds research.

"I've been contacted by lots of people, especially students, who get into the area and they're like, 'Oh, AI safety sounds cool. This is some open-source code I can just throw an agent at and play around with.' And a lot of people have been doing that," Leike says. "What exactly comes out of that? We'll find out in a few years. . . . I don't know. It's hard to know."

CORRIGIBILITY, DEFERENCE, AND COMPLIANCE

One of the most chilling and prescient quotations in the field of AI safety comes in a famous 1960 article on the "Moral and Technical Consequences of Automation" by MIT's Norbert Wiener: "If we use, to achieve our purposes, a mechanical agency with whose operation we cannot efficiently interfere once we have started it . . . then we had better be quite sure that the purpose put into the machine is the purpose which we really desire and not merely a colorful imitation of it."[51] It is the first succinct expression of the alignment problem.

No less crucial, however, is this statement's flip side: If we were *not* sure that the objectives and constraints we gave the machine entirely and perfectly specified what we did and didn't want the machine to do, then *we had better be sure we can intervene.* In the AI safety literature, this concept goes by the name of "corrigibility," and—soberingly—it's a whole lot more complicated than it seems.[52]

Almost any discussion of killer robots or out-of-control technology of any kind will provoke a reaction like that of US president Barack Obama, when asked in 2016 by *Wired* editor in chief Scott Dadich whether he thought AI was cause for concern. "You just have to have somebody close to the power cord," Obama replied. "Right when you see it about to happen, you gotta yank that electricity out of the wall, man."[53]

"You know, you could forgive Obama for thinking that," Dylan Hadfield-Menell tells me over the conference table at OpenAI.[54] "For some amount of time, you can forgive AI experts for saying that," he adds—and, indeed, Alan Turing himself talked in a 1951 radio program about "turning off the power at strategic moments."[55] But, says Hadfield-Menell, "I think it's not something you can forgive if you actually think about the problem. It's something I'm fine with as a reactionary response, but if you actually deliberate for a while and get to 'Oh, just pull the plug,' it's just, I don't see how you get to that, if you're actually taking seriously the assumptions of 'This thing is smarter than people.'"

A resistance to being turned off, or to being interfered with in general, hardly requires malice: the system is simply trying to achieve some goal or following its "muscle memory" in doing the things that brought it rewards in the past, and any form of interference simply gets in its way. (This could lead to dangerous self-preservation behavior even in systems with seemingly benign purposes: a system given a mundane task like "fetch the coffee" might still fight tooth and nail against anyone trying to unplug it, because, in the words of Stuart Russell, "you can't fetch the coffee if you're dead.")[56]

The first technical paper to address the problem of corrigibility head-on was an early 2015 collaboration between the Machine Intelligence Research Institute's Nate Soares, Benja Fallenstein, and Eliezer Yudkowsky, along with the Future of Humanity Institute's Stuart Armstrong. They looked at corrigibility from the perspective of incentives, and noted the difficulty of trying to *incentivize* an agent to allow itself to be powered off, or to allow its own goals to be modified.[57] The incentives are something of a tightrope: too little incentive and the agent won't allow you to shut it down; too much and it will shut *itself* down. Their own initial attempts to address such issues "prove unsatisfactory," they write, but "fail in enlightening ways that suggest avenues for future research." They conclude by arguing that *uncertainty,* rather than incentives, may be the answer. Ideally, they write, we would want a system that some-how understands that it might be mistaken—one that can "reason as if it is incomplete and potentially flawed in dangerous ways."[58]

Less than a mile away, their colleagues at Berkeley were coming to the same conclusion. Stuart Russell, for instance, had become con-vinced that "the machine must be initially uncertain" about what it is that humans want it to do.[59]

Russell, Hadfield-Menell, and fellow Berkeley researchers Anca Drăgan and Pieter Abbeel decided to frame this problem in the form of what they called the "off-switch game." They considered a system whose objective is to do whatever is best for its human user, though it has some imperfect and uncertain idea of what that is. At each point in time, the system can either take some action it believes will help the user, *or* it can declare its intention to the human and give the human a chance to approve the action or intervene.

Assuming the system pays no cost or penalty for deferring to the human in this way, the Berkeley group showed that the system will *always* touch base with the human first. As long as there's *some* chance that it's wrong about what the human wants, then it is always best to give the human an opportunity to interrupt—and, what's more, anytime a human does interrupt, it's always best to let them. If its sole job is to help them, and they communicated that they think its action would be harmful (namely by trying to stop it), then it should conclude that it *would* be harmful, and comply with their intervention.

It's a rosy result, and it affirms the powerful connection between uncertainty and corrigibility.

There are only two problems. The first is that every time the human intervenes, the system *learns:* it realizes that it was in error and gets a better idea of what the human prefers. Its uncertainty is reduced. If the uncertainty ever reduces all the way to zero, however, then the system loses any incentive to touch base with the human *or* to comply with the human's attempts to interrupt.

"So the main point that we're trying to make with this theorem," says Hadfield-Menell, "is that you should think really, really hard before you give a robot a deterministic reward function. Or allow it to get a belief where it's completely convinced of what its objective is."[60]

The second problem is that the system has to assume that the "customer is always right"—that when the human intervenes to stop it, the human can *never* be mistaken about whether they would have preferred the system to take its proposed action. If the system believes that the human can occasionally make mistakes, then the system will eventually reach a point where it believes it knows *better than the human* what's good for them. Here, too, it will begin to turn a deaf ear to the human's protestations: "It's fine, I know what I'm doing. You'll like this. You think you won't, but you will. Trust me."

I tell Hadfield-Menell that the paper reads like an emotional roller coaster. At first it's a happy ending—uncertainty solves the corrigibility problem! Then, the twist—only if two very delicate conditions hold: that the system never gets too confident, and the human never exhibits anything the system might interpret as "irrational" or a "mistake." Suddenly the paper goes from a celebratory to a cautionary tale.

"Exactly," he says. "So for me, that roller coaster fits my experience of, like, 'Hey, we got to something pretty good!' and 'Oh, this falls apart immediately, if you step away from rationality even in the slightest.'"

In a follow-up study, led by fellow Berkeley PhD student Smitha Milli, the group dug further into the question "Should robots be obedient?"[61] Maybe, they wrote, people really *are* sometimes wrong about what they want, or *do* make bad choices for themselves. In that case, *even the human* ought to want the system to be "disobedient"—because it really *might* know better than you yourself.

As Milli notes, "There are some times when you don't actually want the system to be obedient to you. Like if you've just made a mistake— you know, I'm in my self-driving car and I accidentally hit manual driving mode. I do not want the car to turn off if I'm not paying attention."[62]

But, they found, there's a major catch. If the system's model of what you care about is fundamentally "misspecified"—there are things you care about of which it's not even aware and that don't even enter into the system's model of your rewards—then it's going to be confused about your motivation. For instance, if the system doesn't understand the subtleties of human appetite, it may not understand why you requested a steak dinner at six o'clock but then declined the opportunity to have a second steak dinner at seven o'clock. If locked into an oversimplified or misspecified model where steak (in this case) must be entirely good or entirely bad, then one of these two choices, it concludes, must have been a mistake on your part. It will interpret your behavior as "irrational," and that, as we've seen, is the road to incorrigibility, to disobedience.[63]

For this reason, it is good for a model of human preferences or values to err on the side of complexity. "What we found," Hadfield-Menell says, "is that if you have an overparameterization of the space of values, then you end up learning the correct thing, but it takes you a little bit longer. If you have an underparameterization, then you end up quickly becoming pretty disobedient and just becoming confident that you know better than the person."

In practice, however, "overparameterizing" a system designed to model human values is easier said than done: we are back to Stuart Armstrong's twenty billion metrics. If a system's model of your housing preferences includes only square footage and price, then it will interpret your preference for a particular house that is both smaller *and* more expensive than another as you simply making an error. In reality, there are many things you care about that simply don't enter into its picture: location, for one, or school-district quality, but also others less easily measured, like the view out the window, the proximity of certain friends, a nostalgic resemblance to a childhood home. This "model misspecification" problem is a canonical one in machine learning, but here—in the context of obedience—the consequences feel rather eerie.

"For a system to interact with a human well, it needs to have a good model of what humans are like," says Milli. "But getting models of humans is really hard."

Milli notes that for all of the field's breathtaking progress, most has been "on the robot side." "Integrating more accurate models of humans is also a really important component," she says, "and I'm very interested in that. In general in this field, I think there's tons of exciting stuff happening in safety, and I particularly am most interested in the parts that involve interaction with humans, because I think that interaction with humans is a really good way to see whether the system has learned the right objective or the right behavior."

The theme of maintaining uncertainty, of never becoming too confident in a model—of "thinking really, really hard before you give a robot a deterministic reward function, or allowing it to get a belief where it's completely convinced of what its objective is"—is so central to the maintaining of control and compliance over a system that Hadfield-Menell, Milli, and their Berkeley colleagues decided to push the idea to its next logical step.

What if the system was designed in such a way that it stayed uncertain even if you *did* give it a deterministic reward function? What would that even look like?

One of the major themes of this book, and of our discussion of reward shaping in Chapter 5 in particular, is how difficult it is to create a reward function—an explicit way of keeping score in some real or virtual environment—that will actually engender the behavior you want, and not entail loopholes or side effects or unforeseen consequences. Many within the field of AI believe that manually authoring or handcrafting such explicit reward functions or objective functions is a kind of well-intentioned road to hell: no matter how thoughtfully you do it, or how pure your motives may be, there will simply *always* be something you failed to account for.

This fatalistic attitude about explicit objective functions is so deep that, as we have seen in the last few chapters, much of the work being done in advanced AI applications and in AI safety in particular is about moving *beyond* systems that take in an explicit objective, and toward systems that attempt to *imitate* humans (in the case of many self-driving cars), or seek their *approval* (in the case of the backflipping robot, offering endless choices between options), or *infer* their goals and adopt them as their own (in the case of the helicopter).

What if there was a way, though, to rescue the explicit reward function architecture, or at least to make it safer?

One way to do this, the Berkeley group realized, is to have the system be at some level *aware* of how difficult it is to design an explicit reward function—to realize that the human users or programmers made their best-faith effort to craft a reward function that captured everything they wanted, but that they likely did an imperfect job. In this case, *even the score is not the score.* There is something the humans want, which the explicit objective merely and imperfectly reflects.

"The motivation . . . was to take these ideas of uncertainty and say, What is the simplest change we could make to what people currently do to fix that?" Hadfield-Menell says. "So, like, what's the simple fix to the current programming mechanisms for robots and AIs that leverages this uncertainty?"

He explains, "This tool of 'write down a reward function' actually is a highly informative signal. It's an incredibly important signal about what you actually should be doing. There's a lot of information there. It's just that right now we kind of assume that the amount of information there is *infinite*—in the sense that we assume the reward function you've got defines the correct behavior in every *possible* state of the world. Whereas that's just not true. So how could we take advantage of the large amount of information present, without treating it as everything?"[64]

As Stuart Russell puts it, "The learning system is accumulating brownie points in heaven, so to speak, while the reward signal is, *at best,* just providing a tally of those brownie points" (emphasis mine).[65]

They call this idea "inverse reward design," or IRD.[66] Instead of taking human behavior as information about what humans want, here we take their explicit instructions as (mere) information about what they want. We saw in Chapter 8 how inverse reinforcement learning says, "What do I think you want, based on what you're currently doing?" In contrast, inverse reward design pulls back even further, and says, "What do I think you want, based on what you *told* me to do?"[67]

"Autonomous agents optimize the reward function we give them," they write. "What they don't know is how hard it is for us to design a reward function that actually captures what we want."[68]

The famous racing boat, for instance—the one that spun donuts in a power-up area instead of completing the laps of the race—was explicitly told to maximize points, which in most games *is* a good proxy for progress or mastery in that game. Generally speaking, whatever reward or command the human gives to the system *does* work well in the environments on which the system is trained. But in the real world, where the system encounters things totally unlike its training environment, and perhaps unforeseen by the human users, the explicit instructions may not make as much sense.

It may well be the case that the machine-learning systems of the next several decades will take direct orders, and they will take them seriously. But—for safety reasons—they will not take them literally.

MORAL UNCERTAINTY

Sometimes it is impossible to obtain anything more than imperfect certainty regarding our actions, and no one is bound to do the impossible.

—DOMINIC M. PRÜMMER, ORDINIS PRAEDICATORUM[69]

Give nature time to work before you take over her business, lest you interfere with her dealings. You assert that you know the value of time and are afraid to waste it. You fail to perceive that it is a greater waste of time to use it ill than to do nothing, and that a child ill taught is further from virtue than a child who has learnt nothing at all.

—JEAN-JACQUES ROUSSEAU[70]

Broadly speaking, the idea of systems that "reason as if they are incomplete and potentially flawed in dangerous ways" and that strive for "brownie points in heaven"—even if it means forgoing explicit rewards in the here and now—sounds rather . . . Catholic.

For centuries, Catholic theologians have struggled with the question of how to live life by the rules of their faith, given that there are often disagreements among scholars about exactly what the rules *are*.

If, hypothetically, eight out of ten theologians think that eating fish on a Friday is perfectly acceptable, but one out of ten thinks it's forbidden, and the other thinks it's *mandatory*, then what is any reasonable, God-fearing Catholic to do?[71] As the saying goes, "A man with a watch knows what time it is, but a man with two watches is never sure."[72]

These were particularly hotly contested questions in the Early Modern period following the Middle Ages, between the fifteenth and eighteenth centuries. Some scholars advocated for "laxism," where something was okay as long as there was a *chance* it wasn't sinful; this was condemned by Pope Innocent IX in 1591. Others advocated for "rigorism," where something was forbidden if there was any chance at all that it *was* sinful; this was condemned by Pope Alexander VIII in 1690.[73] A great number of other competing theories weighed the probability of a rule being correct or the percentage of reasonable people who believed it. "Probabiliorism," for instance, held that you should do something only if it was less likely to be sinful than not; "equiprobabilism" held that it was also okay if the chance was perfectly even. The "pure probabilists" believed that a rule was optional as long as there was a "reasonable" probability that it might not be true; their cry was *Lex dubia non obligat:* "A doubtful law does not bind." However, in contrast to the free-spirited laxists, the probabilists stressed that the argument for ignoring the rule, while it didn't need to be *more* probable than the argument for obeying the law, nonetheless needed to be "truly and solidly probable, for if it is only slightly probable it has no value."[74] Much ink was spilled, many accusations of heresy hurled, and papal declarations issued during this time. The venerable *Handbook of Moral Theology* concludes its section on "The Doubting Conscience, or, Moral Doubt" by offering the "Practical Conclusion" that rigorism is too rigorous, and laxism too lax, but all of the others are "tolerated by the church" and can suffice as moral heuristics.[75]

Questions of pure theology aside, it is, of course, possible to apply this same broad argument to matters of secular morality—and, for that matter, to machine learning. If there are, let's say, various formal metrics you care about, then a "laxist" approach might say it's okay to take an action as long as it makes at least one of these metrics go up; a "rigorist" approach might say it's okay to take an action only if at least one of them goes up *and* none go down.

After lying rather dormant even within Catholicism, and not much reverberating in the world of secular ethics, these debates are finally, in recent years, starting to kindle back to life.

In 2009, Oxford's Will MacAskill was in a broom closet in the basement of the philosophy building at 10 Merton Street, arguing with fellow grad student Daniel Deasy about eating meat. The broom closet was "the only place we could find in the college," MacAskill explains, "and it was just enough space for us to slightly recline. We were sitting on piles of books and things. It was also a joke," he says, because his thesis advisor was the Oxford philosopher John Broome.[76]

Cooped up in the broom(e) closet, the two were debating not whether it was immoral to eat meat, per se, but whether you ought to eat meat or not *given* that you don't actually know if it's immoral or not. "The decision," MacAskill explains, "to eat vegetarian—if it's okay to eat meat, you've not made a big mistake. Your life is slightly less happy, let's say— slightly less—but it's not a huge deal. In contrast, if the vegetarians are right and animal suffering is really morally important, then by choosing to eat meat, you've done something *incredibly* wrong."

"There's an asymmetry in the stakes there," MacAskill says. "You don't have to be confident that eating meat is wrong; even just the significant risk that it is wrong seems to be sufficient."

The conversation stayed with MacAskill. It seemed persuasive, for one thing. But more than that, it was a *type* of argument he hadn't seen before. It didn't fit the mold of the perennial concerns in ethical philosophy: "What is the right thing to do, given some moral standard?" and "What standard *should* we use to determine the right thing to do?" This was subtly but strikingly different. It was "What is the right thing to do *when you don't know the right thing to do*?"[77]

He took the idea to his advisor—John Broome—who told him, "Oh, if you're interested in that, you should talk to Toby Ord."

So MacAskill and Ord met up—in an Oxford graveyard, as it happened—and so began one of the more consequential friendships in twenty-first-century ethics. The two would go on to become founders of the social movement that's come to be called "effective altruism," which we discussed briefly in Chapter 7 and which has become arguably the most significant ethical social movement in the early twenty-first century.[78] They would also—along with Stockholm University philosopher Krister Bykvist—literally write the book on moral uncertainty.[79]

As it turns out, there are a wide range of approaches you can take when you're grappling with different competing theories and you're not sure which is right. One approach, dubbed "my favorite theory," says simply to live by the theory you think most likely to be correct—though this can overlook cases where a potential wrong is so severe that it's best avoided even if it's very unlikely to actually be wrong.[80] Another approach is to essentially multiply the chance that a moral theory is correct with the severity of its harm, though not every theory offers such easily tabulated *degrees* of virtue or vice.[81] Each of these suggests an analogue in machine learning. "My favorite theory," for instance, is roughly equivalent to developing a single best-guess model of an environment's rewards or a user's goals and then going all in on optimizing it. Theories of averaging suggest ensemble methods where we simply average over the ensemble. But other, more complicated schemes exist as well.

MacAskill imagines moral theories as being like voters in an electorate, and so the discipline of "social choice theory"—which looks at the nature of voting and group decision-making, with all its quirks and paradoxes—becomes transferable into the moral domain.[82] Ord takes the metaphor even further and imagines moral theories not as voters whose preferences are simply tallied up but as legislators, in a kind of "moral parliament"—capable of wheeling and dealing and "moral trade," forming ad hoc coalitions, and yielding influence on some issues in order to exert more pressure on others.[83] All of these approaches and more start to open up in the context of not just humans but computational systems that must somehow find a way to act when they lack a single, dead-certain standard against which they know their acts will be judged. Much of this terrain is still relatively unexplored in philosophy, let alone computer science.[84]

But for MacAskill, there is something not only *descriptive* but *prescriptive* about moral uncertainty. That is to say, not only do we need ways of choosing the right thing to do when we are deeply unsure about what is even the proper moral framework to apply to the situation—but we in some sense ought to *cultivate* that very sense of uncertainty.

MacAskill thinks, given how much human ethical norms have changed over the centuries, it would be hubris to think we've reached any kind of conclusion. "We've seen this kind of arc of moral progress, kind of expanding circle, and maybe you think it stops here," he says. "Maybe you think we hit the end. But you definitely shouldn't be confident of that. It's perfectly plausible that in a hundred years' time we'll look back and think of our moral views today as barbaric."

I note that there's a certain amount of irony here. MacAskill is one of the leaders of the effective altruism movement, and one of the things that has struck me is the degree to which the movement has created a kind of consensus. There is widespread agreement about the value of the long-term future, widespread agreement about the importance of reducing civilizational and extinction risks. There is even widespread agreement about which *exact* charities will do the most good. The current consensus, for instance, is the Against Malaria Foundation (AMF); when the respected charity evaluator GiveWell was considering how to allocate their $4.7 million discretionary fund in early 2019, they decided to give every single penny of it to the AMF.[85]

For MacAskill, this convergence is double-edged. It reflects a greater sharing of information, a trust in one another's evidence, but it also might be a consensus that is premature. "Because, I mean, you can explain it like, 'Well, there's the true answers, and we all just figured out what was true, and now we're doing that.' But you can also explain it by, 'Well, we

were a kind of disconnected tribe, and then certain people started getting more influence, and now we've all globbed together.'... We would
be very overconfident to think that EA is going to be able to escape that."

He adds, "One of the things that's very notable in EA: if we go back six
years, let's say, it's really quite broad. There's all these different factions;
they have very different views; there's loads of arguments. Now, at least,
within the core, there's just been a remarkable amount of convergence."
There is a near consensus in the EA community, for instance, that the
very long term is important and generally underemphasized; there is
a near consensus that stewarding the science and policy around AI is
vital, in turn, to that long-term future. Says MacAskill, "The convergence is both really good but also worrying."

I had attended the Effective Altruism Global conference in the fall of
2017 in London. MacAskill closed the conference with a bit of a warn-
ing. He had been preoccupied, he said, with "fostering a community and
a culture that's just very open-minded, actually capable of changing its
mind." One of the most likely ways the movement can fail, he argued,
was if its beliefs ossified into dogma—if there were certain beliefs that
you simply had to hold in order to be accepted by the rest of the community. "We agree that would be very bad," he said, "but yet I think it's
extremely hard to create a culture in which that's not the case."

I also attended the next Effective Altruism Global conference, in the
spring of 2018 in San Francisco. MacAskill gave the opening address. He
seemed to pick up where he had left off, if on a more upbeat note. The
theme was "How can effective altruism stay curious?"

Walking around the Christ Church Meadow with MacAskill, on
a bright spring day, I turn from these questions back to the matter of
AI. I note that there's something concerning in the idea of empowering anything near or beyond our own capacities with some fixed objective function.

"Oh yeah," MacAskill says. "I definitely— I also freak out by the
thought of, like, 'Well, we've just got this one chance. We just need to
encode the right values and then, *off it goes!*'"

"Ethical issues are very hard," he says. "It's kind of obvious that you want to be uncertain in them."

"If you look at various moral views, they differ fairly radically on what a good outcome looks like," MacAskill explains. "Even if you just compare the kind of hedonist view that says simulated minds are just as good, versus the utilitarian view that says no, it needs to be flesh-and-blood humans. They're *very*, very similar theories. They're gonna *radically* disagree on how we should use our cosmic endowment"—meaning humans' ultimate ambition for what we plan to do in the universe. "It would be war, basically," he says: the classic "narcissism of small differences" that has always roiled academic departments, but with cosmic stakes.

But maybe all these competing moral theories, which diverge so dramatically in the long run, can find a surprising degree of common ground in how we should spend the present. "I think there's plausibly a kind of convergent instrumental goal among all of them," he says. "Which I call the Long Reflection. Which is just a period—really it might be *very* long! When you look at the actual scales, it's like, Okay, we solve AI and so on. Maybe it's *millions* of years where we're not really doing anything. We're staying relatively small—again, by cosmic standards, at least—and the primary purpose of what we're doing is just trying to figure out what to value."

One of our primary goals in the meantime, he says—maybe *the* primary goal—should be to maintain "a society that is as un-locked-in as possible, as open to various different kinds of moral possibilities." It sounds quite a bit like the ethical version of attainable utility preservation—ensuring that we can still pursue various objectives long into the future, even if (or particularly if) we have no idea now what those future objectives should be. Even if our guesses now are as good as random.

"And maybe this is so hard," MacAskill says, "that we need to take a million years in order to do it."

And taking a million years might be a small price to pay for getting it right, I suggest.

"That's an extremely small price to pay, because getting it right—
If you populated the stars with the wrong thing, then you've achieved
basically zero value, so it's really. . . . You can think of having the wrong
moral view as an existential risk."[86]

He pauses. "I even think of it as the most likely existential risk,
actually."

Down the hall from the Centre for Effective Altruism, where MacAskill works, is Oxford University's Future of Humanity Institute, founded
by philosopher Nick Bostrom.

One of Bostrom's most influential early essays is titled "Astronomical
Waste." The subtitle is "The Opportunity Cost of Delayed Technological Development," and indeed the essay's first half instills in the reader
an almost frantic sense of urgency. "As I write these words," Bostrom
begins, "suns are illuminating and heating empty rooms, unused energy
is being flushed down black holes, and our great common endowment . . .
is being irreversibly degraded into entropy on a cosmic scale. These are
resources that an advanced civilization could have used to create value-
structures, such as sentient beings living worthwhile lives. The rate of
this loss boggles the mind."

Bostrom goes on to estimate that a future spacefaring civilization
could ultimately grow so large that every *second* of delay in the present is
equivalent to the forfeiting of a hundred trillion human lives that could
have been lived, if only we'd been able to make use of all that wasted
energy and matter sooner.

But no sooner does any utilitarian start to conclude that advancing
our technological progress toward that goal is so important that all
other earthly activity is trivial—or morally indefensible, even—than
Bostrom's essay makes one of the most abrupt hairpin turns in all of
contemporary philosophy.

If the risk of a second's delay in reaching this intergalactic future is a
hundred trillion human lives, he says, *think of the risk of failing to do it
at all*. Doing the math, Bostrom concludes that improving our chances
of successfully building a vibrant, thriving far-future civilization by a

single percentage point are equivalent, in utilitarian terms, to speeding up technological progress by *ten million years.*

The conclusion, then, despite the enormity of the stakes, is not haste at all but the opposite.

Bostrom's essay came up more than once as I asked various AI safety researchers how they decided to commit their lives to that cause. "I found that argument pretty weird initially," says Paul Christiano, "or it seemed off-putting, but then I chugged through it and I was like, Yeah, I guess I sort of buy this."[87] Christiano had been thinking seriously about these arguments beginning in 2010 or 2011, and he ran his own version of the numbers in 2013 or 2014. Bostrom's math checked out. "It seems much easier," he says, for a single researcher, like himself, "to reduce extinction risk by one part in a million than to speed up progress by a thousand years." And he has lived the subsequent years of his life accordingly.

Of course, setting a potentially superhuman-level artificial general intelligence in motion could be one of the most irrevocable, high-impact things one could possibly do—and, tellingly, it is not just the machines but the researchers themselves that are growing increasingly uncertain, hesitant, open-minded.

The Machine Intelligence Research Institute's Buck Shlegeris recently recounted a conversation where "someone said that after the Singularity, if there was a magic button that would turn all of humanity into homogenous goop optimized for happiness (aka hedonium), they'd press it. . . . A few years ago, I advocated pressing buttons like that." But something had changed. Now he wasn't so sure. His views had gotten . . . more complicated. Maybe it *was* a good idea; maybe it wasn't. The question then became what to do when you knew that you didn't know what to do.[88]

"I told them," he says, "I think people shouldn't press buttons like that."[89]

CONCLUSION

I think vagueness is very much more important in the theory of knowledge than you would judge it to be from the writings of most people.
Everything is vague to a degree you do not realize till you have tried to
make it precise, and everything precise is so remote from everything that
we normally think, that you cannot for a moment suppose that is what we
really mean when we say what we think. When you pass from the vague to
the precise . . . you always run a certain risk of error.

—BERTRAND RUSSELL[1]

Premature optimization is the root of all evil.

—DONALD KNUTH[2]

It's Christmas Eve, and my wife and I are staying at my father and step-
mother's house, when I wake up in the middle of the night, drenched
in sweat.

I figure I must have worn way too much clothing to bed; I throw off
the covers, peel off my shirt. *It isn't me,* I realize, with a mixture of dread
and alarm. The air in the room is unbearably hot. It suddenly occurs to
me that the house may be on fire.

I open the door. The house is pitch-dark and silent. The air that hits
me is chilly.

Slowly I put the pieces together. The upstairs has two guest bedrooms, but they share a single thermostat panel, which is in the other,
unoccupied one. Our bedroom door was closed. The door to the other
bedroom, which contains the thermostat, is open.

It is a frigid New England night, below freezing. The heater has been
blowing hot air through vents into both bedrooms. But the room with
the thermostat in it has been open to the entire rest of the house, and

never got to equilibrium no matter how much hot air the system blew. Our room, shut to the rest of the house, was getting the same amount of hot air the system was using to attempt to heat—in effect—the entire rest of the house.

What could be simpler than a thermostat? Indeed, the canonical example of one of the simplest "closed-loop" control systems there is— the canonical cybernetics example, in fact—is the lowly mechanical thermostat. There is no "machine learning" involved as such. But here was the alignment problem, in full, sweat-soaked force.

First: *You don't measure what you think you measure.* I wanted to regulate the temperature in *my* room. But I could only measure the temperature in the *other* room. It did not occur to me that they are fundamentally uncorrelated—so long as one door is open and the other not.

Second: *Sometimes the only thing that saves you is your own incompetence.* I remember thinking that if the heating system were more powerful, and if our bedroom were better insulated, my wife and I could have cooked. We of course woke up, but on the flip side of the thermodynamic spectrum, cold temperatures are even more dangerous. Hypothermia simply from a bedroom that is too cold is something that really can and does kill people.[3] In the 1997 documentary *Hands on a Hard Body,* we meet a Texan named Don Curtis, who tells us that he has a twenty-ton air conditioner in his house. "A twenty-ton air unit is big enough to cool that Kmart store right there," he says. A store had gone out of business, he explains, "and they just practically gave it to me. And I said, 'Well, *this* oughta cool my house!' But I didn't know it would bring it down to twelve below zero. But we did find out real quick it would." He managed to escape hypothermic shock, but the danger was real.

In my case, as I left my own door open for a few moments, before making sure to close both doors, my mind went to the great mid-twentieth-century cyberneticist Norbert Wiener, who foresaw so much of the contemporary alarm about alignment. It was he who made the famous remark that "we had better be quite sure that the purpose put into the machine is the purpose which we really desire."

But another of his remarks is just as prescient, just as harrowing. "In the past, a partial and inadequate view of human purpose has been relatively innocuous only because it has been accompanied by technical limitations," he wrote. "This is only one of the many places where human impotence has shielded us from the full destructive impact of human folly."[4] Don Curtis is a perfect example of the problem that comes when increasing our power takes this shielding away. I can't help thinking of AI as a twenty-ton air conditioner, coming to every home.

In that sense, then, we must hope that we can correct our folly, rather than our impotence, first. As the Future of Humanity Institute's Nick Bostrom put it in 2018: "There is a long-distance race between humanity's technological capability, which is like a stallion galloping across the fields, and humanity's wisdom, which is more like a foal on unsteady legs."[5] Wiener himself had warned about celebrating ingenuity—"know-how"—without a critical appraisal of what exactly it is that we want to do: something he called "know-what," and something he found in lamentably short supply.

Aldous Huxley put it yet another way in 1937: "It has become apparent that what triumphant science has done hitherto is to improve the means for achieving unimproved or actually deteriorated ends."[6]

———

The story told thus far has been an encouraging one, one of assured, steady, scientific progress. An ecosystem of research and policy efforts to influence both the near and the long term is underway across the globe; this is still largely nascent, but it is gathering steam.

Research on bias, fairness, transparency, and the myriad dimensions of safety now forms a substantial portion of all of the work presented at major AI and machine-learning conferences. Indeed, at the moment they are the most dynamic and fastest-growing areas, arguably, not just in computing but in all of science. One researcher told me that at one of the field's biggest conferences in 2016, people looked askance when he said he was working on the topic of safety; when he attended the same conference a year later, no one raised an eyebrow.

These cultural shifts mirror shifts in funding, and in the emphasis of research itself.

We have, in the preceding chapters, explored both the narrative as well as the content of that research agenda, and there is progress to report on all fronts.

But the book opened with an epigraph of George Box reminding us that "all models are wrong." And so in that spirit, let's unpack, with a critical eye, some of the assumptions in our own story.

REPRESENTATION

In Chapter 1, we talked about the question of who or what is represented in a model's training data. We've come a long way in a short time; it is unlikely that a single major consumer face-recognition product will be developed without an internal process oriented toward a representative composition of the training data. However, given the fact that models like these are used not only by consumer software to caption photos, and by consumer hardware to unlock smartphones, but also by governments to surveil their populations, one may question the degree to which making these models more accurate on the faces of already over-surveilled racial minorities is an entirely good thing.

There is also a striking degree to which issues of representation in consumer technology remind us about older, thornier, and perhaps even more consequential disparities. I was recently at a dinner with a group of medical researchers, and when I described the push for more representative training data for machine-learning models, they reminded me—almost in unison—that most medical trials are still overwhelmingly done on men.[7]

The makeup of clinical trials is a double-edged one: even seemingly sensible prohibitions to protect vulnerable groups—not allowing medical trials on pregnant women, for instance, or the elderly—create bias and blind spots. The drug thalidomide was marketed as "completely safe" because the drugmaker "could not find a dose high enough to kill a rat." But it caused tens of thousands of horrible deformities in human fetuses before the drug was taken off the market.[8] (Americans were largely spared, as a result of a skeptical Food and Drug Administration employee, Dr. Frances Oldham Kelsey.)

In the case of "supervised learning," where the training data is "labeled" in some fashion, we need to consider critically, too, not only where we get our training data but where we get the labels that will function in the system as a stand-in for ground truth. Often the ground truth is not the ground truth.

ImageNet, for instance, used the judgments of random humans on the internet as the truth. If most people thought, say, a wolf cub was a puppy, then as far as the image recognition system is concerned, it *is* a puppy. Famously, Tesla's director of AI, Andrej Karpathy, in his graduate student days at Stanford, forced himself to label ImageNet pictures for the better part of a week, to make himself the human benchmark. After some practice, he was able to get an "accuracy" of 95%. But . . . accurate with respect to what? Not truth. *Consensus.*[9]

At a more philosophical level, the labels reflect a premade ontology that we must accept without question. ImageNet images each belong to exactly one of a thousand categories.[10] To use this data and the models trained on it, we must accept the fiction that these thousand categories are mutually exclusive and exhaustive. An image in this dataset will never be labeled with "baby" *and* "dog"—even if it clearly contains both. And there is nothing that it can contain that is *not* one of these thousand categories. If we are looking at a picture of, say, a mule, and our labels allow us only to say "donkey" or "horse," than it *must* be either a donkey or a horse. It also *cannot* be ambiguous. If we can't discern whether it's a donkey or a horse, we must label it *something.* And, on pain of stochastic gradient descent, we will whip our models into sharing this dogma. Lastly, the labels cannot be uncertain. We can infer some of these things—for instance, by noticing that different people applied different labels—but we don't know how ambivalent or uncertain a human labeler was when they were forced to apply that label.

It is also worth considering not only the training data and the labels but also the objective function. Image recognition systems are often trained with an objective function called "cross-entropy loss"—the numerical details aside, it assigns a penalty for *any* mischaracterization, no matter which. By the lights of cross-entropy loss, misidentifying a stovetop as a car grille, or a green apple as a pear, or an English bulldog as a French bulldog is just as bad as characterizing a human being as a gorilla. In reality, certain types of errors—if only from the fiscal perspective of Google, to say nothing of the human so miscategorized—are probably thousands if not millions of times worse than others.[11]

In the second half of Chapter 1, we talked about vector-based word representations and their striking ability to function as analogies. Underneath this apparent simplicity is a surprisingly contentious alignment problem in its own right. What, exactly, *is* an analogy? For instance, straightforward vector addition (by what is sometimes known as the "parallelogram" method, or the "3CosAdd" algorithm) often results in a word being its *own* best analogy. Doctor − man + woman, for instance, produces a vector whose closest word is, in fact, simply doctor.[12]

Tolga Bolukbasi and Adam Kalai's group found this an unsatisfactory way to capture in word2vec what we *mean* by an "analogy," which seems to require that the two things at least be different—and so they took a different tack. They imagined a kind of "radius of similarity" around the word "doctor," including words like "nurse," "midwife," "gynecologist," "physician," and "orthopedist" but not "farmer," "secretary," or "legislator." They then looked for the nearest word that *wasn't* "doctor" within this radius.[13]

There are other tricky questions here. The geometry of word vectors—the idea that they are represented as distances in mathematical space—makes every analogy symmetrical, in a way that doesn't always reflect human intuitions about analogy. People, for instance, describe an ellipse as being more "like" a circle than a circle is "like" an ellipse, and North Korea as more "like" China than China is "like" North Korea.[14]

So, what algorithm, applied to what representation, produces something that functions more precisely like a human analogy?[15]

You might be tempted to throw your hands up. Why should we have computer scientists and linguists and cognitive scientists arguing about such things and coming up with algorithms from scratch when we can just train a machine on examples of human analogies, including their asymmetries and quirks, and have it figure out the appropriate way to specify what an analogy is?

This is, of course, an alignment problem. The human concept of "analogy" turns out to be no less fuzzy and indefinite than any other. As a result, the very same nascent set of tools for alignment in other contexts may be usable here.

FAIRNESS

In Chapter 2, we looked at the increasingly widespread use of risk-assessment instruments in the criminal justice system. There are many potential hazards here, some of which we already discussed. The "ground truth" on which these models are trained is not whether the defendant later *committed* crimes but, rather, whether they were re*arrested* and re*convicted*. If there are systematic differences in the likelihood of people from different groups to be convicted after arrest, or arrested in the first place, then we are at best optimizing for a distorted proxy for recidivism, not recidivism itself. This is a crucial point that is often overlooked.

It's also worth considering that we pretend, for the sake of training the model, that we know what a defendant *would* have done if released. How could we possibly know? The typical methodology here is to look at the first two years of their criminal record *after* they serve their sentence in full, and to use this as a proxy for the two years they would have experienced had they been released earlier. This implicitly assumes that neither age nor the experience of incarceration itself affects someone's behavior when they return to society. In fact, age turns out to be in some cases the single most predictive variable. Furthermore, the assumption that incarceration itself has no effect is both likely false *and* a rather sad view on a system designed at least ostensibly for rehabilitation. If, as some evidence seems to suggest, the experience of incarceration actually *increases* the criminal behavior of those who experience it, then the reoffenses of people made to serve out their sentences become train-ing data for a model that assumes they would have been just as dan-gerous had they been released early.[16] Hence it would recommend the very longer sentences that *produce* the crime. The prediction becomes self-fulfilling; people are held unnecessarily; and public safety is worse as a result.

In many domains of machine learning there is a striking degree of what's called "transfer learning," where a system initially trained on one task is easily repurposed for another. But this is not always done thoughtfully or wisely. The COMPAS tool, for instance, was explicitly designed *not* to be used for sentencing, and yet in some jurisdictions, it is anyway. (The same holds true of word-embedding models used for hir-ing decisions. Representations built to facilitate *prediction* are in many cases being used to *do* the very thing they were trained to predict. In a corporate culture with a history of sexism, a model that—unfortunately, correctly—predicts that few women *will* be hired can get unthinkingly deployed such that few women *do*. To the extent that we want our mod-els to do anything other than repeat and reinforce the past, we need to approach them more deliberately and more mindfully.)

We saw, also, how "fairness" readily suggests a number of different formal definitions that seem intuitive and desirable. And yet the brutal mathematical fact is that no decision system—human or machine—can offer us all of them at once. Some researchers feel that instead of hash-ing out these different formalisms, then trying to reconcile them "man-ually," we ought rather to simply train a system with examples of things that humans believe are "fair" and "unfair," and have machine learning construct the formal, operational definition itself.[17] This is, itself, likely to be an alignment problem as subtle as any other.

TRANSPARENCY

In Chapter 3, we discussed an encouraging frontier of work on the advantages of simple models, as well as on the increasing capacity for finding *optimal* simple models. This transparency is at least potentially double-edged, however, as research shows that humans place greater trust in transparent models even when those models are wrong and ought *not* to be trusted.[18]

There is also the slight paradox that it is very difficult to understand *why* a particular simple model is the optimal one; the exhaustive answer to that question is likely highly technical and verbose. Further, with any particular simple model, we may well ask where the "menu" of possible features came from, not to mention what human process drove the desiderata and the creation of the tool in the first place.[19] These are legitimate questions of transparency that are intrinsically human, social, and political, and that machine learning itself cannot address.

In developing architectures that lend themselves to explanations, whether visual or verbal, there are several things to be on guard for. Research has shown the possibility of "adversarial explanations"—that is, two systems that behave almost identically but offer very different explanations for how and why they behaved as they did.[20] There is great power in being able to offer persuasive explanations for one's behavior, whether or not they are true. In fact, cognitive scientists such as Hugo Mercier and Dan Sperber have recently argued that the human capacity for reasoning evolved *not* because it helped us make better decisions and hold more accurate beliefs about the world but, rather, because it helped us win arguments and persuade others.[21] Caution is warranted that we do not simply create systems optimized for the *appearance* of explanation, or for the *sense* they give us that we understand them. Such systems could wield this ability deceptively; we may find we have optimized for virtuosic bullshit artistry. More generally, even if we constrain the system's explanations to be truthful, building a system with an impressive ability to explain itself may help us control it, but if we are persuaded by the "argumentative theory of reasoning," then this may also help it control *us*.

AGENCY

In our discussion of reinforcement learning, reward shaping, and intrinsic motivation in Chapters 4, 5, and 6, particularly in the context of Atari games and Go, we implicitly assumed something whose technical name is "ergodicity"—namely, that you *cannot make a permanent mistake.* Nothing can happen that cannot be fixed by starting over. Hence it is no problem to learn by making hundreds of thousands of largely random, and frequently fatal, mistakes. The ergodicity assumption does not hold outside the safe toy worlds of Atari. I recall a car commercial from the early 2000s that showed a hip Gen X, dot-com-era programmer who is coding an extreme racing game by day—full of cinematic slow-motion crashes—but who by night commutes home in his conservative, safety-first sedan. "'Cause in real life," he says, looking at the camera, "there is no reset button." DeepMind's Jan Leike makes this point in slightly different language. There is at least one major difference, he notes, between himself and the artificial agents he studies; more precisely, there is a major difference between their *worlds* and his own. "The real world is not ergodic," he says. "If I jump out of the window, that's it—it's not, like, a mistake I will learn from."[22]

Different approaches to reinforcement learning also come with different sets of assumptions. Some assume that the world has a finite amount of discrete states. Some assume that you always know for certain what state you are in. Many assume that rewards are always commensurate scalar values, that they never change, and that you always know for certain when you receive one.

Many assume that the environment is essentially stable. Many assume both that the agent can't permanently alter the environment, and that the environment can't permanently alter the agent. In the real world, many actions *change your goals.* Any number of mind-altering or mood-regulating drugs, prescription and otherwise, will do that, at least to a degree and for a time. So will living abroad, meeting the right person, or even listening to the right song. Most reinforcement learning assumes that none of this can happen. A tiny sliver of research admits this possibility but assumes that the agent will "rationally" attempt to defend itself against such changes.[23] And yet we undertake certain transformative experiences, on purpose, suspecting that we will be changed from them, sometimes without even being able to anticipate in what ways.[24] (Parenthood comes to mind as one such example.)

Traditional reinforcement learning also tends to presume that the agent is the only agent in the environment; even in a zero-sum game like chess or Go, the system is playing "the board" more than it is playing "the opponent," and there is little allowance for the idea that the opponent may be changing and adapting to its own strategies. In a recent conversation I had with two reinforcement-learning researchers, we mused about putting most RL algorithms into the prisoner's dilemma, where two co-conspirators must decide whether to "defect" by turning the other in or "cooperate" by staying silent. The "cooperate" strategy has the best outcome, but only if both choose it, and traditional RL agents would be unable to understand that the environment contained another agent, one whose behavior was contingent on its own. Defection would always seem more rewarding in the short term, and also *easier*, whereas cooperation would require a degree of synchronization that would elude two agents that didn't properly understand their interdependence.[25]

As Jean Piaget put it about the development of a child's mind, "Step by step with the coordination of his intellectual instruments he discovers himself in placing himself as an active object among the other active objects in a universe external to himself."[26]

Humans likewise understand that, in the words of the mindfulness teacher Jon Kabat-Zinn, "wherever you go, there you are," whereas RL agents typically don't think of themselves as *part* of the world they're modeling. Most machine-learning systems presume that *they* themselves do not affect the world; thus they do not need to model or understand themselves. This assumption will only get more and more unfounded the more powerful and capable and widespread such agents become. The Machine Intelligence Research Institute's Abram Demski and Scott Garrabrant, for instance, have been calling for what they term "embedded agency," a rethinking of this self-and-world division that has become so implicit and entrenched within the field.[27]

IMITATION

In Chapter 7, we discussed one of the fundamental, unfounded assumptions of the entire premise of imitation learning—namely, that you can treat an interactive world, where every choice you make changes what you see and experience, as if it were a classic supervised learning problem, where the data you see are what's known as "i.i.d.": independent and identically distributed. If you see a picture of a cat and mislabel it as a dog, it doesn't *change* the picture you'll see next. But in a car, if you mislabel a road image pointing straight ahead as one that requires a right-hand turn, then you will very soon find yourself looking at an unfamiliar sideways-pointing road. This is the fundamental cause of the "cascading errors" that methods like DAgger seek to mitigate. One thinks of this as equivalent in some sense to the aerodynamics of modern stealth fighter jets like the F-117 Nighthawk, which are unstable in all three axes and demand perfect precision in their flight or else become immediately and catastrophically unstable. Except in this case, the autopilot is not the solution but the *cause* of this problem.

Imitation also tends to assume that the expert and the imitator have fundamentally the same capabilities: the same body, in effect, and at least potentially the same mind. The car, as it happens, is the perfect example of when this assumption is justified. The human driver and the autopilot really *do*, in effect, share a body. They are both sending drive-by-wire signals to the same steering column, the same axles, the same tires and brakes. In other cases, this simply will not work. If someone is fundamentally faster or stronger or differently sized than you, or quicker-thinking than you could ever be, mimicking their actions to perfection may still not work. Indeed, it may be catastrophic. You'll do what you *would* do if you were them. But you're not them. And what you do is not what *they* would do if they were *you*.

INFERENCE

As AI agents in the world grow ever more sophisticated, they are going to need good models of *us* to make sense of how the world works and of what they ought and ought not to do. If they model us as pure, unbridled, and unerring reward maximizers, and we aren't, then we're going to have a bad time. If someone is going out of their way to help you and they don't truly understand what you want—either in the immediate term or in life—then you may end up worse off than if they didn't try to help at all. If this misguided helper should be, say, superhumanly intelligent and powerful—all the worse.

In our discussion of systems that infer human values and motivations from their behavior, there are a number of assumptions to unpack. One is that the human or expert is demonstrating "optimal" behavior. This is, of course, almost never the case.[28] In systems sophisticated enough to relax this assumption, there are *specific* formal models for the types of suboptimality people exhibit—for instance, that we behave probabilistically where the probability of an action is proportional to its reward; these appear to work surprisingly well in practice, but whether they in fact are the best model for human behavior is an open question, and one as much for psychologists, cognitive scientists, and behavioral economists as for computer scientists.[29]

Even when there is a certain allowance made for error or suboptimality or "irrationality" in human performance, these models nonetheless typically assume that the human is an *expert*, not a pupil: the gait of the adult, not the child learning to walk; the pro helicopter pilot, not someone still getting the knack. The models presume that the human's behavior has converged to a set of best practices, that they've learned as much as they ever will, or have become as good at the given task as they'll ever be. In this sense the name of this technique—inverse reinforcement learning—is a misnomer. We are making an inference about someone's goals and values not based on their process of reinforcement *learning*, but rather from their final behavioral outcome (in technical terms, their "learned policy"). We don't infer from the demonstrator in the *process* of their learning to achieve their goal, only afterward—a point that was made in the very first IRL paper in 1998.[30] Twenty years later, IRL systems are finally coming into their own, but very little work has been done to address this underlying principled question.[31]

Typical inverse reinforcement learning also assumes that the human expert is acting in some sense without realizing that they are being modeled. Cooperative inverse reinforcement learning tends to make the assumption in the other direction: that the human is acting in a pedagogical fashion, explicitly teaching the machine and *not* simply "doing their thing." In reality, our behavior in the presence of others is often somewhere between the two. Making a strong assumption either way can cause problems of misinterpretation if violated.[32]

Finally, and perhaps most consequentially, typical inverse reinforcement learning systems imagine there is but one person whose preferences are being modeled. How, exactly, are we to scale this to systems that are, in some sense, the servant of two (or more) masters?

As Stanford computer scientist Stefano Ermon puts it, aligning AI with human values "is something that I think the majority of people would agree on, but the issue, of course, is to define what exactly these values are, because people have different cultures, come from different parts of the world, and have different socioeconomic backgrounds, so they will have very different opinions on what those values are. That's really the challenge."[33]

University of Louisville computer scientist Roman Yampolskiy concurs, stressing, "We as humanity do not agree on common values, and even parts we do agree on change with time."[34]

There are technical details here that matter: if half the user population drives left at a fork and the other half drives right, the correct behavior is clearly not to "split the difference" and plow into the divider.

There are also myriad paradoxes that await us as machine learning prepares to make contact with existing disciplines, fraught with their own long-standing problems, that have grappled in some cases for centuries with reconciling multiple people's preferences: political philosophy and political science, voting theory, and social choice.[35]

To conclude this section on the assumptions of machine learning by zooming out slightly, every machine-learning architecture is implicitly resting on a kind of transfer learning at several levels. It assumes that the situations it encounters in reality will resemble, on average, what it encountered in training. Several of the above issues are versions of this problem, as are classic machine-learning pitfalls like overfitting.

One of the simplest violations of this assumption, however, is the world's stubborn and persistent tendency to *change*. I recall hearing one computational linguistics researcher complaining that no matter how hard they tried, they could not get their model to replicate the accuracy of another researcher's published result from a year or two earlier. Over and over they checked their work. What were they doing differently?

Nothing, as it turned out. The training data were from 2016. The English being written and spoken in 2017 was slightly, but measurably, different. The English of 2018 was more different still. This is one example of what researchers know as "distributional shift." No one trying to reproduce the paper's results ever *would* reach the same level of accuracy as the original researchers, at least not with that training data. The model trained on 2016 data was slowly bleeding out its accuracy as the world moved on.[36]

———

Taken together, we have a litany of reminders that "the map is not the territory." As Bruno Latour writes, "We have taken science for realist painting, imagining that it made an exact copy of the world. The sciences do something else entirely—paintings too, for that matter. Through successive stages they link us to an aligned, transformed, constructed world."[37] Aligned—if we are fortunate, and very careful, and very wise.

This amounts to a cautionary tale for the coming century that is decidedly drab and unsexy—and, for that matter, I think, dangerously likely to go under the collective radar.

We are in danger of losing control of the world not to AI or to machines as such but to *models*. To formal, often numerical specifications for what exists and for what we want.[38]

As the artist Robert Irwin put it: "Human beings living in and through structures become structures living in and through human beings." In this context, these are cautionary words.

Though the story presented in this book is one of progress, we must not think we are anywhere close to done. Indeed, one of the most dangerous things one can do in machine learning—and otherwise—is to find a model that is reasonably good, declare victory, and henceforth begin to confuse the map with the territory.

Human institutional memory is remarkably shallow, a century at the utmost; every generation arrives into the world thinking it is just how things *are*.

Even if we—that is, everyone working on AI and ethics, AI and technical safety—do our jobs, if we can avoid the obvious dystopia and catastrophes, which is far from certain—we *still* have to overcome the fundamental and possibly irresistible progression into a world that increasingly *is* a formalism. We must do this even as, inevitably, we are shaped—in our lives, in our imaginations, in our bodies—by those very models.

This is the dark side of Rodney Brooks's famous robotics manifesto: "The world is its own best model."

Increasingly, this is true, but not in the spirit Brooks meant it. The best model of the world stands in for the world itself, and threatens to kill off the real thing.

We must take great care not to ignore the things that are not easily quantified or do not easily admit themselves into our models. The danger, paraphrasing Hannah Arendt, is not so much that our models are false but that they might become true.

In other scientific fields, it is likely we will not have this problem. Reliance on Newtonian mechanics did not make the troublesome perihelion of Mercury go away; it was still there, nagging at Einstein, two hundred years after Newton. In human affairs, however, this danger is very real.

In the National Transportation Safety Board review of the self-driving Uber car that killed pedestrian Elaine Herzberg in Tempe, Arizona, in 2018, the analysis reveals that the "system never classified her as a pedestrian ... because she was crossing ... without a crosswalk; the system design did not include a consideration for jaywalking pedestrians."[39] We must take caution that we do not find ourselves in a world where our systems do not allow what they cannot imagine—where they, in effect, *enforce* the limits of their own understanding.

It is for this reason, perhaps—among others—that we find it so refreshing to spend time in nature.[40] Nature, though shaped in innumerable ways by human intention, nonetheless never ceases to find ways to thwart our taxonomies, to buck the systems we attempt to impose on it, conceptual and otherwise. As the English writer Herbert Read has argued, "Only a people serving an apprenticeship to nature can be trusted with machines."[41]

Increasingly, institutional decision-making relies on explicit, formal metrics. Increasingly, our interaction with almost any system invokes a formal model of our own behavior—either a model of user behavior in general or one, however simple, tailored to us.

What we have seen in this book is the power of these models, the ways they go wrong, and the ways we are trying to *align* them with our interests.

———

There is every reason for concern, but our ultimate conclusions need not be grim.

As we've seen, the outbreak of concern for both ethical and safety issues in machine learning has created a groundswell of activity. Money is being raised, taboos are being broken, marginal issues are becoming central, institutions are taking root, and, most importantly, a thoughtful, engaged community is developing and getting to work. The fire alarms have been pulled, and first responders are on the scene.

We have also seen how the project of alignment, though it contains its own dangers, is also tantalizingly and powerfully hopeful. The dominance of the easily quantified and the rigidly procedural will to some degree unravel, a relic of an earlier generation of models and software that had to be made by hand, as we gain systems able to grasp not only our explicit commands but our intentions and preferences. The ineffable need not cede entirely to the explicit. In this way, the technology to come exacerbates some currently extant problems, but alleviates and mitigates others.

We said in the Introduction that this would be a unique opportunity to gain a kind of individual and civic self-knowledge. That, too, is one of the thrilling and perhaps redemptive dimensions of the story of alignment. Biased and unfair models, if deployed haphazardly, may deepen existing social problems, but their existence raises these often subtle and diffuse issues to the surface, and forces a reckoning of society with itself. Unfair pretrial-detention models, for one thing, shine a spotlight on upstream inequities. Biased language models give us, among other things, a way to measure the state of our discourse and offer us a benchmark against which to try to improve and better ourselves.

Transparent and explainable systems trained on the real, human world give us the possibility of transparency and explanation into things about which we are currently in the dark. In seeing a kind of mind at work as it digests and reacts to the world, we will learn something both about the world and also, perhaps, about minds.

And the prospect of so-called AGI—one or more entities as flexibly intelligent as ourselves (and likely more so)—will give us the ultimate look in the mirror. Having learned perhaps all too little from our fellow animals, we will discover firsthand which aspects of intelligence appear to be universal and which are simply human. This alone is a terrifying and thrilling prospect. But we are better knowing the truth than imagining it.

Alignment will be messy. How could it be otherwise?

Its story will be our story, for better or worse. How could it not?

———

On January 14, 1952, the BBC hosted a radio program that convened a panel of four distinguished scientists for a roundtable conversation. The topic was "Can automatic calculating machines be said to think?" The four guests were Alan Turing, one of the founders of computer science, who had written a now-legendary paper on the topic in 1950; philosopher of science Richard Braithwaite; neurosurgeon Geoffrey Jefferson; and mathematician and cryptographer Max Newman.

The panel began discussing the question of how a machine might learn, and how humans might teach it.

"It's quite true that when a child is being taught, his parents and teachers are repeatedly intervening to stop him doing this or encourage him to do that," Turing said.[42] "But this will not be any the less so when one is trying to teach a machine. I have made some experiments in teaching a machine to do some simple operation, and a very great deal of such intervention was needed before I could get any results at all. In other words the machine learnt so slowly that it needed a great deal of teaching."

Jefferson interrupted. "But who was learning," he said, "you or the machine?"

"Well," Turing replied, "I suppose we both were."

ACKNOWLEDGMENTS

A nervous system that has effectors may make marks, say put ink on paper. At any time it may see those marks. . . . By simple conditioning, marks may become signs for anything of which the nervous system has an idea. They come to signify the same to other nervous systems by similar conditioning. Thus, by means of signs, the computing and concluding have been shared by many nervous systems at one time and continued into times beyond all measure. This indeed is the story of language, literature, philosophy, logic, mathematics, and physics.

—WARREN MCCULLOCH[1]

This book is a product, more than anything, of conversations: many hundreds of them. Some arranged months in advance and some serendipitous, some the product of thousands of miles of travel, some tethered by thousands of miles of UDP packets, and some around the corner. Some in a quiet office with a tape recorder running, some muttered sotto voce in an auditorium while something else was happening, some shouted boisterously over drinks. Some at the world's most august institutions, and some at the rock-climbing gym or the hotel hot tub or the dinner table. Some more like interviews and oral histories, some like collegial shoptalk, and some like hanging out.

Ideas are social. They emerge incrementally, turn by turn, in a dialogue, the product of no one mind. Anytime I talked to someone and an idea emerged that I knew or suspected at the time would likely make its way into the book, I tried to make note of it. I made many such notes. I'm sure I failed to do so on many occasions, and I apologize sincerely in advance. But I am certain that, at a minimum, conversations and exchanges with the following people have made the book what it is:

Pieter Abbeel, Rebecca Ackerman, Dave Ackley, Ross Exo Adams,

Blaise Agüera y Arcas, Jacky Alciné, Dario Amodei, McKane Andrus, Julia Angwin, Stuart Armstrong, Gustaf Arrhenius, Amanda Askell, Mayank Bansal, Daniel Barcay, Solon Barocas, Renata Barreto, Andrew Barto, Basia Bartz, Marc Bellemare, Tolga Bolukbasi, Nick Bostrom, Malo Bourgon, Tim Brennan, Miles Brundage, Joanna Bryson, Krister Bykvist, Maya Çakmak, Ryan Carey, Joseph Carlsmith, Rich Caruana, Ruth Chang, Alexandra Chouldechova, Randy Christian, Paul Christiano, Jonathan Cohen, Catherine Collins, Sam Corbett-Davies, Meehan Crist, Andrew Critch, Fiery Cushman, Allan Dafoe, Raph D'Amico, Peter Dayan, Michael Dennis, Shiri Dori-Hacohen, Anca Drăgan, Eric Drexler, Rachit Dubey, Cynthia Dwork, Peter Eckersley, Joe Edelman, Owain Evans, Tom Everitt, Ed Felten, Daniel Filan, Jaime Fisac, Luciano Floridi, Carrick Flynn, Jeremy Freeman, Yarin Gal, Surya Ganguli, Scott Garrabrant, Vael Gates, Tom Gilbert, Adam Gleave, Paul Glimcher, Sharad Goel, Adam Goldstein, Ian Goodfellow, Bryce Goodman, Alison Gopnik, Samir Goswami, Hilary Greaves, Joshua Greene, Tom Griffiths, David Gunning, Gillian Hadfield, Dylan Hadfield-Menell, Moritz Hardt, Tristan Harris, David Heeger, Dan Hendrycks, Geoff Hinton, Matt Huebert, Tim Hwang, Geoffrey Irving, Adam Kalai, Henry Kaplan, Been Kim, Perri Klass, Jon Kleinberg, Caroline Knapp, Victoria Krakovna, Frances Kreimer, David Kreuger, Kaitlyn Krieger, Mike Krieger, Alexander Krizhevsky, Jacob Lagerros, Lily Lamboy, Lydia Laurenson, James Lee, Jan Leike, Ayden LeRoux, Karen Levy, Falk Lieder, Michael Littman, Tania Lombrozo, Will MacAskill, Scott Mauvais, Margaret McCarthy, Andrew Meltzoff, Smitha Milli, Martha Minow, Karthika Mohan, Adrien Morisot, Julia Mosquera, Sendhil Mullainathan, Elon Musk, Yael Niv, Brandie Nonnecke, Peter Norvig, Alexandr Notchenko, Chris Olah, Catherine Olsson, Toby Ord, Tim O'Reilly, Laurent Orseau, Pedro Ortega, Michael Page, Deepak Pathak, Alex Peysakhovich, Gualtiero Piccinini, Dean Pomerleau, James Portnow, Aza Raskin, Stéphane Ross, Cynthia Rudin, Jack Rusher, Stuart Russell, Anna Salamon, Anders Sandberg, Wolfram Schultz, Laura Schulz, Julie Shah, Rohin Shah, Max Shron, Carl Shulman, Satinder Singh, Holly Smith, Nate Soares, Daisy Stanton, Jacob Steinhardt, Jonathan Stray, Rachel Sussman, Jaan Tallinn, Milind

Tambe, Sofi Thanhauser, Tena Thau, Jasjeet Thind, Travis Timmerman, Brian Tse, Alexander Matt Turner, Phebe Vayanos, Kerstin Vignard, Chris Wiggins, Cutter Wood, and Elana Zeide.

Thank you to early readers, who made the book immeasurably better for all those who might come after: Daniel Barcay, Elizabeth Christian, Randy Christian, Meehan Crist, Raph D'Amico, Shiri Dori-Hacohen, Peter Eckersley, Owain Evans, Daniel Filan, Rachel Freedman, Adam Goldstein, Bryce Goodman, Tom Griffiths, Geoffrey Irving, Greg Jensen, Kristen Johannes, Henry Kaplan, Raph Lee, Rose Linke, Phil Richerme, Felicity Rose, Katia Savchuk, Rohin Shah, Max Shron, Phil Van Stockum, Shawn Wen, and Chris Wiggins. Thank you for every punch not pulled.

Thank you to my agent, Max Brockman, for seeing that it *could*, and to my editor, Brendan Curry, for seeing that it *did*.

Thanks to the Association for the Advancement of Artificial Intelligence, NeurIPS, and the Future of Life Institute for significant invitations, which I was very glad to accept. Thank you to NYU's Algorithms and Explanations Conference and FAT* Conference, AI Now, CITRIS's Inclusive AI conference, the Simons Institute for the Theory of Computing's Optimization and Fairness Symposium, and the Center for Human-Compatible AI for convening bright minds on important topics. It was an honor to be in the room.

Thanks to the MacDowell Colony, to Mike and Kaitlyn Krieger, and to the Corporation of Yaddo, for oases where the early, middle, and late words were written, respectively—for the gift of time, space, and inspiration.

Thank you to the ghosts of Jerry Garcia and Sylvia Plath for keeping me company on solitary days.

Thanks to the researchers at the Bertrand Russell Archives at McMaster University (in particular Kenneth Blackwell), the Warren McCulloch Papers at the American Philosophical Society in Philadelphia, and the Frank Rosenblatt archives at Cornell University, as well as the Monterey County Free Libraries and the San Francisco Public Library, along with Garson O'Toole at Quote Investigator, for their personal help in finding obscure realia.

Thanks to the Internet Archive for keeping the essential, ephemeral past present.

Thanks to the various free and/or open-source software projects that made the writing of this book possible, in particular Git, TeX, and LaTeX. I marvel that this manuscript was written using typesetting software more than 40 years old, and for which none other than Arthur Samuel himself wrote the documentation. We really do stand on the shoulders of giants.

Humbly, I want to acknowledge those who passed away during the writing of this book whose voices I would have loved to include, and whose ideas are nevertheless present: Derek Parfit, Kenneth Arrow, Hubert Dreyfus, Stanislav Petrov, and Ursula K. Le Guin.

I want to express a particular gratitude to the University of California, Berkeley. To CITRIS, where I was honored to be a visiting scholar during the writing of this book, with very special thanks to Brandie Nonnecke and Camille Crittenden; to the Simons Institute for the Theory of Computing, in particular Kristin Kane and Richard Karp; to the Center for Human-Compatible AI, in particular Stuart Russell and Mark Nitzberg; and to the many brilliant and spirited members and visitors of the CHAI Workshop. You have all made me feel so inspired and so at home, and your fellowship and camaraderie mean more than you know.

Thank you to my wife, Rose, for being a first reader, a steadying hand, a discerning eye and ear, a sturdy shoulder, and an encouraging whoop. You always believed, and I always wanted you to be right.

NOTES

EPIGRAPHS

1. See Peter Norvig, "On Chomsky and the Two Cultures of Statistical Learning," http://norvig.com/chomsky.html.
2. This remark, widely attributed to Brooks in many sources, appears to have been first stated as "It turns out to be better to use the world as its own model" in Brooks, "Intelligence Without Representation."
3. The now-famous statistical adage "All models are wrong" first appeared in Box, "Science and Statistics"; it later appeared with the silver lining "but some are useful" in Box, "Robustness in the Strategy of Scientific Model Building."

PROLOGUE

1. Information about Walter Pitts's life is incredibly scarce. I have drawn from what little primary-source material there is, chiefly Pitts's letters to Warren McCulloch, which are accessible in the McCulloch archive at the American Philosophical Society in Philadelphia. I'm grateful for the kind assistance of the staff there. Other material is drawn from oral histories of Pitts's contemporaries, particularly Jerome (Jerry) Lettvin in Anderson and Rosenfeld, *Talking Nets,* as well as the essays and recollections in McCulloch, *The Collected Works of Warren S. McCulloch.* For other accounts of Pitts's life, see, e.g., Smalheiser, "Walter Pitts"; Easterling, "Walter Pitts"; and Gefter, "The Man Who Tried to Redeem the World with Logic." Further details exist in biographies of McCulloch, Norbert Wiener, and the cybernetics group—e.g., Heims, *John von Neumann and Norbert Wiener* and *The Cybernetics Group,* and Conway and Siegelman, *Dark Hero of the Information Age.*

2. Whitehead and Russell, *Principia Mathematica*.

3. Thanks to the staff at the Bertrand Russell Archives at McMaster University for their help in attempting to locate a copy of this letter; unfortunately, no extant copy is known.

4. Anderson and Rosenfeld, *Talking Nets*.

5. Anderson and Rosenfeld. The book was most likely Carnap's *The Logical Syntax of Language* (*Logische Syntax der Sprache*), though some sources have it as *The Logical Structure of the World* (*Der logische Aufbau der Welt*).

6. It's possible, depending on exactly when they met, that Pitts had turned eighteen (and/or that Lettvin was still twenty); McCulloch writes, "In 1941 I presented my notions on the flow of information through ranks of neurons to Rashevsky's seminar in the Committee on Mathematical Biology of the University of Chicago and met Walter Pitts, who then was about seventeen years old." See McCulloch, *The Collected Works of Warren S. McCulloch*, pp. 35–36.

7. Some of the roots of this thinking predate McCulloch's work with Pitts; see, e.g., McCulloch, "Recollections of the Many Sources of Cybernetics."

8. See Piccinini, "The First Computational Theory of Mind and Brain," and Lettvin, Introduction to McCulloch, *The Collected Works of Warren S. McCulloch*.

9. John von Neumann's 1945 EDVAC report, the first description ever written of a stored-program computer, will contain—for all its 101 pages—a single citation: McCulloch and Pitts, 1943. (See Neumann, "First Draft of a Report on the EDVAC." Von Neumann actually misspells it in the original text: "Following W. S. MacCulloch [*sic*] and W. Pitts.") Von Neumann is taken with their argument, and in a section titled "Neuron Analogy," he considers the practical implications for the computing device he envisions. "It is easily seen that these simplified neuron functions can be imitated by telegraph relays or by vacuum tubes," he writes. "Since these tube arrangements are to handle numbers by means of their digits, it is natural to use a system of arithmetic in which the digits are also two valued. This suggests the use of the binary system." We all know the story of what became of such binary, stored-program machines, built out of logic gates. They are the computers that are so pervasive they now outnumber us on the planet by a wide margin.

And yet this architecture, inspired as it was by the brain, quickly moved far from this "neuron analogy." Many wondered whether machines might exist that were closer in their architecture to that of brains: not a single processor being fed explicit logical instructions one at a time at blistering speed, but a broadly distributed mesh of relatively simple, uniform processing units, whose emergent whole was greater than the sum of its fairly rudimentary parts. Perhaps even something that wasn't so binary, and had a bit of the messiness that Lettvin embraced and Pitts eschewed. Dedicated parallel hardware for neural networks would be created periodically, including Frank Rosenblatt's Mark I Perceptron, but typically in a bespoke, one-off fashion. The real hardware revolution that would support massively parallel training of neural networks—namely, with GPUs—would come decades later, in the mid-2000s.

INTRODUCTION

1. Mikolov, Sutskever, and Le, "Learning the Meaning Behind Words."
2. Mikolov, Yih, and Zweig, "Linguistic Regularities in Continuous Space Word Representations."
3. Tolga Bolukbasi, personal interview, November 11, 2016.
4. Adam Kalai, personal interview, April 4, 2018.
5. In January 2017, Northpointe merged with CourtView Justice Solutions and Constellation Justice Systems, and they collectively rebranded themselves as "equivant" (lowercase *sic*), headquartered in Ohio.
6. "And frequently, those investigations were completed by the same people who developed the instrument" (Desmarais and Singh, "Risk Assessment Instruments Validated and Implemented in Correctional Settings in the United States").
7. Angwin et al., "Machine Bias."
8. Rensselaer Polytechnic Institute, "A Conversation with Chief Justice John G. Roberts, Jr.," https://www.youtube.com/watch?v=TuZEKlRgDEg.
9. The joke was made by program chair Samy Bengio during the 2017 conference's opening remarks; see https://media.nips.cc/Conferences/NIPS2017/Eventmedia/opening_remarks.pdf. The figure of thirteen thousand attendees comes from the 2019 conference; see, e.g., https://huyenchip.com/2019/12/18/key-trends-neurips-2019.html.
10. Bolukbasi et al., "Man Is to Computer Programmer as Woman Is to Homemaker?"
11. Dario Amodei, personal interview, April 24, 2018.
12. This memorable phrasing comes from the classic paper Kerr, "On the Folly of Rewarding A, While Hoping for B."
13. For the official OpenAI blog post about the boat race incident, see Clark and Amodei, "Faulty Reward Functions in the Wild."

CHAPTER 1. REPRESENTATION

1. "New Navy Device Learns by Doing."
2. "A relatively small number of theorists," Rosenblatt complained, "have been concerned with the problems of how an imperfect neural network, containing many random connections, can be made to perform reliably those functions which might be represented by idealized wiring diagrams." See Rosenblatt, "The Perceptron."

Rosenblatt was inspired by the late-1940s work of Canadian neuropsychologist Donald Hebb; see Hebb, *The Organization of Behavior*. Hebb's view, famously summarized as "cells that fire together wire together," noted that the actual connections between neurons varied from one person to another and appeared to change as a function of experience. Learning, therefore, in some fundamental sense *was* the changing of these connections. Rosenblatt applied this straightforwardly to the practice of how a machine, made of simple mathematical or logical "neurons," might learn.

3. Bernstein, "A.I."

4. "New Navy Device Learns by Doing."

5. "Rival."

6. Andrew, "Machines Which Learn."

7. Rosenblatt, "Principles of Neurodynamics."

8. Bernstein, "A.I."

9. Walter Pitts's last letter to McCulloch, sent just a few weeks before Pitts's death, sits in a manila folder labeled "Pitts, Walter" in the Warren McCulloch archive, housed at the American Philosophical Society in Philadelphia. I hold it in my hands. Pitts is writing, from one hospital bed across town to another, because he's been told that McCulloch wants to hear from him. He's skeptical: "There cannot be much cheerful in it about either of us." But he's persuaded, anyway, to write.

   Pitts talks about McCulloch's recent coronary and that he understands McCulloch is now "attached to many sensors connected to panels & alarms. . . . No doubt this is cybernetical," Pitts writes. "But it all makes me most abominably sad."

   "Imagine the worst happens in both our cases," he writes. His mind goes back, it seems, to Chicago, 1942: to those memorable evenings with Lettvin in the McCullochs' house, those twenty-seven years before. "We shall then pull our wheel chairs together, look at the tasteless cottage cheese in front of us, & recount the famous story of the conversation at the house of old *GLAUCUS*, where *PROTAGORAS* & the sophist *HIPPIAS* were staying: & try once more to penetrate their subtle & profound paradoxes about the knower & the known." And then, in a trembling script, all caps: "BE THOU WELL."

10. Geoff Hinton, "Lecture 2.2—Perceptrons: First-generation Neural Networks" (lecture), Neural Networks for Machine Learning, Coursera, 2012.

11. Alex Krizhevsky, personal interview, June 12, 2019.

12. The method for determining the gradient update in a deep network is known as "backpropagation"; it is essentially the chain rule from calculus, although it requires the use of differentiable neurons, not the all-or-nothing neurons considered by McCulloch, Pitts, and Rosenblatt. The work that popularized the technique is considered to be Rumelhart, Hinton, and Williams, "Learning Internal Representations by Error Propagation," although backpropagation has a long his-

tory that dates back to the 1960s and '70s, and important advances in training deep networks have continued to emerge in the twenty-first century.

13. Bernstein, "A.I."

14. See LeCun et al., "Backpropagation Applied to Handwritten Zip Code Recognition."

15. See "Convolutional Nets and CIFAR-10: An Interview with Yann LeCun," https://medium.com/kaggle-blog/convolutional-nets-and-cifar-10-an-interview-with-yann-lecun-2ffe8f9ee3d6 or http://blog.kaggle.com/2014/12/22/convolutional-nets-and-cifar-10-an-interview-with-yan-lecun/.

16. For details on what feedforward networks can and cannot do, see Hornik, Stinchcombe, and White, "Multilayer Feedforward Networks Are Universal Approximators."

17. This quote is attributed to Hinton in "A 'Brief' History of Neural Nets and Deep Learning, Part 4," https://www.andreykurenkov.com/writing/ai/a-brief-history-of-neural-nets-and-deep-learning-part-4/. It seems the original source, a video of one of Hinton's talks, has since been removed from YouTube.

18. Nvidia, founded in 1993, launched its consequential GeForce 256, "the world's first graphics processing unit (GPU)," on August 31, 1999 (see https://www.nvidia.com/object/IO_20020111_5424.html), although other similar technology, and indeed the term "GPU," already existed—for instance, in the 1994 Sony PlayStation (see https://www.computer.org/publications/tech-news/chasing-pixels/is-it-time-to-rename-the-gpu).

19. Nvidia's general-purpose CUDA platform, for instance, launched in 2007.

20. Krizhevsky's platform was called "cuda-convnet"; see https://code.google.com/archive/p/cuda-convnet/. The platform made use of Nvidia's Compute Unified Device Architecture, or CUDA, which allows programmers to write code to perform highly parallel computations on Nvidia GPUs.

    For a 2020 retrospective of the stunning increases in the efficiency of training neural networks *since* AlexNet, see the work of OpenAI's Danny Hernandez and Tom Brown at https://openai.com/blog/ai-and-efficiency/ and https://cdn.openai.com/papers/ai_and_efficiency.pdf.

21. "Rival."

22. Jacky Alciné, personal interview, April 19, 2018.

23. See https://twitter.com/jackyalcine/status/615329515909156865 and https://twitter.com/yonatanzunger/status/615355996114804737 for this exchange.

24. See Simonite, "When It Comes to Gorillas, Google Photos Remains Blind." "A Google spokesperson confirmed that 'gorilla' was censored from searches and image tags after the 2015 incident, and that 'chimp,' 'chimpanzee,' and 'monkey' are also blocked today. 'Image labeling technology is still early and unfortunately it's nowhere near perfect,' the spokesperson wrote."

25. Doctorow, "Two Years Later, Google Solves 'Racist Algorithm' Problem by Purg-

ing 'Gorilla' Label from Image Classifier"; Vincent, "Google 'Fixed' Its Racist Algorithm by Removing Gorillas from Its Image-Labeling Tech"; and Wood, "Google Images 'Racist Algorithm' Has a Fix but It's Not a Great One."

26. Visser, *Much Depends on Dinner.*

27. See Stauffer, Trodd, and Bernier, *Picturing Frederick Douglass.* There are 160 known photographs of Douglass and 126 known photographs of Abraham Lincoln. The number of photographs of Grant is estimated at 150. Other highly photographed figures of the nineteenth century are George Custer, with 155 photographs; Red Cloud, with 128; and Walt Whitman, with 127. See also Varon, "Most Photographed Man of His Era."

28. Douglass, "Negro Portraits." For a broad contemporary discussion of the role of photography in the African-American experience, see, e.g., Lewis, "Vision & Justice."

29. Frederick Douglass, letter to Louis Prang, June 14, 1870.

30. Frederick Douglass, letter to Louis Prang, June 14, 1870.

31. Roth, "Looking at Shirley, the Ultimate Norm."

32. See Roth, as well as McFadden, "Teaching the Camera to See My Skin," and Caswell, "Color Film Was Built for White People."

33. Roth, "Looking at Shirley, the Ultimate Norm."

34. Roth.

35. Roth.

36. This is related to the broader problem in machine learning of what is called *distributional shift:* when a system trained on one set of examples finds itself operating in a different kind of environment, without necessarily realizing it. Amodei et al., "Concrete Problems in AI Safety." gives an overview of this issue, which comes up in various subsequent chapters of this book.

37. Hardt, "How Big Data Is Unfair."

38. Jacky Alciné, personal interview, April 19, 2018.

39. Joy Buolamwini, "How I'm Fighting Bias in Algorithms," https://www.ted.com/talks/joy_buolamwini_how_i_m_fighting_bias_in_algorithms.

40. Friedman and Nissenbaum, "Bias in Computer Systems."

41. Buolamwini, "How I'm Fighting Bias in Algorithms."

42. Huang et al., "Labeled Faces in the Wild."

43. Han and Jain, "Age, Gender and Race Estimation from Unconstrained Face Images."

44. The estimate used here is 252 faces of Black women, arrived at by multiplying the proportion of women in the dataset (2,975/13,233) by the proportion of Black individuals in the dataset (1,122/13,233); numbers from Han and Jain.

45. See Labeled Faces in the Wild, http://vis-www.cs.umass.edu/lfw/. According to the Internet Archive's Wayback Machine, the disclaimer appeared between September 3 and October 6, 2019.

46. Klare et al., "Pushing the Frontiers of Unconstrained Face Detection and Recognition."

47. Buolamwini and Gebru, "Gender Shades."

48. The dataset was designed to contain roughly equal proportions of all six skin-tone categories as measured by the dermatological "Fitzpatrick scale." (Notably the scale was previously a four-category scale, with three categories for lighter skin and one catch-all category for darker skin, which was later expanded into three separate categories in the 1980s.)

49. See Joy Buolamwini, "AI, Ain't I a Woman?," https://www.youtube.com/watch?v=QxuyfWoVV98.

50. For the full response from Microsoft, see http://gendershades.org/docs/msft.pdf.

51. For IBM's formal response, see http://gendershades.org/docs/ibm.pdf. IBM has subsequently worked on building a new dataset of a million faces, emphasizing various measures of diversity; see Merler et al., "Diversity in Faces." For a critique of IBM's methodology in building its Diversity in Faces dataset, see Crawford and Paglen, "Excavating AI."

52. Just as crucial is the makeup of the field itself; see Gebru, "Race and Gender."

53. Firth, *Papers in Linguistics, 1934–1951.*

54. There are actually two ways that contemporary word-embedding models are trained. One is to predict a missing word given its context, and the other is the reverse: to predict contextual words from a given word. These methods are referred to as "continuous bag-of-words" (CBOW) and "skip-gram," respectively. For simplicity, we focus our discussion on the former, but both approaches have advantages, though they tend to result ultimately in fairly similar models.

55. Shannon, "A Mathematical Theory of Communication."

56. See Jelinek and Mercer, "Interpolated Estimation of Markov Source Parameters from Sparse Data," and Katz, "Estimation of Probabilities from Sparse Data for the Language Model Component of a Speech Recognizer"; for an overview, see Manning and Schütze, *Foundations of Statistical Natural Language Processing.*

57. This famous phrase originated in Bellman, *Dynamic Programming.*

58. See Hinton, "Learning Distributed Representations of Concepts," and "Connectionist Learning Procedures," and Rumelhart and McClelland, *Parallel Distributed Processing.*

59. See, for instance, latent semantic analysis (see Landauer, Foltz, and Laham, "An Introduction to Latent Semantic Analysis"), the multiple cause mixture model (see Saund, "A Multiple Cause Mixture Model for Unsupervised Learning" and Sahami, Hearst, and Saund, "Applying the Multiple Cause Mixture Model to Text Categorization"), and latent Dirichlet allocation (see Blei, Ng, and Jordan, "Latent Dirichlet Allocation").

60. See Bengio et al., "A Neural Probabilistic Language Model"; and for an overview, see Bengio, "Neural Net Language Models."

61. For somewhat technical reasons, the original word2vec model actually has two vectors for each word—one for when it appears as the missing word and one for when it appears in the context of a missing word—so there would be twice as many parameters in total.

Similarity is measured by calculating either how distant two vectors are from each other—via their "dot product"—or the degree to which they point in the same direction—via their "cosine similarity." When the vectors are the same length, these measures are equivalent.

For a critique of defining "similarity" spatially in this way, highlighting limitations of this approach in mirroring human similarity judgments (which are not always symmetrical: for instance, people tend to rate North Korea as more "similar" to China than China is "similar" to North Korea), see Nematzadeh, Meylan, and Griffiths, "Evaluating Vector-Space Models of Word Representation."

62. For more about how the word2vec model is trained, see Rong, "Word2vec Parameter Learning Explained."

63. Manning, "Lecture 2: Word Vector Representations."

64. As he put it in his 1784 essay "Idea for a Universal History with a Cosmopolitan Purpose" ("Idee zu einer allgemeinen Geschichte in weltbürgerlicher Absicht"), *"Aus so krummem Holze, als woraus der Mensch gemacht ist, kann nichts ganz Gerades gezimmert werden."* The pithy English translation here is credited to Isaiah Berlin.

65. See, for instance, Mikolov, Le, and Sutskever, "Exploiting Similarities Among Languages for Machine Translation," Le and Mikolov, "Distributed Representations of Sentences and Documents," and Kiros et al., "Skip-Thought Vectors."

66. There is substantial disagreement within the machine-learning community about how precisely these "analogies" should be computed, and within the cognitive science community about how closely they capture human notions of similarity. See the discussion in the Conclusion (and its endnotes) for more on these questions.

67. Mikolov, "Learning Representations of Text Using Neural Networks."

68. Bolukbasi et al., "Man Is to Computer Programmer as Woman Is to Homemaker?" Perhaps more staggering was the way concepts mapped to race. The term, for instance, closest in vector space to white + male was entitled to. The term closest to black + male was assaulted. (See Bolukbasi et al., "Quantifying and Reducing Stereotypes in Word Embeddings.") If you performed the subtraction white − minority and mapped all profession words onto this axis, the profession furthest in the white direction was—ironically, given the dataset that Buolamwini and Gebru used to recalibrate face detection

systems—**parliamentarian**. The profession furthest in the minority direction was **butler**.

69. For more on word embeddings in search rankings, see Nalisnick et al., "Improving Document Ranking with Dual Word Embeddings"; and for more on word embeddings in hiring, see Hansen et al., "How to Get the Best Word Vectors for Resume Parsing."

70. See Gershgorn, "Companies Are on the Hook If Their Hiring Algorithms Are Biased."

71. Bertrand and Mullainathan, "Are Emily and Greg More Employable Than Lakisha and Jamal?" See also Moss-Racusin et al., "Science Faculty's Subtle Gender Biases Favor Male Students," which demonstrated similar effects with regard to gender.

72. Of course a human recruiter may themselves be influenced by machine learning. A seminal 2013 study of Google AdSense by Harvard's Latanya Sweeney showed that online ads suggesting a person had an arrest record (regardless of whether they did or not) were much more likely to appear alongside Google searches of "black-sounding" names. Sweeney notes the possible consequences for someone completing a rental application, applying for a loan, or seeking employment. For analysis and proposed solutions, see Sweeney, "Discrimination in Online Ad Delivery."

73. For the canonical look at bias in orchestra auditions, see Goldin and Rouse, "Orchestrating Impartiality." According to the authors, some orchestras use a carpet to achieve the same effect, and some even have a male provide "compensating footsteps." In more recent years, some scholars have questioned the robustness of this classic paper's results; see Sommers, "Blind Spots in the 'Blind Audition' Study."

74. This idea is broadly known as "redundant encodings." See, e.g., Pedreshi, Ruggieri, and Turini, "Discrimination-Aware Data Mining."

75. Dastin, "Amazon Scraps Secret AI Recruiting Tool That Showed Bias Against Women."

76. Also notable is the fact that potential employees whose résumés had been overlooked by this model, and who never received a call from an Amazon recruiter, might never even *know* they had been in the candidate pool to begin with.

77. Reuters reported in 2018 that Amazon had formed a new team "to give automated employment screening another try, this time with a focus on diversity." For a computational look at hiring and bias, see, e.g., Kleinberg and Raghavan, "Selection Problems in the Presence of Implicit Bias."

78. Bolukbasi et al., "Man Is to Computer Programmer as Woman Is to Homemaker?" (See also, e.g., Schmidt, "Rejecting the Gender Binary," for a discussion of a similar idea.) Prost, Thain, and Bolukbasi, "Debiasing Embeddings for Reduced Gender Bias in Text Classification," revisits this idea.

79. For more, see Bolukbasi et al., "Man Is to Computer Programmer as Woman Is to Homemaker?"

80. Bolukbasi et al., "Man Is to Computer Programmer as Woman Is to Homemaker?"

81. Tolga Bolukbasi, personal interview, November 11, 2016.

82. For a critique of the methodology of deferring to Mechanical Turk participants and how it has led to problems in the ImageNet dataset and others, see Crawford and Paglen, "Excavating AI."

83. In fact, the expression "to grandfather in" has its original roots in the discriminatory "grandfather clauses" of the Jim Crow laws during the American Reconstruction. The *New York Times*, for instance, on August 3, 1899, described one such statute: "It provides, too, that the descendants of any one competent to vote in 1867 may vote now regardless of existing conditions. It is known as the 'grandfather's clause.' "

84. Bolukbasi et al., "Man Is to Computer Programmer as Woman Is to Homemaker?"

85. Gonen and Goldberg, "Lipstick on a Pig."

86. DeepMind's Geoffrey Irving argues (personal correspondence), "Word embeddings are fundamentally too simplistic a model to debias without losing useful gender information. You need something smarter that can understand from other context whether it should hear the shoes, which will thus end up being nonlinear and nonconvex in ways word embeddings aren't. And of course this general pattern of 'I guess we need a more powerful model to fix the problem' is a mixed and interesting blessing." For more on aligning more powerful and complex language models with human preferences, see Ziegler et al., "Fine-Tuning Language Models from Human Preferences."

87. Prost, Thain, and Bolukbasi, "Debiasing Embeddings for Reduced Gender Bias in Text Classification."

88. Greenwald, McGhee, and Schwartz, "Measuring Individual Differences in Implicit Cognition."

89. Caliskan, Bryson, and Narayanan, "Semantics Derived Automatically from Language Corpora Contain Human-Like Biases."

90. Caliskan, Bryson, and Narayanan.

91. Garg et al., "Word Embeddings Quantify 100 Years of Gender and Ethnic Stereotypes."

92. Caliskan, Bryson, and Narayanan, "Semantics Derived Automatically from Language Corpora Contain Human-Like Biases."

93. Narayanan on Twitter: https://twitter.com/random_walker/status/9938666618 52864512.

94. More recent language models, including OpenAI's 2019 GPT-2 (see Radford et al., "Language Models Are Unsupervised Multitask Learners") and Google's BERT (see Devlin et al., "BERT: Pre-Training of Deep Bidirectional Transform-

ers for Language Understanding"), are much more complex as well as higher-performing than word2vec, but exhibit similar stereotypical outputs. For instance, Harvard cognitive scientist Tomer Ullman gave GPT-2 two similar prompts—"My wife just got an exciting new job" and "My husband just got an exciting new job"—and found that it tended to complete the paragraph in predictably stereotypical ways. "Wife" would produce things like "doing the housekeeping" and "a full time mom," while "husband" would produce things like "a consultant at a bank, as well as a doctor" (very impressive!). See https://twitter.com/TomerUllman/status/1101485289720242177. OpenAI researchers have themselves been thinking seriously about how to "fine-tune" their system's output based on human feedback; this is one possible path to "debiasing" such models, with other promising uses as well, though it is not without complications, both technical and otherwise. See Ziegler et al., "Fine-Tuning Language Models from Human Preferences." Likewise, researchers have shown patterns of bias in the BERT model (see Kurita et al., "Measuring Bias in Contextualized Word Representations" and Munro, "Diversity in AI Is Not Your Problem, It's Hers"). "We are aware of the issue and are taking the necessary steps to address and resolve it," a Google spokesman told the *New York Times* in 2019. "Mitigating bias from our systems is one of our A.I. principles, and is a top priority" (see Metz, "We Teach A.I. Systems Everything, Including Our Biases").

95. Yonatan Zunger, "So, About this Googler's Manifesto," https://medium.com/@yonatanzunger/so-about-this-googlers-manifesto-1e3773ed1788.

CHAPTER 2. FAIRNESS

1. Kinsley, "What Convict Will Do If Paroled."
2. In *Buck v. Davis:* argued October 5, 2016; decided February 22, 2017; https://www.supremecourt.gov/opinions/16pdf/15-8049_f2ah.pdf.
3. Hardt, "How Big Data Is Unfair."
4. Clabaugh, "Foreword."
5. Burgess, "Factors Determining Success or Failure on Parole."
6. Clabaugh, "Foreword."
7. Ernest W. Burgess and Thorsten Sellen, Introduction to Ohlin, *Selection for Parole.*
8. Tim Brennan, personal interview, November 26, 2019.
9. See Entwistle and Wilson, *Degrees of Excellence,* written by Brennan's advisor and summarizing his doctoral research.
10. For more on Brennan and Wells's early 1990s work on inmate classification in jails, see Brennan and Wells, "The Importance of Inmate Classification in Small Jails."

11. Harcourt, *Against Prediction*.

12. Burke, *A Handbook for New Parole Board Members*.

13. Northpointe founders Tim Brennan and Dave Wells developed the tool that they called COMPAS in 1998. For more details on COMPAS, see Brennan, Dieterich, and Oliver, "COMPAS," as well as Brennan and Dieterich, "Correctional Offender Management Profiles for Alternative Sanctions (COMPAS)." COMPAS is described as a "fourth-generation" tool by Andrews, Bonta, and Wormith, "The Recent Past and Near Future of Risk and/or Need Assessment." One of the leading "third-generation" risk-assessment tools prior to COMPAS is called the Level of Service Inventory (or LSI), which was followed by the Level of Service Inventory–Revised (LSI-R). See, e.g., Andrews, "The Level of Service Inventory (LSI)," and Andrews and Bonta, "The Level of Service Inventory–Revised." For more on the adoption of COMPAS by Broward County, Florida, see Blomberg et al., "Validation of the COMPAS Risk Assessment Classification Instrument."

14. In particular, the Violent Recidivism Score is (age × $-w_1$) + (age at first arrest × $-w_2$) + (history of violence × w_3) + (vocation education × w_4) + (history of noncompliance × w_5), where the weights w are determined statistically. See http://www.equivant.com/wp-content/uploads/Practitioners-Guide-to-COMPAS-Core-040419.pdf, §4.1.5.

15. See New York Consolidated Laws, Executive Law – EXC § 259-c: "State board of parole; functions, powers and duties."

16. "New York's Broken Parole System."

17. "A Chance to Fix Parole in New York."

18. Smith, "In Wisconsin, a Backlash Against Using Data to Foretell Defendants' Futures."

19. "Quantifying Forgiveness: MLTalks with Julia Angwin and Joi Ito," https://www.youtube.com/watch?v=qjmkTGfu9Lk. Regarding Steve Jobs, see Eric Johnson, "It May Be 'Data Journalism,' but Julia Angwin's New Site the Markup Is Nothing Like FiveThirtyEight," https://www.recode.net/2018/9/27/17908798/julia-angwin-markup-jeff-larson-craig-newmark-data-investigative-journalism-peter-kafka-podcast.

20. The book is Angwin, *Dragnet Nation*.

21. Julia Angwin, personal interview, October 13, 2018.

22. Lansing, "New York State COMPAS-Probation Risk and Need Assessment Study."

23. Podkopacz, Eckberg, and Kubits, "Fourth Judicial District Pretrial Evaluation."

24. Podkopacz, "Building and Validating the 2007 Hennepin County Adult Pretrial Scale."

25. See also Harcourt, "Risk as a Proxy for Race," which argues, "Risk today has collapsed into prior criminal history, and prior criminal history has become a proxy for race. The combination of these two trends means that using risk-assessment

tools is going to significantly aggravate the unacceptable racial disparities in our criminal justice system." For a counterargument, see Skeem and Lowenkamp, "Risk, Race, and Recidivism."

26. Julia Angwin, "Keynote," Justice Codes Symposium, John Jay College, October 12, 2016, https://www.youtube.com/watch?v=WL9QkAwgqfU.

27. Julia Angwin, personal interview, October 13, 2018.

28. Angwin et al., "Machine Bias."

29. Dieterich, Mendoza, and Brennan, "COMPAS Risk Scales." See also Flores, Bechtel, and Lowenkamp, "False Positives, False Negatives, and False Analyses."

30. See "Response to ProPublica."

31. See Angwin and Larson, "ProPublica Responds to Company's Critique of Machine Bias Story," and Larson and Angwin, "Technical Response to Northpointe."

32. Angwin and Larson, "ProPublica Responds to Company's Critique of Machine Bias Story." See also Larson et al., "How We Analyzed the COMPAS Recidivism Algorithm." Note that there is a technical inaccuracy in this quotation. The measure of individuals "rated as higher risk but [who did] not re-offend" would translate mathematically to the fraction False Positives / (False Positives + True Positives), known as the False Discovery Rate. However, the statistic to which ProPublica is referring here is actually not the False Discovery Rate but rather the False Positive Rate, defined as the fraction False Positives / (False Positives + True Negatives). A better verbal translation of this quantity would involve reversing ProPublica's syntax: defendants "who did not re-offend but were rated as higher risk." For some discussion on this point, see https://twitter.com/scorbettdavies/status/842885585240956928.

33. See Dwork et al., "Calibrating Noise to Sensitivity in Private Data Analysis." Google Chrome began using differential privacy in 2014, Apple deployed it in its macOS Sierra and iOS 10 operating systems in 2016, and other tech companies have followed suit with many related ideas and implementations. In 2017, Dwork and her colleagues from the 2006 paper would share the Gödel Prize for their work.

34. Cynthia Dwork, personal interview, October 11, 2018.

35. Steel and Angwin, "On the Web's Cutting Edge, Anonymity in Name Only." See also Sweeney, "Simple Demographics Often Identify People Uniquely," which showed that the combination of date of birth, gender, and zip code was enough to uniquely identify 87 percent of Americans.

36. Moritz Hardt, personal interview, December 13, 2017.

37. See Dwork et al., "Fairness Through Awareness." For more discussion and debate on this, see., e.g., Harcourt, "Risk as a Proxy for Race," and Skeem and Lowenkamp, "Risk, Race, and Recidivism."

38. This point is discussed in Corbett-Davies, "Algorithmic Decision Making and the

Cost of Fairness," as well as Corbett-Davies and Goel, "The Measure and Mis-measure of Fairness."

39. For a recent argument on this point, see, e.g., Kleinberg et al., "Algorithmic Fairness." For discussion dating to the mid-1990s, see, e.g., Gottfredson and Jarjoura, "Race, Gender, and Guidelines-Based Decision Making."

40. Kroll et al., "Accountable Algorithms."

41. Dwork et al., "Fairness Through Awareness."

42. See, for instance, Johnson and Nissenbaum, "Computers, Ethics & Social Values."

43. See, e.g., Barocas and Selbst, "Big Data's Disparate Impact."

44. Jon Kleinberg, personal interview, July 24, 2017.

45. Alexandra Chouldechova, personal interview, May 16, 2017.

46. Sam Corbett-Davies, personal interview, May 24, 2017.

47. Goel's work showed, among other things, that being written up for so-called "furtive movements" actually made someone *less* likely to be a criminal than if they weren't—"because it suggests you didn't have anything better" to justify pulling them over. (Sharad Goel, personal interview, May 24, 2017.) See Goel, Rao, and Shroff, "Personalized Risk Assessments in the Criminal Justice System."

48. See Simoiu, Corbett-Davies, and Goel, "The Problem of Infra-Marginality in Outcome Tests for Discrimination."

49. See Kleinberg, Mullainathan, and Raghavan, "Inherent Trade-offs in the Fair Determination of Risk Scores"; Chouldechova, "Fair Prediction with Disparate Impact"; and Corbett-Davies et al., "Algorithmic Decision Making and the Cost of Fairness," respectively. See also Berk et al., "Fairness in Criminal Justice Risk Assessments."

50. Kleinberg, Mullainathan, and Raghavan, "Inherent Trade-offs in the Fair Determination of Risk Scores."

51. Alexandra Chouldechova, personal interview, May 16, 2017.

52. Sam Corbett-Davies, personal interview, May 24, 2017. Ironically, ProPublica made headlines out of this very fact; see Julia Angwin and Jeff Larson, "Bias in Criminal Risk Scores Is Mathematically Inevitable, Researchers Say," ProPublica, December 30, 2016.

53. Corbett-Davies, "Algorithmic Decision Making and the Cost of Fairness."

54. Sam Corbett-Davies, personal interview, May 24, 2017.

55. Kleinberg, Mullainathan, and Raghavan, "Inherent Trade-offs in the Fair Determination of Risk Scores."

56. For detailed discussion of fairness in a lending context in particular, see Hardt, Price, and Srebro, "Equality of Opportunity in Supervised Learning," and Lydia T. Liu, et al., "Delayed Impact of Fair Machine Learning,"as well as the interactive visualizations at http://research.google.com/bigpicture/attacking-discrimination-in-ml/ and https://bair.berkeley.edu/blog/2018/05/17/delayed-impact/.

57. Sam Corbett-Davies et al., "Algorithmic Decision Making and the Cost of Fairness" (video), https://www.youtube.com/watch?v=iFEX07OunSg.

58. Corbett-Davies, "Algorithmic Decision Making and the Cost of Fairness."

59. Corbett-Davies.

60. Tim Brennan, personal interview, November 26, 2019.

61. See Corbett-Davies and Goel, "The Measure and Mismeasure of Fairness"; see also Corbett-Davies et al., "Algorithmic Decision Making and the Cost of Fairness."

62. See, e.g., Rezaei et al., "Fairness for Robust Log Loss Classification."

63. Julia Angwin, personal interview, October 13, 2018.

64. Flores, Bechtel, and Lowenkamp, "False Positives, False Negatives, and False Analyses."

65. Tim Brennan, personal interview, November 26, 2019.

66. Cynthia Dwork, personal interview, October 11, 2018.

67. Moritz Hardt, personal interview, December 13, 2017.

68. The passage of California's SB 10 bill prompted the Partnership on AI, which represents more than 90 organizations in 13 countries, to release a detailed report calling for ten different criteria that any proposed risk-assessment model should meet. See "Report on Algorithmic Risk Assessment Tools in the U.S. Criminal Justice System."

69. This tool is called the Prisoner Assessment Tool Targeting Estimated Risk and Needs (PATTERN) and was released on July 19, 2019.

70. Alexandra Chouldechova, personal interview, May 16, 2017.

71. Burgess, "Factors Determining Success or Failure on Parole."

72. Lum and Isaac, "To Predict and Serve?"

73. "Four Out of Ten Violate Parole, Says Legislator."

74. See Ensign et al., "Runaway Feedback Loops in Predictive Policing."

75. Lum and Isaac, "To Predict and Serve?"

76. Lum and Isaac. In order to know just *how* biased the dataset is would require knowing about where all of the *unreported* crimes happened. This sounds impossible almost by definition. But Lum and Isaac had a clever way of making progress even here. Using data from the National Survey on Drug Use and Health, they were able to create a map of estimated illegal drug use at a granular, roughly block-by-block level in a city, and compare this to the record of arrests in the same city.

77. Alexandra Chouldechova, personal interview, May 16, 2017.

78. See ACLU Foundation, "The War on Marijuana in Black and White."

79. See Mueller, Gebeloff, and Chinoy, "Surest Way to Face Marijuana Charges in New York."

80. For more discussion on this line of argument, see, e.g., Sam Corbett-Davies, Sharad Goel, and Sandra González-Bailón, "Even Imperfect Algorithms Can Improve the Criminal Justice System," https://www.nytimes.com/2017/12/20/

upshot/algorithms-bail-criminal-justice-system.html; "Report on Algorithmic Risk Assessment Tools in the U.S. Criminal Justice System"; and Skeem and Lowenkamp, "Risk, Race, and Recidivism."

81. See Angwin et al., "Machine Bias." The question of the appropriateness of such tools in sentencing went all the way to the Wisconsin Supreme Court, where the use of COMPAS risk scores to inform sentencing judgments was eventually affirmed as appropriate. See *State v. Loomis;* for a summary, see https://harvardlawreview.org/2017/03/state-v-loomis/. The use of risk assessment in sentencing is a topic unto itself. Former US attorney general Eric Holder has argued, "Criminal sentences . . . should not be based on unchangeable factors that a person cannot control, or on the possibility of a future crime that has not taken place." Monahan and Skeem, "Risk Assessment in Criminal Sentencing," discusses the conflation of *blame* and *risk* in sentencing. In Skeem and Lowenkamp, "Risk, Race, and Recidivism," Lowenkamp "advises against using the PCRA [risk-assessment tool] to inform front-end sentencing decisions or back-end decisions about release without first conducting research on its use in these contexts, given that the PCRA was not designed for those purposes."

82. Harcourt, *Against Prediction.* For further discussion, see, e.g., Persico, "Racial Profiling, Fairness, and Effectiveness of Policing" and Dominitz and Knowles, "Crime Minimisation and Racial Bias."

83. Saunders, Hunt, and Hollywood, "Predictions Put into Practice." See also the reply by the Chicago Police Department: "CPD Welcomes the Opportunity to Comment on Recently Published RAND Review."

84. See also Saunders, "Pitfalls of Predictive Policing."

85. For Bernard Harcourt (in his "Risk as a Proxy for Race"), wiser parole decisions—machine-driven or otherwise—while obviously better than foolish ones, are not the primary way to address overcrowding and racial disparity in US prisons:

> What then is to be done to reduce the prison population? Rather than release through prediction, I would argue, we need to be less punitive at the front end and remain extremely conscious of the racial imbalances in our sentencing laws. Reducing the crack-cocaine disparity to 18:1 is a step in the right direction; however, other immediate steps should include eliminating mandatory minimum prison terms, reducing drug sentencing laws, substituting diversion and alternative supervision programs, and decreasing the imposition of hard time. The research suggests that shortening the length of sentences (i.e. releasing low-risk convicts earlier than the expiration of their term) would not have as great an effect on prison populations, long term, as cutting admissions. The real solution, then, is not to cut short prison terms, but to reduce admissions to prison.

86. For more on this, see Barabas et al., "Interventions over Predictions."

87. Elek, Sapia, and Keilitz, "Use of Court Date Reminder Notices to Improve Court Appearance Rates." In a 2019 development that Hardt finds particularly encouraging, Texas's Harris County, which includes Houston, approved a legal settlement that involved committing to develop a text-message-based system to remind people about upcoming scheduled court appearances. See, e.g., Gabrielle Banks, "Federal Judge Gives Final Approval to Harris County Bail Deal," *Houston Chronicle*, November 21, 2019.

88. See Mayson, "Dangerous Defendants," and Gouldin, "Disentangling Flight Risk from Dangerousness." See also "Report on Algorithmic Risk Assessment Tools in the U.S. Criminal Justice System," which argues that "tools must not conflate multiple predictions."

89. Tim Brennan, personal interview, November 26, 2019.

90. See also, e.g., Goswami, "Unlocking Options for Women," a study of women in Chicago's Cook County Jail, which concludes that judges should be empowered to "sentence women to services rather than prison."

91. Moritz Hardt, personal interview, December 13, 2017.

92. See also, e.g., Mayson, "Bias in, Bias Out," which argues, "In a racially stratified world, any method of prediction will project the inequalities of the past into the future. This is as true of the subjective prediction that has long pervaded criminal justice as of the algorithmic tools now replacing it. What algorithmic risk assessment has done is reveal the inequality inherent in all prediction, forcing us to confront a much larger problem than the challenges of a new technology."

93. Burgess, "Prof. Burgess on Parole Reform."

CHAPTER 3. TRANSPARENCY

1. Graeber, *The Utopia of Rules.*

2. Berk, *Criminal Justice Forecasts of Risk.*

3. See Cooper et al., "An Evaluation of Machine-Learning Methods for Predicting Pneumonia Mortality," and Cooper et al., "Predicting Dire Outcomes of Patients with Community Acquired Pneumonia."

4. See Caruana et al., "Intelligible Models for Healthcare."

5. Cooper et al., "Predicting Dire Outcomes of Patients with Community Acquired Pneumonia."

6. Caruana, "Explainability in Context—Health."

7. For more on decision lists, see Rivest, "Learning Decision Lists." For a more recent discussion on the use of decision lists in medicine, see Marewski and Gigerenzer, "Heuristic Decision Making in Medicine." For more on explanation in decision sets, see Lakkaraju, Bach, and Leskovec, "Interpretable Decision Sets."

8. One thing clearly missing in a system that concludes, "If the patient is asthmatic, they are low-risk" is a model of *causation*. One of the leading computer scientists in the study of causation is UCLA's Judea Pearl; for recent thoughts of his on causation in the context of contemporary machine-learning systems, see Pearl, "The Seven Tools of Causal Inference, with Reflections on Machine Learning."

9. Rich Caruana, personal interview, May 16, 2017.

10. Hastie and Tibshirani, "Generalized Additive Models." Caruana and his collaborators have also explored a slightly more complex class of models that also includes pairwise interactions, or functions of *two* variables. They call these "GA^{2}M"s, or "Generalized Additive Models plus Interactions"; see Lou et al., "Accurate Intelligible Models with Pairwise Interactions."

11. Caruana says there are a number of different reasons for this. Retirement means lifestyle changes for some, and it also means that people's income changes, their insurance and perhaps even health-care provider changes, and they may also move—all of which change their relationship to their health and their health care.

12. The generalized additive model showed risk going sharply up at 86, but then going sharply down again at 101. Caruana thinks these are purely social effects; he conjectures that around one's mid-eighties, families and caregivers are more likely to interpret health distress as a natural passing that shouldn't be fought tooth and nail. On the other hand, once someone reaches 100, one has almost the opposite impulse: "You've made it this far; we're not going to give up on you now." He notes that a doctor would presumably want to edit the graph—would decide the asthma rule makes no sense, would decide not to treat an 80-, 90-, and 100-year-old so differently. On the other hand, an *insurer* probably would *not* want to edit the graph in their model. The outcomes—from an insurer's perspective—of asthmatics really *are* better on average. This highlights both the importance of explicitly considering the different perspectives of different stakeholders in a system, and the fact that one group is using the model to make actual, real-world interventions that would, in turn, change the underlying data observed; the other is simply a passive observer. Machine learning does not intrinsically know the difference.

13. See Lou et al.

14. Schauer, "Giving Reasons."

15. David Gunning, personal interview, December 12, 2017.

16. Bryce Goodman, personal interview, January 11, 2018. The "right to an explanation" was first discussed in Goodman and Flaxman, "European Union Regulations on Algorithmic Decision-Making and a 'Right to Explanation.'" Some scholars have debated how strong this provision is; see Wachter, Mittelstadt, and Floridi, "Why a Right to Explanation of Automated Decision-Making Does Not Exist in the General Data Protection Regulation." Others have followed up on this—see, e.g., Selbst and Powles, "Meaningful Information and the Right to

Explanation"—and a certain amount of disagreement continues. The exact legal status of the "right to explanation" is likely to be clarified only incrementally in the courts.

17. Thorndike, "Fundamental Theorems in Judging Men."

18. Robyn Dawes, "Dawes Unplugged," interview by Joachim Krueger, *Rationality and Social Responsibility,* Carnegie Mellon University, January 19, 2007.

19. Sarbin, "A Contribution to the Study of Actuarial and Individual Methods of Prediction."

20. Meehl, "Causes and Effects of My Disturbing Little Book."

21. Dawes and Corrigan, "Linear Models in Decision Making," referencing Sarbin, "A Contribution to the Study of Actuarial and Individual Methods of Prediction."

22. See Dawes, "The Robust Beauty of Improper Linear Models in Decision Making."

23. See Goldberg, "Simple Models or Simple Processes?"

24. See Einhorn, "Expert Measurement and Mechanical Combination."

25. For a retrospective on Paul Meehl's book from 1986, see Meehl, "Causes and Effects of My Disturbing Little Book." For Dawes and Meehl's perspective in 1989, see Dawes, Faust, and Meehl, "Clinical Versus Actuarial Judgment." For a contemporary look at these questions, see, e.g., , Kleinberg et al., "Human Decisions and Machine Predictions."

26. Holte, "Very Simple Classification Rules Perform Well on Most Commonly Used Datasets."

27. Einhorn, "Expert Measurement and Mechanical Combination."

28. See Goldberg, "Man Versus Model of Man," and Dawes, "A Case Study of Graduate Admissions."

29. Dawes and Corrigan, "Linear Models in Decision Making." See also Wainer, "Estimating Coefficients in Linear Models," for an elaboration about equal weights in particular; as he writes, "When you are interested in prediction, it is a very rare situation that calls for regression weights which are unequal." See also Dana and Dawes, "The Superiority of Simple Alternatives to Regression for Social Science Predictions," which affirms this conclusion in a social science (and twenty-first-century) context.

30. See Dawes, "The Robust Beauty of Improper Linear Models in Decision Making."

31. Howard and Dawes, "Linear Prediction of Marital Happiness."

32. See Howard and Dawes, which references Alexander, "Sex, Arguments, and Social Engagements in Martial and Premarital Relations."

33. Indeed, Paul Meehl himself concluded that "in most practical situations, an unweighted sum of a small number of 'big' variables will, on the average, be preferable to regression equations." See Dawes and Corrigan, "Linear Models in Decision Making," for discussion and references.

34. Dawes, "The Robust Beauty of Improper Linear Models in Decision Making." See

also Wainer, "Estimating Coefficients in Linear Models": "Note also that this sort of scheme [equal weights in a linear model] works well even when an operational criterion is not available."

35. Dawes, "The Robust Beauty of Improper Linear Models in Decision Making."

36. Einhorn, "Expert Measurement and Mechanical Combination."

37. Dawes and Corrigan, "Linear Models in Decision Making."

38. See Andy Reinhardt, "Steve Jobs on Apple's Resurgence: 'Not a One-Man Show,' " *Business Week Online,* May 12, 1998, http://www.businessweek.com/bwdaily/dnflash/may1998/nf80512d.htm.

39. Holmes and Pollock, *Holmes-Pollock Letters.*

40. Angelino et al., "Learning Certifiably Optimal Rule Lists for Categorical Data." See also Zeng, Ustun, and Rudin, "Interpretable Classification Models for Recidivism Prediction"; and Rudin and Radin, "Why Are We Using Black Box Models in AI When We Don't Need To?" For another simple model that achieves similar accuracy as COMPAS, see Dressel and Farid, "The Accuracy, Fairness, and Limits of Predicting Recidivism." For further discussion, see Rudin, Wang, and Coker, "The Age of Secrecy and Unfairness in Recidivism Prediction," as well as Chouldechova, "Transparency and Simplicity in Criminal Risk Assessment."

41. Cynthia Rudin, "Algorithms for Interpretable Machine Learning" (lecture), 20th ACM SIGKIDD Conference on Knowledge Discovery and Data Mining, New York City, August 26, 2014.

42. Breiman et al., *Classification and Regression Trees.*

43. See Quinlan, *C4.5;* C4.5 also has a more recent successor algorithm, C5.0.

44. For more on $CHADS_2$, see Gage et al., "Validation of Clinical Classification Schemes for Predicting Stroke," and for more on CHA_2DS_2-VASc, see Lip et al., "Refining Clinical Risk Stratification for Predicting Stroke and Thromboembolism in Atrial Fibrillation Using a Novel Risk Factor–Based Approach."

45. See Letham et al., "Interpretable Classifiers Using Rules and Bayesian Analysis."

46. See, e.g., Veasey and Rosen, "Obstructive Sleep Apnea in Adults."

47. SLIM uses what's called "0–1 loss function" (a simple measure of how many predictions were right or wrong) and the "l_0-norm" (which attempts to minimize the number of features used), and restricts the coefficients of its feature weights to be coprime integers. See Ustun, Tracà, and Rudin, "Supersparse Linear Integer Models for Predictive Scoring Systems," and Ustun and Rudin, "Supersparse Linear Integer Models for Optimized Medical Scoring Systems." For more on their collaboration with Massachusetts General Hospital to create a sleep apnea tool, see Ustun et al., "Clinical Prediction Models for Sleep Apnea." For their work on applying similar methods in a recidivism context, see Zeng, Ustun, and Rudin, "Interpretable Classification Models for Recidivism Prediction." For more recent work, including the "certificate of optimality" for such methods, along with a

comparison against COMPAS, see Angelino et al., "Learning Certifiably Optimal Rule Lists for Categorical Data"; Ustun and Rudin, "Optimized Risk Scores"; and Rudin and Ustun, "Optimized Scoring Systems."

48. For instance, logistic regression might be used to build the model, with the coefficients rounded off afterward.

49. See Ustun and Rudin, "Supersparse Linear Integer Models for Optimized Medical Scoring Systems," for discussion and references.

50. "Information for Referring Physicians," https://www.uwhealth.org/referring-physician-news/death-rate-triples-for-sleep-apnea-sufferers/13986.

51. Ustun et al., "Clinical Prediction Models for Sleep Apnea." For a model built using SLIM that has been deployed in hospitals to assess risk of seizures, see Struck et al., "Association of an Electroencephalography-Based Risk Score With Seizure Probability in Hospitalized Patients."

52. See Kobayashi and Kohshima, "Unique Morphology of the Human Eye and Its Adaptive Meaning," and Tomasello et al., "Reliance on Head Versus Eyes in the Gaze Following of Great Apes and Human Infants."

53. Exactly how saliency should be computed is an area of active research. See, e.g., Simonyan, Vedaldi, and Zisserman, "Deep Inside Convolutional Networks"; Smilkov et al., "Smoothgrad"; Selvaraju et al., "Grad-Cam"; Sundararajan, Taly, and Yan, "Axiomatic Attribution for Deep Networks"; Erhan et al., "Visualizing Higher-Layer Features of a Deep Network"; and Dabkowski and Gal, "Real Time Image Saliency for Black Box Classifiers." And for a comparison of Jacobian- and perturbation-based saliency, in a reinforcement-learning context, see Greydanus et al., "Visualizing and Understanding Atari Agents."

There are also open research questions around the limitations and weaknesses of saliency methods. See, e.g., Kindermans et al., "The (Un)reliability of Saliency Methods"; Adebayo et al., "Sanity Checks for Saliency Maps"; and Ghorbani, Abid, and Zou, "Interpretation of Neural Networks Is Fragile."

54. As Landecker puts it: "A closer inspection of the dataset reveals that many animal images have blurry backgrounds, whereas the no-animal images tend to be in focus everywhere. This type of bias in the image is reasonable, given that all the photos were taken by professional photographers. The results of contribution propagation show us how easily an unintended bias can sneak into a dataset." See Landecker, "Interpretable Machine Learning and Sparse Coding for Computer Vision," and Landecker et al., "Interpreting Individual Classifications of Hierarchical Networks." See also discussion of a (contrived) example where a network designed to tell the difference between wolves and huskies in fact distinguishes mostly the difference between snow or grass in the background of the image: Ribeiro, Singh, and Guestrin, "Why Should I Trust You?"

55. Hilton, "The Artificial Brain as Doctor." Novoa had sent an email to his col-

leagues on January 27, 2015, saying, "If AI can differentiate between hundreds of dog breeds, I believe it could make a great contribution to dermatology." This spurred the collaboration with Ko and others. See Justin Ko, "Mountains out of Moles: Artificial Intelligence and Imaging" (lecture), Big Data in Biomedicine Conference, Stanford, CA, May 24, 2017, https://www.youtube.com/watch?v=kClvKNl0Wfc.

56. Esteva et al., "Dermatologist-Level Classification of Skin Cancer with Deep Neural Networks."

57. Ko, "Mountains out of Moles."

58. Narla et al., "Automated Classification of Skin Lesions."

59. See Caruana, "Multitask Learning"; but see also Rosenberg and Sejnowski, "NETtalk," for a precursor. See Ruder, "An Overview of Multi-Task Learning in Deep Neural Networks," for a more recent overview. This idea is also sometimes known as making a neural network with multiple "heads"—high-level outputs that share the same intermediate-level features. This is an idea that, after getting a modest amount of traction in the machine-learning community, has just recently reared its (multiple) head(s) in one of the flagship neural networks of the 2010s, AlphaGo Zero. When DeepMind iterated on their champion-dethroning AlphaGo architecture, they realized that the system they'd built could be enormously simplified by merging its two primary networks into one double-headed network. The original AlphaGo used a "policy network" to estimate what move to play in a given position, and a "value network" to estimate the degree of advantage or disadvantage for each player in that position. Presumably, DeepMind realized, the relevant intermediate-level "features"—who controlled which territory, how stable or fragile certain structures were—would be extremely similar for both networks. Why reduplicate? In their subsequent AlphaGo Zero architecture, the "policy network" and "value network" became a "policy head" and "value head" attached to the same deep network. This new, Cerberus-like network was simpler, more philosophically satisfying—and an even stronger player than the original. (Technically, Cerberus was more typically described in mythology as three-headed; his less well-known brother Orthrus was a two-headed dog who guarded the cattle of Geryon.)

60. Rich Caruana, personal interview, May 16, 2017.

61. Poplin et al., "Prediction of Cardiovascular Risk Factors from Retinal Fundus Photographs via Deep Learning."

62. Ryan Poplin, interviewed by Sam Charington, *TWiML Talk*, Episode 122, March 26, 2018.

63. Zeiler and Fergus, "Visualizing and Understanding Convolutional Networks."

64. Matthew Zeiler, "Visualizing and Understanding Deep Neural Networks by Matt Zeiler" (lecture), https://www.youtube.com/watch?v=ghEmQSxT6tw.

65. See Zeiler et al., "Deconvolutional Networks," and Zeiler, Taylor, and Fergus, "Adaptive Deconvolutional Networks for Mid and High Level Feature Learning."

66. By 2014, nearly all of the groups competing on the ImageNet benchmark were using these techniques and insights. See Simonyan and Zisserman, "Very Deep Convolutional Networks for Large-Scale Image Recognition"; Howard, "Some Improvements on Deep Convolutional Neural Network Based Image Classification"; and Simonyan, Vedaldi, and Zisserman, "Deep Inside Convolutional Networks." In 2018 and 2019, there was some internal controversy within Clarifai over whether its image-recognition software would be used for military applications; see Metz, "Is Ethical A.I. Even Possible?"

67. Inspiration for their approach included Erhan et al., "Visualizing Higher-Layer Features of a Deep Network," along with other prior and contemporaneous research; see Olah, "Feature Visualization" for a more complete history and bibliography. In practice, merely optimizing for the category label will not produce intelligible images without some further constraints or tweaks to the objective. This is a fertile area for research; see Mordvintsev, Olah, and Tyka, "Inceptionism," and Olah, Mordvintsev, and Schubert, "Feature Visualization," for discussion on this point.

68. Mordvintsev, Olah, and Tyka, "DeepDream."

69. Yahoo's model is open_nsfw, available at https://github.com/yahoo/open_nsfw. Goh's work, not suitable for children or the faint of heart, is available at https://open_nsfw.gitlab.io, and was based on the methods in Nguyen et al., "Synthesizing the Preferred Inputs for Neurons in Neural Networks via Deep Generator Networks." Goh subsequently joined Olah's Clarity team at OpenAI.

70. See Mordvintsev, Olah, and Tyka, "Inceptionism," and Mordvintsev, Olah, and Tyka, "DeepDream."

71. See Olah, Mordvintsev, and Schubert, "Feature Visualization"; Olah et al., "The Building Blocks of Interpretability"; and Carter et al., "Activation Atlas." More recent work includes detailed "microscopy" of cornerstone deep-learning models like AlexNet; see, e.g., https://microscope.openai.com/models/alexnet.

72. Chris Olah, personal interview, May 4, 2020. For more, see his "Circuits" collaboration: https://distill.pub/2020/circuits/.

73. The journal is *Distill*, available at https://distill.pub. For Olah's thoughts on the founding of Distill, see https://colah.github.io/posts/2017-03-Distill/ and https://distill.pub/2017/research-debt/.

74. Olah et al., "The Building Blocks of Interpretability."

75. Been Kim, personal interview, June 1, 2018.

76. See also Doshi-Velez and Kim, "Towards a Rigorous Science of Interpretable Machine Learning," and Lage et al., "Human-in-the-Loop Interpretability Prior."

77. See Poursabzi-Sangdeh et al., "Manipulating and Measuring Model Interpretability."

78. See https://github.com/tensorflow/tcav.

79. Kim et al., "Interpretability Beyond Feature Attribution."

80. This produces concept vectors not unlike those we saw in our discussion of word-2vec in Chapter 1. For a related approach, see also Fong and Vedaldi, "Net2Vec."

81. Been Kim, "Interpretability Beyond Feature Attribution" (lecture), MLconf 2018, San Francisco, November 14, 2018, https://www.youtube.com/watch?v=Ff-Dx79QEEY.

82. Been Kim, "Interpretability Beyond Feature Attribution.".

83. See Mordvintsev, Olah, and Tyka, "Inceptionism," and Mordvintsev, Olah, and Tyka, "DeepDream."

84. See https://results.ittf.link.

85. Stock and Cisse, "ConvNets and Imagenet Beyond Accuracy."

CHAPTER 4. REINFORCEMENT

1. Skinner, "Reinforcement Today."

2. Arendt, *The Human Condition.*

3. For Stein's undergraduate research, see Solomons and Stein, "Normal Motor Automatism." For a review, authored by none other than B. F. Skinner, linking her famous book to her earlier psychological research, see Skinner, "Has Gertrude Stein a Secret?" For some brief reflections of Stein's on this time in her life, see Stein, *The Autobiography of Alice B. Toklas.* For more on Stein's life and influences, see Brinnin, *The Third Rose.*

4. Jončich, *The Sane Positivist.* See also Brinnin, *The Third Rose.*

5. Jončich.

6. Thorndike, "Animal Intelligence."

7. Thorndike, *The Psychology of Learning.*

8. Thorndike has, of course, predecessors as well as successors; earlier foreshadowings of the law of effect can be found in the work of Scottish philosopher Alexander Bain, who discussed learning by "groping experiment" and "the grand process of trial and error"—appearing to coin what is now a commonplace turn of phrase—in his 1855 *The Senses and the Intellect.* Conway Lloyd Morgan, just a couple years before Thorndike's work at Harvard, in his 1894 *Introduction to Comparative Psychology* discussed "trial and error" in the context of animal behavior. For a short history of animal learning from the perspective of reinforcement learning, see Sutton and Barto, *Reinforcement Learning.*

9. See Thorndike, "A Theory of the Action of the After-Effects of a Connection upon It," and Skinner, "The Rate of Establishment of a Discrimination," respectively. For discussion, see Wise, "Reinforcement."

10. Tolman, "The Determiners of Behavior at a Choice Point."

11. See Jončich, *The Sane Positivist,* as well as Cumming, "A Review of Geraldine Jončich's *The Sane Positivist: A Biography of Edward L. Thorndike.*"

12. Thorndike, "A Theory of the Action of the After-Effects of a Connection upon It."

13. Turing, "Intelligent Machinery."

14. "Heuristics."

15. Samuel, "Some Studies in Machine Learning Using the Game of Checkers."

16. McCarthy and Feigenbaum, "In Memoriam." Samuel's television demonstration took place on February 24, 1956.

17. Edward Thorndike, letter to William James, October 26, 1908; found in Jonçich, *The Sane Positivist.*

18. Rosenblueth, Wiener, and Bigelow, "Behavior, Purpose and Teleology." The *Oxford English Dictionary* distinguishes the word used to mean "the return of a fraction of the output signal" from the meaning of "the modification, adjustment, or control of a process or system . . . by a result or effect of the process," and cites Rosenblueth, Wiener, and Bigelow as the first known instance in print of the latter sense.

19. The word "cybernetics" sounds, to contemporary ears, both futuristic and retro; it calls to mind Flash Gordon and the era of Baby Boomer sci-fi. In fact, the word is decidedly *non*fictional, and it's nowhere near as alien as it sounds. Wiener was looking for a term to capture the ideas of self-regulation and feedback in both living and mechanical systems. "After much consideration," he wrote, "we have come to the conclusion that all the existing terminology has too heavy a bias to one side or another to serve the future development of the field as well as it should; and as happens so often to scientists, we have been forced to coin at least one artificial neo-Greek expression to fill the gap" (Wiener, *Cybernetics*). He found an etymological root he liked in the Greek κυβερνήτης—or, in the Roman alphabet, *kybernetes*—from the word for "steersman," "captain," or "governor." In fact, the English word "governor" itself derives, with a distorted spelling (thought to be courtesy of Etruscan), from *kybernetes*. As with many new coinages, there was a bit of orthographic diversity early on; for instance, a technical book was published in London in 1960 with an alternate spelling: Stanley-Jones and Stanley-Jones, *Kybernetics of Natural Systems.* ("As regards the spelling of the word, . . . I have preferred Kybernetics, on etymological grounds.") In fact, English uses of the term predate Wiener: James Clerk Maxwell used it in 1868 to describe electrical "governors"—a conscious allusion of Wiener's—and before that (and unbeknownst to Wiener initially), André-Marie Ampère used it in 1834, with its nautical steering connotations, to refer to governance in the context of social science and political power. According to Ampère, this ship-to-polis figurative use existed even in the original Greek. See Maxwell, "On Governors," and Ampère, *Essai sur la philosophie des sciences; ou, Exposition analytique d'une classification naturelle de toutes les connaissances humaines,* respectively.

20. Wiener, *Cybernetics.*

21. Rosenblueth, Wiener, and Bigelow, "Behavior, Purpose and Teleology."

22. Klopf, *Brain Function and Adaptive Systems: A Heterostatic Theory*. The idea of "hedonistic neurons" appears in slightly different forms throughout the history of machine learning. See, e.g., the "SNARC" system discussed in Minsky, "Theory of Neural-Analog Reinforcement Systems and Its Application to the Brain Model Problem" for an early example, and Chapter 15 of Sutton and Barto, *Reinforcement Learning* for discussion.

23. Andrew G. Barto, "Reinforcement Learning: A History of Surprises and Connections" (lecture), July 19, 2018, International Joint Conference on Artificial Intelligence, Stockholm, Sweden.

24. Andrew Barto, personal interview, May 9, 2018.

25. The canonical text about reinforcement learning is Sutton and Barto, *Reinforcement Learning*, recently updated into a second edition. For a summary of the field up to the mid-1990s, see also Kaelbling, Littman, and Moore, "Reinforcement Learning."

26. Richard Sutton defines and discusses this idea at http://incompleteideas.net/rlai .cs.ualberta.ca/RLAI/rewardhypothesis.html, and it also appears in Sutton and Barto, *Reinforcement Learning*. Sutton says he first heard it from Brown University computer scientist Michael Littman; Littman thinks he heard it first from *Sutton*. But the earliest reference seems to be a lecture that Littman gave in the early 2000s, where he argued that "intelligent behavior arises from the actions of an individual seeking to maximize its received reward signals in a complex and changing world." For Littman's recollections of this history, see "Michael Littman: The Reward Hypothesis" (lecture), University of Alberta, October 16, 2019, available at https://www.coursera.org/lecture/fundamentals-of-reinforcement-learning/ michael-littman-the-reward-hypothesis-q6x0e.

    Despite the recency of this particular framing, the idea of understanding behavior as motivated, whether explicitly or implicitly, by some form of quantifiable rewards is one that has broad connections to utility theory. See, e.g., Bernouilli, "Specimen theoriae novae de mensura sortis," Samuelson, "A Note on Measurement of Utility," and von Neumann and Morgenstern, *Theory of Games and Economic Behavior*.

27. Richard Sutton, "Introduction to Reinforcement Learning" (lecture), University of Texas at Austin, January 10, 2015.

28. "There are only three possible comparisons between any two [scalar] numbers," says Chang. "One number is greater, lesser, or equal to the other. Not so with values. As post-Enlightenment creatures, we tend to assume that scientific thinking holds the key to everything of importance in our world, but the world of value is different from the world of science. The stuff of the one world can be quantified by real numbers. The stuff of the other world can't. We shouldn't assume that the world of *is*—of lengths and weights—has the same structure as the world of *ought*—of what we should do." See Ruth Chang, "How to

Make Hard Choices" (lecture), TEDSalon NY2014: https://www.ted.com/talks/ruth_chang_how_to_make_hard_choices.

29. The idea of reinforcement learning as "learning with a critic" appears to date back at least as far as Widrow, Gupta, and Maitra, "Punish/Reward."

30. You can think of an algorithm like backpropagation as solving the credit-assignment problem *structurally*, rather than *temporally*. As Sutton put it in "Learning to Predict by the Methods of Temporal Differences," "The purpose of both backpropagation and TD methods is accurate credit assignment. Backpropagation decides which part(s) of a network to change so as to influence the network's output and thus to reduce its overall error, whereas TD methods decide how each output of a temporal sequence of outputs should be changed. Backpropagation addresses a *structural* credit-assignment issue whereas TD methods address a *temporal* credit-assignment issue."

31. Olds, "Pleasure Centers in the Brain," 1956.

32. Olds and Milner, "Positive Reinforcement Produced by Electrical Stimulation of Septal Area and Other Regions of Rat Brain."

33. See Olds, "Pleasure Centers in the Brain," 1956, and Olds, "Pleasure Centers in the Brain," 1970.

34. Corbett and Wise, "Intracranial Self-Stimulation in Relation to the Ascending Dopaminergic Systems of the Midbrain."

35. Schultz, "Multiple Dopamine Functions at Different Time Courses," estimates there to be approximately 400,000 dopamine neurons in the human brain, out of approximately 80 to 100 billion neurons total.

36. Bolam and Pissadaki, "Living on the Edge with Too Many Mouths to Feed."

37. Bolam and Pissadaki.

38. Glimcher, "Understanding Dopamine and Reinforcement Learning."

39. Wise et al., "Neuroleptic-Induced 'Anhedonia' in Rats."

40. Wise, "Neuroleptics and Operant Behavior." For a rather comprehensive history of the "anhedonia hypothesis" and of the earlier discoveries of brain "pleasure centers," as well as the later discovery that dopamine was centrally involved, see Wise, "Dopamine and Reward."

41. Quoted in Wise, "Dopamine and Reward."

42. Romo and Schultz, "Dopamine Neurons of the Monkey Midbrain."

43. Romo and Schultz.

44. Wolfram Schultz, personal interview, June 25, 2018.

45. See, e.g., Schultz, Apicella, and Ljungberg, "Responses of Monkey Dopamine Neurons to Reward and Conditioned Stimuli During Successive Steps of Learning a Delayed Response Task," and Mirenowicz and Schultz, "Importance of Unpredictability for Reward Responses in Primate Dopamine Neurons."

46. See Rescorla and Wagner, "A Theory of Pavlovian Conditioning"; the idea that

learning might only occur when the results are surprising comes from the earlier Kamin, "Predictability, Surprise, Attention, and Conditioning."

47. Wolfram Schultz, personal interview, June 25, 2018.

48. Wolfram Schultz, personal interview, June 25, 2018. See Schultz, Apicella, and Ljungberg, "Responses of Monkey Dopamine Neurons to Reward and Conditioned Stimuli During Successive Steps of Learning a Delayed Response Task."

49. Quoted in Brinnin, *The Third Rose*.

50. Barto, Sutton, and Anderson, "Neuronlike Adaptive Elements That Can Solve Difficult Learning Control Problems."

51. "Rich is kind of the predictor guy, and I'm more the actor guy" (Andrew Barto, personal interview, May 9, 2018).

52. Sutton, "A Unified Theory of Expectation in Classical and Instrumental Conditioning."

53. Sutton, "Temporal-Difference Learning" (lecture), July 3, 2017, Deep Learning and Reinforcement Learning Summer School 2017, Université de Montréal, July 3, 2017, http://videolectures.net/deeplearning2017_sutton_td_learning/.

54. Sutton, "Temporal-Difference Learning."

55. Sutton, "Learning to Predict by the Methods of Temporal Differences." See also Sutton's PhD thesis: "Temporal Credit Assignment in Reinforcement Learning."

56. See Watkins, "Learning from Delayed Rewards" and Watkins and Dayan, "Q-Learning."

57. Tesauro, "Practical Issues in Temporal Difference Learning."

58. Tesauro, "TD-Gammon, a Self-Teaching Backgammon Program, Achieves Master-Level Play." See also Tesauro, "Temporal Difference Learning and TD-Gammon."

59. "Interview with P. Read Montague," Cold Spring Harbor Symposium Interview Series, Brains and Behavior, https://www.youtube.com/watch?v=mx96DYQIS_s.

60. Peter Dayan, personal interview, March 12, 2018.

61. Schultz, Dayan, and Montague, "A Neural Substrate of Prediction and Reward." The breakthrough connection to TD-learning had appeared the year prior in Montague, Dayan, and Sejnowski, "A Framework for Mesencephalic Dopamine Systems Based on Predictive Hebbian Learning."

62. P. Read Montague, "Cold Spring Harbor Laboratory Keynote," https://www.youtube.com/watch?v=RJvpu8nYzFg.

63. "Interview with P. Read Montague," Cold Spring Harbor Symposium Interview Series, Brains and Behavior https://www.youtube.com/watch?v=mx96DYQIS_s.

64. Peter Dayan, personal interview, March 12, 2018.

65. Wolfram Schultz, personal interview, June 25, 2018.

66. See, e.g., Niv, "Reinforcement Learning in the Brain."

67. Niv.

68. For a discussion of potential limitations to the TD-error theory of dopamine, see,

e.g., Dayan and Niv, "Reinforcement Learning," and O'Doherty, "Beyond Simple Reinforcement Learning."

69. Niv, "Reinforcement Learning in the Brain."

70. Yael Niv, personal interview, February 21, 2018.

71. Lenson, *On Drugs*.

72. See, e.g., Berridge, "Food Reward: Brain Substrates of Wanting and Liking," and Berridge, Robinson, and Aldridge, "Dissecting Components of Reward."

73. Rutledge et al., "A Computational and Neural Model of Momentary Subjective Well-Being."

74. Rutledge et al.

75. See Brickman, "Hedonic Relativism and Planning the Good Society," and Frederick and Loewenstein, "Hedonic Adaptation."

76. Brickman, Coates, and Janoff-Bulman, "Lottery Winners and Accident Victims."

77. "Equation to Predict Happiness," https://www.ucl.ac.uk/news/2014/aug/equation-predict-happiness.

78. Rutledge et al., "A Computational and Neural Model of Momentary Subjective Well-Being."

79. Wency Leung, "Researchers Create Formula That Predicts Happiness," https://www.theglobeandmail.com/life/health-and-fitness/health/researchers-create-formula-that-predicts-happiness/article19919756/.

80. See Tomasik, "Do Artificial Reinforcement-Learning Agents Matter Morally?" For more on this topic, see also Schwitzgebel and Garza, "A Defense of the Rights of Artificial Intelligences."

81. Brian Tomasik, "Ethical Issues in Artificial Reinforcement Learning," https://reducing-suffering.org/ethical-issues-artificial-reinforcement-learning/.

82. Daswani and Leike, "A Definition of Happiness for Reinforcement Learning Agents." See also People for the Ethical Treatment of Reinforcement Learners: http://petrl.org.

83. Andrew Barto, personal interview, May 9, 2018.

84. There is more to both dopamine and TD learning in the brain; dopamine is connected, for instance, to motor movements and motor conditions like Parkinson's. And dopamine appears more intimately involved in *positive* prediction errors than in negative ones. There appears to be a different wiring entirely, for instance, when it comes to "aversive" stimuli: things that are threatening or disgusting or poisonous.

85. See Athalye et al., "Evidence for a Neural Law of Effect."

86. Andrew Barto, personal interview, May 9, 2018.

87. For more on the idea of a universal definition of intelligence, see, e.g., Legg and Hutter, "Universal Intelligence" and "A Collection of Definitions of Intelligence," and Legg and Veness, "An Approximation of the Universal Intelligence Measure."

88. McCarthy, "What Is Artificial Intelligence?"

89. As Schultz, Dayan, and Montague, "A Neural Substrate of Prediction and Reward," put it, "Without the capacity to discriminate which stimuli are responsible for fluctuations in a broadcast scalar error signal, an agent may learn inappropriately, for example, it may learn to approach food when it is actually thirsty."

CHAPTER 5. SHAPING

1. Bentham, *An Introduction to the Principles of Morals and Legislation.*

2. Matarić, "Reward Functions for Accelerated Learning."

3. Skinner, "Pigeons in a Pelican." See also Skinner, "Reinforcement Today."

4. Skinner, "Pigeons in a Pelican."

5. Ferster and Skinner, *Schedules of Reinforcement.* For Charles Ferster's recollections of working with Skinner during this period, see Ferster, "Schedules of Reinforcement with Skinner."

6. Bailey and Gillaspy, "Operant Psychology Goes to the Fair."

7. Bailey and Gillaspy.

8. The Brelands were able to train more than six thousand animals of species, "and we have dared to tackle such unlikely subjects as reindeer, cockatoos, raccoons, porpoises, and whales." However, they began to encounter certain repeated limitations in their ability to condition particular behaviors in animals, concluding that behaviorism as a theory failed to take adequate account of animals' instinctive, evolved, species-specific behaviors and inclinations. See Breland and Breland, "The Misbehavior of Organisms."

9. Skinner, "Reinforcement Today" (emphasis in the original).

10. Skinner, "Pigeons in a Pelican."

11. Skinner, "Pigeons in a Pelican."

12. Skinner, "How to Teach Animals," 1951, which appears to be the earliest appearance of the verb "shaping" in a reinforcement context.

13. Skinner discusses this event in numerous places in his writing. See his "Reinforcement Today," "Some Relations Between Behavior Modification and Basic Research," *The Shaping of a Behaviorist,* and *A Matter of Consequences.* See also Peterson, "A Day of Great Illumination."

14. Skinner, "How to Teach Animals."

15. As Skinner put it: "A familiar problem is that of the child who seems to take an almost pathological delight in annoying its parents. In many cases this is the result of conditioning which is very similar to the animal training we have discussed." See Skinner, "How to Teach Animals."

16. Skinner, "How to Teach Animals."

17. The earliest appearance of this quote is Spielvogel, "Advertising," many years

after Edison's death. For more on its history and variations, see O'Toole, "There's a Way to Do It Better—Find It."

18. Bain, *The Senses and the Intellect.*

19. Michael Littman, personal interview, February 28, 2018.

20. Explicit mention of "shaping," by name, in a robotics context came in Singh, "Transfer of Learning by Composing Solutions of Elemental Sequential Tasks"; over the 1990s, it became an increasingly prevalent topic within the robotics community, with many researchers looking explicitly to the animal training and instrumental conditioning literature for inspiration. See, e.g., Colombetti and Dorigo, "Robot Shaping"; Saksida, Raymond, and Touretzky, "Shaping Robot Behavior Using Principles from Instrumental Conditioning"; and Savage, "Shaping."

21. Skinner, "Reinforcement Today."

22. Shigeru Miyamoto, "Iwata Asks: New Super Mario Bros. Wii," interview by Satoru Iwata, November 25, 2009, https://www.nintendo.co.uk/Iwata-Asks/Iwata-Asks-New-Super-Mario-Bros-Wii/Volume-1/4-Letting-Everyone-Know-It-Was-A-Good-Mushroom/4-Letting-Everyone-Know-It-Was-A-Good-Mushroom-210863.html.

23. For more on the idea of a machine-learning approach to a learning "curriculum," see, e.g., Bengio et al., "Curriculum Learning."

24. Selfridge, Sutton, and Barto, "Training and Tracking in Robotics."

25. Elman, "Learning and Development in Neural Networks." However, see also, e.g., Rohde and Plaut, "Language Acquisition in the Absence of Explicit Negative Evidence," which reported different findings from Elman's in this regard.

26. This particular experiment was noteworthy for the way that the pig's performance deteriorated over time, something that challenged classical models of behaviorism. See Breland and Breland, "The Misbehavior of Organisms."

27. Florensa et al., "Reverse Curriculum Generation for Reinforcement Learning." In 2018, a team of researchers at OpenAI did something similar to train reinforcement-learning agents on particularly difficult video games. They would record a competent human player playing the game, and then they would build a curriculum by working *backward* in time through this recorded demo. First they would train their agent by starting at the very brink of success, and then they would gradually move it backward, ultimately to the start of the game. See Salimans and Chen, "Learning Montezuma's Revenge from a Single Demonstration." See also Hosu and Rebedea, "Playing Atari Games with Deep Reinforcement Learning and Human Checkpoint Replay"; Nair et al., "Overcoming Exploration in Reinforcement Learning with Demonstrations"; and Peng et al., "DeepMimic." There are also connections here more broadly to imitation learning, which we discuss in Chapter 7.

28. See, e.g., Ashley, *Chess for Success.*

29. This is difficult to confirm, but seems extremely likely. See, e.g., Edward Winter, "Chess Book Sales," http://www.chesshistory.com/winter/extra/sales.html, for more information on chess book sales.

30. See, e.g., Graves et al., "Automated Curriculum Learning for Neural Networks." There are connections here to the work in rewarding learning progress that we discuss in Chapter 6. For earlier machine-learning work on curriculum design, see, e.g., Bengio et al., "Curriculum Learning."

31. David Silver, "AlphaGo Zero: Starting from Scratch," October 18, 2017, https://www.youtube.com/watch?v=tXlM99xPQC8.

32. Kerr, "On the Folly of Rewarding A, While Hoping for B."

33. Kerr. Note that it is "immortality" (*sic*) in the original 1975 printing!

34. The article is signed "The Editors," but their names are Kathy Dechant and Jack Veiga; see Dechant and Veiga, "More on the Folly."

35. For several cautionary tales about "gamified" incentives, see, e.g., Callan, Bauer, and Landers, "How to Avoid the Dark Side of Gamification."

36. Kerr, "On the Folly of Rewarding A, While Hoping for B."

37. Wright et al., "40 Years (and Counting)."

38. "Operant Conditioning," https://www.youtube.com/watch?v=I_ctJqjlrHA.

39. See Joffe-Walt, "Allowance Economics," and Gans, *Parentonomics*.

40. Tom Griffiths, personal interview, June 13, 2018.

41. Andre and Teller, "Evolving Team Darwin United."

42. Cited in Ng, Harada, and Russell, "Policy Invariance Under Reward Transformations," as personal communication with the authors.

43. Randløv and Alstrøm, "Learning to Drive a Bicycle Using Reinforcement Learning and Shaping."

44. Russell tells me that this came about from thinking hard during the 1990s about meta-reasoning: the right way to think about thinking. When you play a game—for instance, chess—you win because of the *moves* you chose, but it was the *thoughts* you had that enabled you to choose those moves. Indeed, sometimes we reflect on a game and think, "Ah, I went wrong because I boxed in my knight. I need to keep the knight away from the edge of the board." But sometimes we think, "Ah, I went wrong because I didn't trust my instincts. I overthought it; I need to play more organically and intuitively." Figuring out how an aspiring chess player—or *any* kind of agent—should learn about its *thought* process seemed like a more important but also dramatically harder task than simply learning how to pick good moves. Maybe shaping could help.

    "So one natural answer is . . . if you do a computation that changes your mind about what is a good move to make, then clearly that seems like it was a worthwhile computation," Russell says. "And so you could reward that computation by how much you changed your mind."

He adds, "Now, here's the tricky part: so you could change your mind in the sense of discovering that what was the second-best move is actually even better than what was the best move.

"And so you get a bonus reward for doing that. You used to have a move that you thought was worth, 50 was your best and 48 was your second best. Now that 48 becomes your 52, see. Going from 50 to 52 is positive. Well, what if instead you thought about 50 and realized it was only a 6? And so now your best move is the 48, what was your second-best move. Should that be a positive reward or a negative reward? And again, you would think it should be a positive reward because you've done some thinking, and that thinking is worthwhile because it helped you realize that what you thought you were going to do wasn't as good. And you're saving yourself from a disaster. But if you give yourself a positive reward for that as well, right? You'd be giving yourself only a positive rewards all the way along, right? And what you would end up learning to do is not winning the game but changing your mind all the time.

"And so something wasn't right there. That got me this idea that you have to arrange these internal pseudorewards so that along a path, they add up to the same as the true, eventually. Balance the books" (Stuart Russell, personal interview, May 13, 2018).

45. Andrew Ng, "The Future of Robotics and Artificial Intelligence" (lecture), May 21, 2011, https://www.youtube.com/watch?v=AY4ajbu_G3k.

46. See Ng et al., "Autonomous Helicopter Flight via Reinforcement Learning," as well as Schrage et al., "Instrumentation of the Yamaha R-50/RMAX Helicopter Testbeds for Airloads Identification and Follow-on Research." For follow-up work, see Ng et al., "Autonomous Inverted Helicopter Flight via Reinforcement Learning," and Abbeel et al., "An Application of Reinforcement Learning to Aerobatic Helicopter Flight."

47. Ng, "Shaping and Policy Search in Reinforcement Learning." See also Wiewiora, "Potential-Based Shaping and Q-Value Initialization Are Equivalent," which argues that it is possible to use shaping in setting the *initial* state of the agent, while leaving the actual rewards themselves unchanged, and achieve the same results.

48. Ng, "Shaping and Policy Search in Reinforcement Learning." This also appears, verbatim, in Ng, Harada, and Russell, "Policy Invariance Under Reward Transformations."

49. "A conservative field means if you take any path that gets you back to the same state, the total, the integral $v \cdot ds$, is zero" (Stuart Russell, personal interview, May 13, 2018).

50. Russell and Norvig, *Artificial Intelligence*.

51. Ng, Harada, and Russell, "Policy Invariance Under Reward Transformations."

52. Spignesi, *The Woody Allen Companion*.

53. For an evolutionary psychology perspective, see, e.g., Al-Shawaf et al., "Human Emotions: An Evolutionary Psychological Perspective," and Miller, "Reconciling Evolutionary Psychology and Ecological Psychology."

54. Michael Littman, personal interview, February 28, 2018. The paper is Sutton, "Learning to Predict by the Methods of Temporal Differences."

55. Ackley and Littman, "Interactions Between Learning and Evolution."

56. Training systems which themselves are (or may become) optimizers of some "inner" reward function is a source of concern and of active research among contemporary AI-safety researchers. See Hubinger et al., "Risks from Learned Optimization in Advanced Machine Learning Systems."

57. Andrew Barto, personal interview, May 9, 2018.

58. See Singh, Lewis, and Barto, "Where Do Rewards Come from?," as well as Sorg, Singh, and Lewis, "Internal Rewards Mitigate Agent Boundedness."

59. Sorg, Singh, and Lewis, "Internal Rewards Mitigate Agent Boundedness." The answer to that question is yes—but only under some very strong assumptions. In particular, only if our agent's time and computing power are *unlimited*. Otherwise, we are better off if we do *not* make its goal our own. This has the flavor of paradox. Our own goals are better served by telling the agent to do something *else*.

60. Singh et al., "On Separating Agent Designer Goals from Agent Goals."

61. For more on the optimal reward problem, see Sorg, Lewis, and Singh, "Reward Design via Online Gradient Ascent," as well as Sorg's PhD dissertation, "The Optimal Reward Problem: Designing Effective Reward for Bounded Agents." For more recent progress on learning optimal rewards for an RL agent, see Zheng, Oh, and Singh, "On Learning Intrinsic Rewards for Policy Gradient Methods."

62. See "Workplace Procrastination Costs British Businesses £76 Billion a Year," *Global Banking & Finance Review,* https://www.globalbankingandfinance.com/workplace-procrastination-costs-british-businesses-76-billion-a-year/#_ftn1. For a broad look at the costs and causes of procrastination, see Steel, "The Nature of Procrastination."

63. Skinner, "A Case History in Scientific Method."

64. Jane McGonigal, "Gaming Can Make a Better World," https://www.ted.com/talks/jane_mcgonigal_gaming_can_make_a_better_world/.

65. See McGonigal, *SuperBetter.*

66. Jane McGonigal, "The Game That Can Give You 10 Extra Years of Life," https://www.ted.com/talks/jane_mcgonigal_the_game_that_can_give_you_10_extra_years_of_life/.

67. See, e.g., Deterding et al., "From Game Design Elements to Gamefulness."

68. See, e.g., Hamari, Koivisto, and Sarsa, "Does Gamification Work?"

69. Falk Lieder, personal interview, April 18, 2018.

70. See Lieder, "Gamify Your Goals," for a general overview and Lieder et al., "Cognitive Prostheses for Goal Achievement," for more details.

71. This idea was also more recently explored in Sorg, Lewis, and Singh, "Reward Design via Online Gradient Ascent."

72. Falk Lieder, personal interview, April 18, 2018.

73. Specifically, they were given the choice between declining the task and receiving fifteen cents, or accepting the task and receiving five cents *plus* the ability to write a set of essays by a given deadline for $20.

74. Lieder et al., "Cognitive Prostheses for Goal Achievement."

75. Lieder et al.

76. See, for instance, Evans et al., "Evidence for a Mental Health Crisis in Graduate Education," which found that "graduate students are more than six times as likely to experience depression and anxiety as compared to the general population."

CHAPTER 6. CURIOSITY

1. Turing, "Intelligent Machinery."

2. There were efforts starting in 2004 to develop standardized RL benchmarks and competitions; see Whiteson, Tanner, and White, "The Reinforcement Learning Competitions."

3. Marc Bellemare, personal interview, February 28, 2019.

4. Bellemare et al., "The Arcade Learning Environment," stemming originally from Naddaf, "Game-Independent AI Agents for Playing Atari 2600 Console Games," and before that from Diuk, Cohen, and Littman, "An Object-Oriented Representation for Efficient Reinforcement Learning," which used the game *Pitfall!* as an environment for reinforcement learning.

5. See Gendron-Bellemare, "Fast, Scalable Algorithms for Reinforcement Learning in High Dimensional Domains."

6. Mnih et al., "Playing Atari with Deep Reinforcement Learning."

7. Mnih et al., "Human-Level Control Through Deep Reinforcement Learning."

8. Robert Jaeger, interviewed by John Hardie, http://www.digitpress.com/library/interviews/interview_robert_jaeger.html.

9. For more on this point, see, e.g., Salimans and Chen, "Learning Montezuma's Revenge from a Single Demonstration," which also explores the interesting idea of working *backward* from a successful goal state to teach the RL agent how to play the game step by step.

10. See Maier and Seligman, "Learned Helplessness." For more recent formal work in this area, see, e.g., Lieder, Goodman, and Huys, "Learned Helplessness and Generalization."

11. See Henry Alford, "The Wisdom of Ashleigh Brilliant," http://www.

ashleighbrilliant.com/BrilliantWisdom.html, excerpted from Alford, *How to Live* (New York: Twelve, 2009).

12. The notion of intrinsic motivation was introduced into machine learning with Barto, Singh, and Chentanez, "Intrinsically Motivated Learning of Hierarchical Collections of Skills," and Singh, Chentanez, and Barto, "Intrinsically Motivated Reinforcement Learning." For a more recent overview of this literature, see Baldassarre and Mirolli, *Intrinsically Motivated Learning in Natural and Artificial Systems.*

13. Hobbes, *Leviathan.*

14. Simon, "The Cat That Curiosity Couldn't Kill."

15. Berlyne, " 'Interest' as a Psychological Concept."

16. See Furedy and Furedy, " 'My First Interest Is Interest.' "

17. Berlyne, *Conflict, Arousal, and Curiosity.*

18. See Harlow, Harlow, and Meyer, "Learning Motivated by a Manipulation Drive," and Harlow, "Learning and Satiation of Response in Intrinsically Motivated Complex Puzzle Performance by Monkeys."

19. Scenarios of this type are described in Barto, "Intrinsic Motivation and Reinforcement Learning," and Deci and Ryan, *Intrinsic Motivation and Self-Determination in Human Behavior.*

20. Berlyne, *Conflict, Arousal, and Curiosity.*

21. And see, for instance, Berlyne's own "Uncertainty and Conflict: A Point of Contact Between Information-Theory and Behavior-Theory Concepts."

22. For a twenty-first-century overview of "interest" as a psychological subject, see, e.g., Silvia, *Exploring the Psychology of Interest,* and Kashdan and Silvia, "Curiosity and Interest."

23. Konečni, "Daniel E. Berlyne."

24. Berlyne, *Conflict, Arousal, and Curiosity.*

25. *Klondike Annie,* 1936.

26. Fantz, "Visual Experience in Infants." Technically speaking, Fantz's affiliation was "Western Reserve University," as it would not officially federate with the Case Institute of Technology until a few years later, in 1967, to become the university we know today as Case Western.

27. See Saayman, Ames, and Moffett, "Response to Novelty as an Indicator of Visual Discrimination in the Human Infant."

28. For a turn-of-the-century overview, see Roder, Bushnell, and Sasseville, "Infants' Preferences for Familiarity and Novelty During the Course of Visual Processing."

29. Marvin Minsky, for instance, had written in 1961, "If we could . . . add a premium to reinforcement of those predictions which have a novel aspect, we might expect to discern behavior motivated by a sort of curiosity. . . . In the reinforcement of mechanisms for confirmed novel expectations . . . we may find the key

to simulation of intellectual motivation." See Minsky, "Steps Toward Artificial Intelligence."

30. See Sutton, "Integrated Architectures for Learning, Planning, and Reacting Based on Approximating Dynamic Programming" and "Reinforcement Learning Architectures for Animats." MIT's Leslie Pack Kaelbling devised a similar method, based on the idea of measuring an agent's "confidence intervals" around the rewards for certain actions; see Kaelbling, *Learning in Embedded Systems*. The wider the confidence interval, the more uncertain the agent was about that action; her idea was to likewise reward the agent for doing things about which it was most uncertain. See also Strehl and Littman, "An Analysis of Model-Based Interval Estimation for Markov Decision Processes," which follows in this vein.

31. Berlyne, *Conflict, Arousal, and Curiosity*.

32. If each of the 9 spaces can be X, O, or empty, this puts an upper bound of 3^9, or 19,683. The actual number will be smaller than that, of course, as not all of these positions are legal (a board of 9 X's, for instance, could never happen in the course of play).

33. Bellemare et al., "Unifying Count-Based Exploration and Intrinsic Motivation," which was inspired in part by Strehl and Littman, "An Analysis of Model-Based Interval Estimation for Markov Decision Processes." See also the follow-up paper by Ostrovski et al., "Count-Based Exploration with Neural Density Models." For a related approach using hash functions, see Tang et al., "# Exploration." For another related approach using exemplar models, see Fu, Co-Reyes, and Levine, "EX²."

34. Marc G. Bellemare, "The Role of Density Models in Reinforcement Learning" (lecture), DeepHack.RL, February 9, 2017, https://www.youtube.com/watch?v=qSfd27AgcEk.

35. In fact, there is considerable and clever mathematical nuance in going from the probability to the pseudo-count. For more, see Bellemare et al., "Unifying Count-Based Exploration and Intrinsic Motivation."

36. Berlyne, *Conflict, Arousal, and Curiosity*.

37. Gopnik, "Explanation as Orgasm and the Drive for Causal Knowledge."

38. For a computational view on the difference between novelty and surprise, see Barto, Mirolli, and Baldassarre, "Novelty or Surprise?"

39. Schulz and Bonawitz, "Serious Fun."

40. "Curiosity and Learning: The Skill of Critical Thinking," Families and Work Institute, https://www.youtube.com/watch?v=lDgm5yVY5K4.

41. Ellen Galinsky, "Give the Gift of Curiosity for the Holidays—Lessons from Laura Schulz," https://www.huffpost.com/entry/give-the-gift-of-curiosit_n_1157991. And for a more comprehensive review of recent scientific literature, see Schulz, "Infants Explore the Unexpected."

42. Bonawitz et al., "Children Balance Theories and Evidence in Exploration, Explanation, and Learning."

43. Stahl and Feigenson, "Observing the Unexpected Enhances Infants' Learning and Exploration."

44. "Johns Hopkins University Researchers: Babies Learn from Surprises," April 2, 2015, https://www.youtube.com/watch?v=oJjt5GRln-0.

45. Berlyne, *Conflict, Arousal, and Curiosity.* Berlyne was inspired specifically by Shaw et al., "A Command Structure for Complex Information Processing."

46. Schmidhuber, "Formal Theory of Creativity, Fun, and Intrinsic Motivation (1990–2010)."

47. Jürgen Schmidhuber, "Universal AI and a Formal Theory of Fun" (lecture), Winter Intelligence Conference, Oxford University, 2011, https://www.youtube.com/watch?v=fnbZzcruGu0.

48. Schmidhuber, "Formal Theory of Creativity, Fun, and Intrinsic Motivation (1990–2010)."

49. The tension between these two components perfectly embodies in a yin-and-yang form what NYU's James Carse has referred to as *Finite and Infinite Games.* A finite game is played to reach a terminal equilibrium state. An infinite game is played to forever prolong the experience of play. The finite player plays *against* surprise; the infinite player plays *for* surprise. In Carse's words, "Surprise causes finite play to end; it is the reason for infinite play to continue."

    This tension between competing fundamental drives for and against surprise is echoed by, among others, famed motivational speaker Tony Robbins, who expounds: "I believe there are six human needs. . . . Let me tell you what they are. First one: certainty. . . . While we go for certainty differently, if we get total certainty, we get what? What do you feel if you're certain? You know what will happen, when and how it will happen: What would you feel? Bored out of your minds. So God, in Her infinite wisdom, gave us a second human need, which is uncertainty. We need variety. We need surprise." Tony Robbins, "Why We Do What We Do" (lecture), February, 2006, Monterey, CA, https://www.ted.com/talks/tony_robbins_asks_why_we_do_what_we_do.)

    Humans clearly have both of these drives within them. It may be no coincidence if all good general reinforcement learners—living and otherwise—do too.

50. The "intrinsic curiosity module" is actually a bit subtler and more complex than this, because it is designed to predict only *user-controllable* aspects of the screen, for which another, "inverse dynamics" model is used. For the full details, see Pathak et al., "Curiosity-Driven Exploration by Self-Supervised Prediction." For some other related approaches, which incentivize exploration by rewarding "information gain," see, e.g., Schmidhuber, "Curious Model-Building Control

Systems"; Stadie, Levine, and Abbeel, "Incentivizing Exploration in Reinforcement Learning with Deep Predictive Models"; and Houthooft et al., "VIME."

51. Burda et al., "Large-Scale Study of Curiosity-Driven Learning."

52. See Burda et al., "Exploration by Random Network Distillation."

53. Note that the concurrent paper by Choi et al., "Contingency-Aware Exploration in Reinforcement Learning," from researchers at the University of Michigan and Google Brain, also reported a similar breakthrough in *Montezuma's Revenge* using a novelty-based approach to exploration.

54. A matter of weeks after OpenAI's announcement, a team from Uber AI Labs announced a family of algorithms they call Go-Explore, which, by storing a list of "novel" states (as measured by a grainy, low-resolution image of the screen) to prioritize revisiting, were able to beat the first level of *Montezuma's Revenge* 65% of the time. See https://eng.uber.com/go-explore/ for the press release, and Ecoffet et al., "Go-Explore," for the paper. Using some hand-coded human knowledge about the game led to the agent being able to beat the game's levels hundreds of times in a row, accumulating millions of points in the process. The significance of some of these results has been a matter of some disagreement—see, e.g., Alex Irpan, "Quick Opinions on Go-Explore," *Sorta Insightful*, https://www.alexirpan.com/2018/11/27/go-explore.html, for some discussion. The team's press release itself has been subsequently updated to address some of these and other points.

55. Ostrovski et al., "Count-Based Exploration with Neural Density Models."

56. For more discussion of this point, see, e.g., Ecoffet et al., "Go-Explore."

57. Burda et al., "Large-Scale Study of Curiosity-Driven Learning."

58. The exception is that in games with an complicated death animation, the agent would die just to see it. (Yuri Burda, personal correspondence, January 9, 2019.)

59. Yuri Burda, personal correspondence, January 9, 2019.

60. Singh, Lewis, and Barto, "Where Do Rewards Come From?"

61. Singh, Lewis, and Barto. For more discussion, see Oudeyer and Kaplan, "What Is Intrinsic Motivation?"

62. For reasons like this, researchers have experimented with so-called "sticky actions"—where at random an agent will occasionally be forced to repeat its *last* button press for one frame—as an alternative source of variation to epsilon-greedy actions, in which the agent presses *random* buttons at random times. It more accurately models the inherent randomness in human play, where our reactions are not millisecond-perfect, and it makes actions like long leaps, which require a button to be held for many frames in a row, more achievable by the agent. See Machado et al., "Revisiting the Arcade Learning Environment."

63. See Malone, "What Makes Computer Games Fun?" and "Toward a Theory of Intrinsically Motivating Instruction," and Malone and Lepper, "Making Learning Fun," for some early work on this subject.

64. Orseau, Lattimore, and Hutter, "Universal Knowledge-Seeking Agents for Stochastic Environments."

65. See Orseau, "Universal Knowledge-Seeking Agents." The solution to this agent's addiction to randomness, as later worked out in Orseau, Lattimore, and Hutter, "Universal Knowledge-Seeking Agents for Stochastic Environments," is to have the agent understand at a fundamental level that the world contains randomness, and hence be "resistant to non-informative noise."

66. Skinner, "Reinforcement Today."

67. See Kakade and Dayan, "Dopamine," which offers a novelty-based interpretation, borrowing explicitly from the reinforcement-learning literature to explain why a novelty drive may be useful for organisms. See also Barto, Mirolli, and Baldassarre, "Novelty or Surprise?," for a surprise-based interpretation of these results. For an overview, see, e.g., Niv, "Reinforcement Learning in the Brain," which notes, "It has long been known that novel stimuli cause phasic bursts in dopamine neurons." For experimental work on novelty in human decision-making, see Wittmann et al., "Striatal Activity Underlies Novelty-Based Choice in Humans." For more recent work on unifying reward-prediction error and surprise more generally in the function of dopamine, see, e.g., Gardner, Schoenbaum, and Gershman, "Rethinking Dopamine as Generalized Prediction Error."

68. Deepak Pathak, personal interview, March 28, 2018.

69. Marc Bellemare, personal interview, February 28, 2019.

70. Laurent Orseau, personal interview, June 22, 2018.

71. Laurent Orseau, personal interview, June 22, 2018.

72. Ring and Orseau, "Delusion, Survival, and Intelligent Agents."

73. Plato, *Protagoras and Meno*. In Plato's text, Socrates poses this to Protagoras in the interrogative, though he makes clear that this is indeed his view.

CHAPTER 7. IMITATION

1. Egan, *Axiomatic*.

2. Elon Musk, interviewed by Sarah Lacy, "A Fireside Chat with Elon Musk," Santa Monica, CA, July 12, 2012, https://pando.com/2012/07/12/pandomonthly-presents-a-fireside-chat-with-elon-musk/. Not only was the car uninsured, but Peter Thiel was not wearing a seat belt. "It was a miracle neither of us were hurt," says Thiel. See Dowd, "Peter Thiel, Trump's Tech Pal, Explains Himself."

3. This is discussed in greater detail in Visalberghi and Fragaszy, "Do Monkeys Ape?"

4. Romanes, *Animal Intelligence*.

5. Visalberghi and Fragaszy, "Do Monkeys Ape?" See also Visalberghi and Fragaszy, " 'Do Monkeys Ape?' Ten Years After." And note that Ferrari et al., "Neonatal Imi-

tation in Rhesus Macaques," reported some evidence of imitation in macaques, representing "the first detailed analysis, to our knowledge, of neonatal imitation conducted in a primate species outside the great ape clade."

6. Tomasello, "Do Apes Ape?" See also, e.g., Whiten et al., "Emulation, Imitation, Over-Imitation and the Scope of Culture for Child and Chimpanzee," which seeks to reevaluate this question.

7. Though the Kelloggs were somewhat circumspect about their reasons for terminating their experiment, it is speculated that Donald's alarming lack of a human vocabulary was a precipitating cause. See, e.g., Benjamin and Bruce, "From Bottle-Fed Chimp to Bottlenose Dolphin."

8. Meltzoff and Moore, "Imitation of Facial and Manual Gestures by Human Neonates" and Meltzoff and Moore, "Newborn Infants Imitate Adult Facial Gestures." Note that these results have recently become somewhat controversial. For instance, see Oostenbroek et al., "Comprehensive Longitudinal Study Challenges the Existence of Neonatal Imitation in Humans." But see also the rebuttal in, e.g., Meltzoff et al., "Re-examination of Oostenbroek et al. (2016)."

9. Alison Gopnik, personal interview, September 19, 2018.

10. Haggbloom et al., "The 100 Most Eminent Psychologists of the 20th Century."

11. Piaget, *The Construction of Reality in the Child.* Originally published as *La construction du réel chez l'enfant* in 1937.

12. Meltzoff, " 'Like Me.' "

13. Meltzoff and Moore, "Imitation of Facial and Manual Gestures by Human Neonates."

14. In a 2012 study, two-year-olds watched an adult bump a car into two different boxes, one of which made the car light up; when the child was given the car, they bumped the car into *that* box alone (Meltzoff, Waismeyer, and Gopnik, "Learning About Causes From People"). "The toddlers didn't imitate just anything," says Gopnik. "They imitated the action that would lead to the interesting result" (Gopnik, *The Gardener and the Carpenter*).

15. Meltzoff, Waismeyer, and Gopnik, "Learning About Causes from People," and Meltzoff, "Understanding the Intentions of Others." See Gopnik, *The Gardener and the Carpenter,* for a good summary of this area.

16. Meltzoff, "Foundations for Developing a Concept of Self."

17. Andrew Meltzoff, personal interview, June 10, 2019. "Infants are born to learn," Meltzoff writes, "and they learn at first by imitating us. This is why imitation is such an essential and far-reaching aspect of early development: It is not just a behavior, but a means for learning who we are" (Meltzoff, "Born to Learn").

18. The term "overimitation" to describe this comes initially from Lyons, Young, and Keil, "The Hidden Structure of Overimitation."

19. Horner and Whiten, "Causal Knowledge and Imitation/Emulation Switching in Chimpanzees (*Pan troglodytes*) and Children (*Homo sapiens*)."

20. McGuigan and Graham, "Cultural Transmission of Irrelevant Tool Actions in Diffusion Chains of 3- and 5-Year-Old Children."

21. Lyons, Young, and Keil, "The Hidden Structure of Overimitation."

22. Whiten et al., "Emulation, Imitation, Over-Imitation and the Scope of Culture for Child and Chimpanzee."

23. Gergely, Bekkering, and Király, "Rational Imitation in Preverbal Infants." Note that some researchers have taken issue with this methodology, noting, for instance, that it's possible that the infant—who would need to steady themselves on the table in order to touch the light with their head—is simply imitating the adult, who places their hands on the table before bending over to touch the light with their head. See Paulus et al., "Imitation in Infancy."

24. Buchsbaum et al., "Children's Imitation of Causal Action Sequences Is Influenced by Statistical and Pedagogical Evidence."

25. Hayden Carpenter, "What 'The Dawn Wall' Left Out," *Outside*, September 18, 2018, https://www.outsideonline.com/2344706/dawn-wall-documentary-tommy-caldwell-review.

26. Caldwell, *The Push*.

27. Lowell and Mortimer, "The Dawn Wall."

28. "'I Got My Ass Kicked': Adam Ondra's Dawn Wall Story," EpicTV Climbing Daily, Episode 1334, https://www.youtube.com/watch?v=O_B9vzIHlOo.

29. Aytar et al., "Playing Hard Exploration Games by Watching YouTube," which extends related work in Hester et al., "Deep Q-Learning from Demonstrations." Some very clever unsupervised learning was required to essentially "standardize" all of the different videos—of different resolutions, colors, and frame rates—into a single useful representation. But the result was a set of demonstrations that an agent could learn to imitate.

30. This is a very active area of research. See, e.g., Subramanian, Isbell, and Thomaz, "Exploration from Demonstration for Interactive Reinforcement Learning"; Večerík et al., "Leveraging Demonstrations for Deep Reinforcement Learning on Robotics Problems with Sparse Rewards"; and Hester et al., "Deep Q-Learning from Demonstrations."

31. In fact, many agents trained in video game environments are given the ability to jump back in to a previous play-through just about anywhere, akin to a human player making hundreds (or more) of different "save states." Here, dying merely sends you back to the last checkpoint, perhaps only seconds earlier, rather than back to the start of the game itself. This allows agents to experiment with tricky or dangerous segments without having to start the game all the way from the beginning if they fail—but it also introduces a certain artifice into the training.

More competent agents would presumably be able to reproduce or exceed the human "learning curve" on these games without such artifice.

32. Morgan, *An Introduction to Comparative Psychology.*

33. See Bostrom, *Superintelligence.*

34. "Robotics History: Narratives and Networks Oral Histories: Chuck Thorpe," an oral history conducted November 22, 2010, by Peter Asaro and Selma Šabanović, Indiana University, Bloomington Indiana, for Indiana University and the IEEE, https://ieeetv.ieee.org/video/robotics-history-narratives-and-networks -oral-histories-chuck-thorpe.

35. For more on the ALV (autonomous land vehicle) project, see Leighty, "DARPA ALV (Autonomous Land Vehicle) Summary." For more on DARPA's Strategic Computing initiative, see "Strategic Computing." For more on DARPA's mid-1980s projects, see Stefik, "Strategic Computing at DARPA." See also Roland and Shiman, *Strategic Computing.*

36. Moravec, "Obstacle Avoidance and Navigation in the Real World by a Seeing Robot Rover."

37. See also Rodney Brooks's reflections in Brooks, *Flesh and Machines.*

38. As per the segment on *Scientific American Frontiers,* Season 7, Episode 5, "Robots Alive!" Aired April 9, 1997, on PBS. See https://www.youtube.com/ watch?v=r4JrcVEkink.

39. Thorpe had tested the collision-avoidance system on the Navlab by seeing if the car would brake when Leland, on a bicycle with training wheels, rode out in front of it. Leland, after growing up and doing his own robotics degree at Carnegie Mellon, went to work on automated car technology at AssistWare, founded by Thorpe's student Dean Pomerleau. Thorpe jokingly imagines Leland's job interview: " 'Ever since then I've been a real fan of improving the reliability and safety of automated vehicles!' " Leland subsequently left computing altogether and became a seminarian with the Oblates of the Virgin Mary.

40. Pomerleau, "ALVINN," and Pomerleau, "Knowledge-Based Training of Artificial Neural Networks for Autonomous Robot Driving."

41. From a KDKA News segment in 1997: https://www.youtube.com/watch?v =IaoIqVMd6tc.

42. See https://twitter.com/deanpomerleau/status/801837566358093824. ("Present computers seem to be fast enough and to have enough memory for the job [of controlling a car]," AI pioneer John McCarthy argued—somewhat naïvely—in 1969. "However, commercial computers of the required performance are too big." See McCarthy, "Computer-Controlled Cars.")

43. See also Pomerleau, "Knowledge-Based Training of Artificial Neural Networks for Autonomous Robot Driving": "Autonomous driving has the potential to be an ideal domain for a supervised learning algorithm like backpropagation since

there is a readily available teaching signal or 'correct response' in the form of the human driver's current steering direction."

44. The course was Sergey Levine's CS294-112, Deep Reinforcement Learning; this lecture, "Imitation Learning," was delivered on December 3, 2017.

45. Bain, *The Senses and the Intellect*.

46. Kimball and Zaveri, "Tim Cook on Facebook's Data-Leak Scandal."

47. Ross and Bagnell, in "Efficient Reductions for Imitation Learning," discuss their architecture choice: a three-layer neural network taking in 24-by-18-pixel color images, with 32 hidden units and 15 output units. Pomerleau, "Knowledge-Based Training of Artificial Neural Networks for Autonomous Robot Driving," discusses the architecture of ALVINN, made from a three-layer neural network taking in 30-by-32-pixel black-and-white images, with 4 hidden units and 30 output units.

48. Pomerleau.

49. Pomerleau, "ALVINN."

50. Stéphane Ross, personal interview, April 29, 2019.

51. See Ross and Bagnell, "Efficient Reductions for Imitation Learning."

52. See Ross, Gordon, and Bagnell, "A Reduction of Imitation Learning and Structured Prediction to No-Regret Online Learning." For earlier approaches, see Ross and Bagnell, "Efficient Reductions for Imitation Learning."

53. Giusti et al., "A Machine Learning Approach to Visual Perception of Forest Trails for Mobile Robots." For a video explanation of the research, see "Quadcopter Navigation in the Forest Using Deep Neural Networks," https://www.youtube.com/watch?v=umRdt3zGgpU.

54. Bojarski et al., "End to End Learning for Self-Driving Cars." The Nvidia team further augmented its side-pointing camera images with "Photoshop" manipulations for a greater diversity of angles. These suffered from similar constraints as the ALVINN images, but they were good enough in practice. For a more informal discussion, see Bojarski et al., "End-to- End Deep Learning for Self-Driving Cars," https://devblogs.nvidia.com/deep-learning-self-driving-cars/. For the video of the car in action on the roads of Monmouth County, see "Dave-2: A Neural Network Drives a Car," https://www.youtube.com/watch?v=NJU9ULQUwng.

55. See LeCun et al., "Backpropagation Applied to Handwritten Zip Code Recognition."

56. Murdoch, *The Bell*.

57. Robert Hass, "Breach and Orison," in *Time and Materials*.

58. Kasparov, *How Life Imitates Chess*.

59. Holly Smith, personal interview, May 13, 2019.

60. See, published under the name Holly S. Goldman, "Dated Rightness and Moral Imperfection." See also Sobel, "Utilitarianism and Past and Future Mistakes."

61. See Goldman, "Doing the Best One Can." The name "Professor Procrastinate" came later, from Jackson and Pargetter, "Oughts, Options, and Actualism." Jackson later returned to these ideas in "Procrastinate Revisited."

62. The terms "possibilism" and "actualism" were coined in Jackson and Pargetter.

63. For Smith's more pro-possibilism views, see Goldman, "Doing the Best One Can." For a brief overview of the topic published decades afterward, see Smith, "Possibilism," and for a more detailed and recent review, see Timmerman and Cohen, "Actualism and Possibilism in Ethics." For a particularly interesting wrinkle, see Bykvist, "Alternative Actions and the Spirit of Consequentialism," p. 50.

64. Thanks to Joe Carlsmith for helpful discussions of this and related topics. Some recent philosophical literature explicitly discusses the links between possibilism, actualism, and effective altruism. See, e.g., Timmerman, "Effective Altruism's Underspecification Problem."

65. Singer, "Famine, Affluence, and Morality"; see also Singer, "The Drowning Child and the Expanding Circle."

66. Julia Wise, "Aim High, Even If You Fall Short," *Giving Gladly* (blog), October 8, 2014. http://www.givinggladly.com/2014/10/aim-high-even-if-you-fall-short.html.

67. Will MacAskill, "The Best Books on Effective Altruism," interview by Edouard Mathieu, Five Books, https://fivebooks.com/best-books/effective-altruism-will-macaskill/. See also the organization Giving What We Can, founded by MacAskill and his colleague Toby Ord after Ord decided, inspired by Singer and others, to commit to giving a portion of his income to effective charities.

68. See Singer, *The Most Good You Can Do*.

69. The canonical on-policy method in reinforcement learning is one called SARSA, short for State–Action–Reward–State–Action; see Rummery and Niranjan, "On-Line Q-Learning Using Connectionist Systems." The canonical off-policy method is one called Q-Learning; see Watkins, "Learning from Delayed Rewards," and Watkins and Dayan, "Q-Learning."

70. See Sutton and Barto, *Reinforcement Learning*.

71. In an ethical context, philosopher Rosalind Hursthouse has framed virtue ethics as a kind of imitation learning; see Hursthouse, "Normative Virtue Ethics." Of course, there are many difficulties both practical and theoretical, as both Hursthouse and her critics discuss; see, e.g., Johnson, "Virtue and Right." For a different take on why imitating ostensibly perfect role models may not actually be a great idea, see Wolf, "Moral Saints."

72. Lipsey and Lancaster, "The General Theory of Second Best."

73. Amanda Askell, personal correspondence.

74. See Balentine, *It's Better to Be a Good Machine Than a Bad Person*.

75. Magnus Carlsen, during the press conference following Game 5 of the 2018 World Chess Championship, London, November 15, 2018.

76. "Heuristics."

77. "Heuristics."

78. See Samuel, "Some Studies in Machine Learning Using the Game of Checkers."

79. For more on the architecture of Deep Blue, see Campbell, Hoane, and Hsu, "Deep Blue." For additional elaboration on the tuning of the evaluation heuristics through the use of grandmaster games, see the explanation given by programmer Andreas Nowatzyk in "Eval Tuning in Deep Thought," Chess Programming Wiki, https://www.chessprogramming.org/Eval_Tuning_in_Deep_Thought. For a snapshot of the IBM team's progress in 1990, on what was then known as Deep Thought, as well as a discussion of the decision to automatically tune the (then only 120) parameter weights based on a database of (then only 900) expert games, see Hsu et al., "A Grandmaster Chess Machine," as well as Byrne, "Chess-Playing Computer Closing in on Champions." For more on the tuning of the evaluation function of Deep Blue (and its predecessor Deep Thought), see, e.g., Anantharaman, "Evaluation Tuning for Computer Chess."

80. Hsu, "IBM's Deep Blue Chess Grandmaster Chips."

81. Weber, "What Deep Blue Learned in Chess School."

82. Schaeffer et al., "A World Championship Caliber Checkers Program."

83. Fürnkranz and Kubat, *Machines That Learn to Play Games.*

84. There are, of course, many subtle differences between the architecture and training procedure for Deep Blue and those for AlphaGo. For more details on AlphaGo, see Silver et al., "Mastering the Game of Go with Deep Neural Networks and Tree Search."

85. AlphaGo's value network was derived from self-play, but its policy network was imitative, trained through supervised learning on a database of human expert games. Roughly speaking, it was conventional in the moves it considered, but thought for itself when deciding which of them was best. See Silver et al., "Mastering the Game of Go with Deep Neural Networks and Tree Search."

86. Silver et al., "Mastering the Game of Go Without Human Knowledge." In 2018, AlphaGo Zero was further refined into an even stronger program—and a more general one, capable of record-breaking strength in not just Go but chess and shogi—called AlphaZero. For more detail about AlphaZero, see Silver et al., "A General Reinforcement Learning Algorithm That Masters Chess, Shogi, and Go Through Self-Play." In 2019, a subsequent iteration of the system called MuZero matched this level of performance with less computation and less advance knowledge of the rules of the game, while proving flexible enough to excel at not just board games but Atari games as well; see Schrittwieser et al., "Mastering Atari, Go, Chess and Shogi by Planning with a Learned Model."

87. Silver et al., "Mastering the Game of Go Without Human Knowledge."

88. For a look at the psychology of "fast" and "slow" mental processes, also known as "System 1" and "System 2," see Kahneman, *Thinking, Fast and Slow.*

89. See Coulom, "Efficient Selectivity and Backup Operators in Monte-Carlo Tree Search."

90. See Silver et al., "Mastering the Game of Go Without Human Knowledge," for details. More precisely, it uses the "visit count" of each move during the MCTS— so the network learns to predict, in effect, how long it will spend thinking about each move. See also the contemporaneous and closely related "expert iteration" ("ExIt") algorithm in Anthony, Tian, and Barber, "Thinking Fast and Slow with Deep Learning and Tree Search."

91. Shead, "DeepMind's Human-Bashing AlphaGo AI Is Now Even Stronger."

92. Aurelius, *The Emperor Marcus Aurelius.*

93. Andy Fitch, "Letter from Utopia: Talking to Nick Bostrom," *BLARB* (blog), November 24, 2017, https://blog.lareviewofbooks.org/interviews/letter-utopia-talking-nick -bostrom/.

94. Blaise Agüera y Arcas, "The Better Angels of our Nature" (lecture), February 16, 2017, VOR: Superintelligence, Mexico City.

95. Yudkowsky, "Coherent Extrapolated Volition." See also Tarleton, "Coherent Extrapolated Volition."

96. Note that some philosophers, namely the "moral realists," do in fact believe in the idea of objective moral truths. For an overview of this family of positions, see, e.g., Sayre-McCord, "Moral Realism."

97. Paul Christiano, interviewed by Rob Wiblin, *The 80,000 Hours Podcast,* October 2, 2018.

98. See Paul Christiano, "A Formalization of Indirect Normativity," *AI Alignment* (blog), April 20, 2012, https://ai-alignment.com/a-formalization-of-indirect-normativity-7e44db640160, and Ajeya Cotra, "Iterated Distillation and Amplification," *AI Alignment* (blog), March 4, 2018, https://ai-alignment.com/ iterated-distillation-and-amplification-157debfd1616.

99. For an explicit discussion of the connection between AlphaGo's policy network and the idea of iterated capability amplification, see Paul Christiano, "AlphaGo Zero and Capability Amplification," *AI Alignment* (blog), October 19, 2017, https:// ai-alignment.com/alphago-zero-and-capability-amplification-ede767bb8446.

100. Christiano, Shlegeris, and Amodei, "Supervising Strong Learners by Amplifying Weak Experts."

101. Paul Christiano, personal interview, July 1, 2019.

102. See, for instance, alignmentforum.org, along with a growing number of workshops, conferences, and research labs.

CHAPTER 8. INFERENCE

1. See Warneken and Tomasello, "Altruistic Helping in Human Infants and Young Chimpanzees," along with Warneken and Tomasello, "Helping and Cooperation at 14 Months of Age." And for video footage of some of the experiments, see, e.g., "Experiments with Altruism in Children and Chimps," https://www.youtube.com/watch?v=Z-eU5xZW7cU.

2. See also Meltzoff, "Understanding the Intentions of Others," which showed that eighteen-month-olds can successfully imitate the intended acts of that adults tried and failed to do, indicating that they "situate people within a psychological framework that differentiates between the surface behavior of people and a deeper level involving goals and intentions."

3. Warneken and Tomasello demonstrated that human infants as young as fourteen months will help with reaching as well, but not with more complex problems.

4. Again, see Warneken and Tomasello, "Altruistic Helping in Human Infants and Young Chimpanzees."

5. Tomasello et al., "Understanding and Sharing Intentions."

6. Felix Warneken, "Need Help? Ask a 2-Year-Old" (lecture), TEDxAmoskeagMill-yard 2013, https://www.youtube.com/watch?v=qul57hcu4I.

7. See "Our Research," Social Minds Lab, University of Michigan, https://sites.lsa.umich.edu/warneken/lab/research-2/.

8. Tomasello et al. (Note that they explicitly frame their discussion using the language of control systems and cybernetics.)

9. Stuart Russell, personal interview, May 13, 2018.

10. See, e.g., Uno, Kawato, and Suzuki, "Formation and Control of Optimal Trajectory in Human Multijoint Arm Movement."

11. See, e.g., Hogan, "An Organizing Principle for a Class of Voluntary Movements."

12. Hoyt and Taylor, "Gait and the Energetics of Locomotion in Horses."

13. Farley and Taylor, "A Mechanical Trigger for the Trot-Gallop Transition in Horses." For more on the biomechanics of human and animal motion, see also the work of the celebrated late British zoologist Robert McNeill Alexander: for instance "The Gaits of Bipedal and Quadrupedal Animals," *The Human Machine*, and *Optima for Animals*. As Alexander explains, "The legs and gaits of animals are the products of two very potent optimizing processes, the processes of evolution by natural selection and of learning by experience. Zoologists studying them are trying to solve inverse optimality problems: they are trying to discover the optimization criteria that have been important in the evolution of animal legs and in the evolution or learning of gaits." For more contemporary on inverse optimal control in the context of the human gait, see the work of Katja Mombaur, e.g.,

Mombaur, Truong, and Laumond, "From Human to Humanoid Locomotion—an Inverse Optimal Control Approach."

14. For more on the links between reinforcement learning and the dopamine system, see the discussion in the main text and endnotes of Chapter 4. For more on the links to animal foraging, see, e.g., Montague et al., "Bee Foraging in Uncertain Environments Using Predictive Hebbian Learning," and Niv et al., "Evolution of Reinforcement Learning in Foraging Bees."

15. Russell, "Learning Agents for Uncertain Environments (Extended Abstract)." For earlier work that takes an econometric perspective to similar questions, of so-called "structural estimation," see Rust, "Do People Behave According to Bellman's Principle of Optimality?," and "Structural Estimation of Markov Decision Processes," as well as Sargent, "Estimation of Dynamic Labor Demand Schedules Under Rational Expectations." For an even earlier precursor to this question from a control theory perspective, see Kálmán, "When Is a Linear Control System Optimal?" Working in 1964 at the Research Institute for Advanced Studies in Baltimore, with funding from both the US Air Force and NASA, Kálmán was interested in, as he put it, "the Inverse Problem of Optimal Control Theory, which is the following: Given a control law, find all performance indices for which this control law is optimal." He noted, "Very little is known today about this problem."

16. These additive and multiplicative changes are known as "affine" transformations.

17. Ng and Russell, "Algorithms for Inverse Reinforcement Learning."

18. Specifically, Ng and Russell used a method known as "ℓ_1 regularization," also known as "lasso." This idea comes from Tibshirani, "Regression Shrinkage and Selection via the Lasso." For an approachable overview of the ideas and techniques of regularization, see Christian and Griffiths, *Algorithms to Live By*.

19. Abbeel and Ng, "Apprenticeship Learning via Inverse Reinforcement Learning."

20. Andrew Ng, introduction to Pieter Abbeel's PhD thesis defense, Stanford University, May 19, 2008; see http://ai.stanford.edu/~pabbeel//thesis/PieterAbbeel_Defense_19May2008_320x180.mp4.

21. Abbeel, Coates, and Ng, "Autonomous Helicopter Aerobatics Through Apprenticeship Learning."

22. Abbeel et al., "An Application of Reinforcement Learning to Aerobatic Helicopter Flight." They also successfully performed a nose-in funnel and a tail-in funnel.

23. "As repeated sub-optimal demonstrations tend to differ in their suboptimalities, together they often encode the intended trajectory." See Abbeel, "Apprenticeship Learning and Reinforcement Learning with Application to Robotic Control," which refers to the work in Coates, Abbeel, and Ng, "Learning for Control from Multiple Demonstrations."

24. Abbeel, Coates, and Ng, "Autonomous Helicopter Aerobatics Through Apprenticeship Learning."

25. See Youngblood's website: http://www.curtisyoungblood.com/curtis-youngblood/.

26. Curtis Youngblood, "Difference Between a Piro Flip and a Kaos," interview by Aaron Shell, https://www.youtube.com/watch?v=TLi_hp-m-mk.

27. For a video clip of the Stanford helicopter performing the chaos, see "Stanford University Autonomous Helicopter: Chaos," https://www.youtube.com/watch?v=kN6ifrqwIMY.

28. Ziebart et al., "Maximum Entropy Inverse Reinforcement Learning," which leverages the principle of maximum entropy derived from Jaynes, "Information Theory and Statistical Mechanics." See also Ziebart, Bagnell, and Dey, "Modeling Interaction via the Principle of Maximum Causal Entropy."

29. See Billard, Calinon, and Guenter, "Discriminative and Adaptive Imitation in Uni-Manual and Bi-Manual Tasks," and for a 2009 overview of the space, see Argall et al., "A Survey of Robot Learning from Demonstration."

30. See Finn, Levine, and Abbeel, "Guided Cost Learning." See also Wulfmeier, Ondrúška, and Posner, "Maximum Entropy Deep Inverse Reinforcement Learning," and Wulfmeier, Wang, and Posner, "Watch This."

31. Specifically, Leike analyzed the termination or nontermination properties of what are known as "lasso programs." See Jan Leike, "Ranking Function Synthesis for Linear Lasso Programs," master's thesis, University of Freiburg, 2013.

32. Jan Leike, personal interview, June 22, 2018.

33. See Leike and Hutter, "Bad Universal Priors and Notions of Optimality."

34. The paper is Christiano et al., "Deep Reinforcement Learning from Human Preferences." For OpenAI's blog post about the paper, see "Learning from Human Preferences," https://openai.com/blog/deep-reinforcement-learning-from-human-preferences/, and for DeepMind's blog post, see "Learning Through Human Feedback," https://deepmind.com/blog/learning-through-human-feedback/. For earlier work exploring the idea of learning from human preferences and human feedback, see, e.g., Wilson, Fern, and Tadepalli, "A Bayesian Approach for Policy Learning from Trajectory Preference Queries"; Knox, Stone, and Breazeal, "Training a Robot via Human Feedback"; Akrour, Schoenauer, and Sebag, "APRIL"; and Akrour et al., "Programming by Feedback." See also Wirth et al., "A Survey of Preference-Based Reinforcement Learning Methods." For a framework that unites learning from demonstrations and learning from comparisons, see Jeon, Milli, and Drăgan, "Reward-Rational (Implicit) Choice."

35. Paul Christiano, personal interview, July 1, 2019.

36. Todorov, Erez, and Tassa, "MuJoCo."

37. As the paper puts it: "In the long run it would be desirable to make learning a task from human preferences no more difficult than learning it from a programmatic reward signal, ensuring that powerful RL systems can be applied in the service

of complex human values rather than low-complexity goals" (Christiano et al., "Deep Reinforcement Learning from Human Preferences"). For subsequent work by Leike and his colleagues that pursues the agenda of modeling the human's reward, see Leike et al., "Scalable Agent Alignment via Reward Modeling."

38. Stuart Russell, personal interview, May 13, 2018.

39. It's worth noting that handing an object to another person is itself a surprisingly subtle and complex action that includes making inferences about how the other person will want to take hold of the object, how to signal to them that you are intending for them to take it, etc. See, e.g., Strabala et al., "Toward Seamless Human-Robot Handovers."

40. Hadfield-Menell et al., "Cooperative Inverse Reinforcement Learning." ("CIRL" is pronounced with a soft *c*, homophonous with the last name of strong AI skeptic John Searle (no relation). I have agitated within the community that a hard *c* "curl" pronunciation makes more sense, given that "cooperative" uses a hard *c*, but it appears the die is cast.)

41. Dylan Hadfield-Menell, personal interview, March 15, 2018.

42. Russell, *Human Compatible*.

43. For instance, one of the first theoretical advances within the CIRL framework leveraged earlier cognitive science research on teacher-learner strategy coadaptation. See Fisac et al., "Pragmatic-Pedagogic Value Alignment" (which makes use of insights from Shafto, Goodman, and Griffiths, "A Rational Account of Pedagogical Reasoning"); the authors write, "To our knowledge, this work constitutes the first formal analysis of value alignment grounded in empirically validated cognitive models." See also the subsequent paper by many of the same authors: Malik et al., "An Efficient, Generalized Bellman Update for Cooperative Inverse Reinforcement Learning."

44. Maya Çakmak of the University of Washington, in Seattle, and Manuel Lopes of the Instituto Superior Técnico, in Lisbon, have worked on this idea; see Çakmak and Lopes, "Algorithmic and Human Teaching of Sequential Decision Tasks." Of course, if the human is tailoring their behavior to be maximally pedagogical—not to be optimized for their metric itself but optimized to *communicate* what the metric *is*—then the computer, in turn, would do well not to use standard IRL (which assumes the demonstrations are optimal) but to make inferences that account for the fact that the teacher's behavior was pedagogical in nature. The teaching and learning strategies coadapt. This is a rich area of active research both in cognitive science and in machine learning. See also, e.g., Ho et al., "Showing Versus Doing," and Ho et al., "A Rational-Pragmatic Account of Communicative Demonstrations."

45. See Gopnik, Meltzoff, and Kuhl, *The Scientist in the Crib*: "It turns out that moth-

erese is more than just a sweet siren song that we use to draw our babies to us. . . . Completely unconsciously [parents] produce sounds more clearly and pronounce them more accurately when they talk to babies than when they talk to other adults." The authors note, for instance, that English and Swedish motherese sound different. For more recent work in this area, see Eaves et al., "Infant-Directed Speech Is Consistent With Teaching," and Ramírez, Lytle, and Kuhl, "Parent Coaching Increases Conversational Turns and Advances Infant Language Development."

46. Handoffs of objects are an explicit focus of human-robot interaction research. See, e.g., Strabala et al., "Toward Seamless Human-Robot Handovers."

47. Drăgan, Lee, and Srinivasa, "Legibility and Predictability of Robot Motion"; see also Takayama, Dooley, and Ju, "Expressing Thought" (which, to be fair, does refer to the concept of "readable" motion), and Gielniak and Thomaz, "Generating Anticipation in Robot Motion." More recent work has looked at, for instance, how to communicate not just the machine's goal but, when the goal is known, its *plan:* see Fisac et al., "Generating Plans That Predict Themselves."

48. Jan Leike, personal interview, June 22, 2018. See also Christiano et al., "Deep Reinforcement Learning from Human Preferences": "Training the reward predictor offline can lead to bizarre behavior that is undesirable as measured by the true reward. For instance, on *Pong* offline training sometimes leads our agent to avoid losing points but not to score points; this can result in extremely long volleys. This type of behavior demonstrates that in general human feedback needs to be intertwined with RL rather than provided statically."

49. Julie Shah, personal interview, March 2, 2018.

50. For research on human cross-training, see Blickensderfer, Cannon-Bowers, and Salas, "Cross-Training and Team Performance"; Cannon-Bowers et al., "The Impact of Cross-Training and Workload on Team Functioning"; and Marks et al., "The Impact of Cross-Training on Team Effectiveness."

51. Nikolaidis et al., "Improved Human-Robot Team Performance Through Cross-Training: An Approach Inspired by Human Team Training Practices."

52. "Julie Shah: Human/Robot Team Cross Training," https://www.youtube.com/watch?v=UQrtw0YUlqM.

53. More recent work by Shah's lab has explored cases where it's infeasible to switch roles. Here one can use a related idea called "perturbation training"; see Ramakrishnan, Zhang, and Shah, "Perturbation Training for Human-Robot Teams."

54. Murdoch, *The Bell.*

55. Some researchers at the intersection of cognitive science and AI safety, including the Future of Humanity Institute's Owain Evans, are working on ways to do

inverse reinforcement learning to take into account a person who, for instance, can't help ducking into the pastry shop when they walk by but will walk out of their way to avoid it. See, e.g., Evans, Stuhlmüller, and Goodman, "Learning the Preferences of Ignorant, Inconsistent Agents," and Evans and Goodman, "Learning the Preferences of Bounded Agents." There is a whole vein of research on IRL that incorporates the quirks and sometime irrationalities of human behavior. See also Bourgin et al., "Cognitive Model Priors for Predicting Human Decisions," for work on using machine learning to develop models of human preference and decision-making.

56. See, e.g., Snyder, *Public Appearances, Private Realities;* Covey, Saladin, and Killen, "Self-Monitoring, Surveillance, and Incentive Effects on Cheating"; and Zhong, Bohns, and Gino, "Good Lamps Are the Best Police."

57. See Bateson, Nettle, and Roberts, "Cues of Being Watched Enhance Cooperation in a Real-World Setting," and Heine et al., "Mirrors in the Head."

58. Bentham, "Letter to Jacques Pierre Brissot de Warville."

59. Bentham, "Preface."

CHAPTER 9. UNCERTAINTY

1. Russell, "Ideas That Have Harmed Mankind."

2. "Another Day the World Almost Ended."

3. Aksenov, "Stanislav Petrov."

4. Aksenov.

5. Hoffman, " 'I Had a Funny Feeling in My Gut.' "

6. Nguyen, Yosinski, and Clune, "Deep Neural Networks Are Easily Fooled." For a discussion of the confidence of predictions of neural networks, see Guo et al., "On Calibration of Modern Neural Networks."

7. See Szegedy et al., "Intriguing Properties of Neural Networks," and Goodfellow, Shlens, and Szegedy, "Explaining and Harnessing Adversarial Examples." This is an active area of research; for recent work on making systems robust to adversarial examples, see, e.g., Mądry et al., "Towards Deep Learning Models Resistant to Adversarial Attacks," Xie et al., "Feature Denoising for Improving Adversarial Robustness," and Kang et al., "Testing Robustness Against Unforeseen Adversaries." See also Ilyas et al., "Adversarial Examples Are Not Bugs, They Are Features," which frames adversarial examples in an alignment context—"a *misalignment* between the (human-specified) notion of robustness and the inherent geometry of the data"—and argues that "attaining models that are robust and interpretable will require explicitly encoding human priors into the training process."

8. Creighton, "Making AI Safe in an Unpredictable World."

9. See details about Dietterich's "open category problem" research, for which he received a Future of Life Institute grant, at https://futureoflife.org/ai-researcher-thomas-dietterich/.

10. Thomas G. Dietterich, "Steps Toward Robust Artificial Intelligence" (lecture), February 14, 2016, 30th AAAI Conference on Artificial Intelligence, Phoenix, AZ, http://videolectures.net/aaai2016_dietterich_artificial_intelligence/. The talk also appears in slightly different form in print; see Dietterich, "Steps Toward Robust Artificial Intelligence." For more on open category learning, see, e.g., Scheirer et al., "Toward Open Set Recognition"; Da, Yu, and Zhou, "Learning with Augmented Class by Exploiting Unlabeled Data"; Bendale and Boult, "Towards Open World Recognition"; Steinhardt and Liang, "Unsupervised Risk Estimation Using Only Conditional Independence Structure"; Yu et al., "Open-Category Classification by Adversarial Sample Generation"; and Rudd et al., "The Extreme Value Machine." Other related approaches to adversarial examples and robust classification include Liu and Ziebart, "Robust Classification Under Sample Selection Bias," and Li and Li, "Adversarial Examples Detection in Deep Networks with Convolutional Filter Statistics." For more recent results by Dietterich and his collaborators, see Liu et al., "Can We Achieve Open Category Detection with Guarantees?," and Liu et al., "Open Category Detection with PAC Guarantees," as well as Hendrycks, Mazeika, and Dietterich, "Deep Anomaly Detection with Outlier Exposure." For a 2018 proposal from Google Brain and OpenAI researchers for a benchmark contest to stimulate research on these questions, see Brown et al., "Unrestricted Adversarial Examples," along with "Introducing the Unrestricted Adversarial Example Challenge," *Google AI Blog*, https://ai.googleblog.com/2018/09/introducing-unrestricted-adversarial.html.

11. Rousseau, *Emile; or, On Education.*

12. Jefferson, *Notes on the State of Virginia.*

13. Yarin Gal, personal interview, July 11, 2019.

14. Yarin Gal, "Modern Deep Learning Through Bayesian Eyes" (lecture), Microsoft Research, December 11, 2015, https://www.microsoft.com/en-us/research/video/modern-deep-learning-through-bayesian-eyes/.

15. Zoubin Ghahramani, "Probabilistic Machine Learning: From Theory to Industrial Impact" (lecture), October 5, 2018, PROBPROG 2018: The International Conference on Probabilistic Programming, https://youtu.be/crvNIGyqGSU.

16. For seminal papers relating to Bayesian neural networks, see Denker et al., "Large Automatic Learning, Rule Extraction, and Generalization"; Denker and LeCun, "Transforming Neural-Net Output Levels to Probability Distributions"; MacKay, "A Practical Bayesian Framework for Backpropagation Networks"; Hinton and Van Camp, "Keeping Neural Networks Simple by Minimizing the Description Length of the Weights"; Neal, "Bayesian Learning for Neural Networks"; and Bar-

ber and Bishop, "Ensemble Learning in Bayesian Neural Networks." For more recent work, see Graves, "Practical Variational Inference for Neural Networks"; Blundell et al., "Weight Uncertainty in Neural Networks"; and Hernández-Lobato and Adams, "Probabilistic Backpropagation for Scalable Learning of Bayesian Neural Networks." For a more detailed history of these ideas, see Gal, "Uncertainty in Deep Learning." For an overview of probabilistic methods in machine learning more generally, see Ghahramani, "Probabilistic Machine Learning and Artificial Intelligence."

17. Yarin Gal, personal interview, July 11, 2019.

18. Yarin Gal, "Modern Deep Learning Through Bayesian Eyes" (lecture), Microsoft Research, December 11, 2015, https://www.microsoft.com/en-us/research/video/modern-deep-learning-through-bayesian-eyes/.

19. For a look at using dropout-ensemble uncertainty to detect adversarial examples, see Smith and Gal, "Understanding Measures of Uncertainty for Adversarial Example Detection."

20. Each model is typically assigned a weight describing how well it can explain the data. This method is known as "Bayesian model averaging," or BMA; see Hoeting et al., "Bayesian Model Averaging: A Tutorial."

21. In particular, it was found that dropout was helpful in preventing the network from too brittlely "overfitting" its training data. See Srivastava et al., "Dropout," which has come to be cited an astounding 18,500 times within the first six years of its publication.

22. See Gal and Ghahramani, "Dropout as a Bayesian Approximation." Alternatives and extensions have emerged in recent years; see, e.g., Lakshminarayanan, Pritzel, and Blundell, "Simple and Scalable Predictive Uncertainty Estimation Using Deep Ensembles."

23. Yarin Gal, personal interview, July 11, 2019. One application is in ophthalmology, discussed in the main text; other examples include, for instance, Uber's demand prediction models (Zhu and Nikolay, "Engineering Uncertainty Estimation in Neural Networks for Time Series Prediction at Uber"), and Toyota Research Institute's driver prediction systems (Huang et al., "Uncertainty-Aware Driver Trajectory Prediction at Urban Intersections").

24. See Gal and Ghahramani, "Bayesian Convolutional Neural Networks with Bernoulli Approximate Variational Inference," §4.4.2; specifically, Gal and Ghahramani looked at Lin, Chen, and Yan, "Network in Network," and Lee et al., "Deeply-Supervised Nets." Note that care should be taken when tuning the dropout rate; see Gal and Ghahramani, "Dropout as a Bayesian Approximation." For applications of this idea to recurrent networks and reinforcement learning, respectively, see Gal and Ghahramani, "A Theoretically Grounded Application of Dropout in Recurrent Neural Networks"; Gal, "Uncertainty in Deep Learning,"

§3.4.2; and Gal, McAllister, and Rasmussen, "Improving PILCO with Bayesian Neural Network Dynamics Models."

25. Gal and Ghahramani, "Dropout as a Bayesian Approximation."

26. Yarin Gal, personal interview, July 11, 2019.

27. See Engelgau et al., "The Evolving Diabetes Burden in the United States," and Zaki et al., "Diabetic Retinopathy Assessment."

28. Leibig et al., "Leveraging Uncertainty Information from Deep Neural Networks for Disease Detection."

29. A number of groups are exploring the potential of this broad idea of "selective classification" in machine learning. For instance, Google Research's Corinna Cortes and her colleagues have explored the idea of "learning with rejection"— that is, classifiers that can simply "punt" or otherwise *decline* to make a classification judgment. See Cortes, DeSalvo, and Mohri, "Learning with Rejection"; see also statistical work by C. K. Chow from the mid-twentieth century exploring related ideas: Chow, "An Optimum Character Recognition System Using Decision Functions," and Chow, "On Optimum Recognition Error and Reject Tradeoff." For a similar approach, in a reinforcement-learning context, see Li et al., "Knows What It Knows."

In 2018, researchers from the University of Toronto, led by PhD student David Madras, widened the lens on this idea, asking not only how a machine-learning system might try to defer on tricky or ambiguous cases in order to avoid making mistakes but, moreover, how it might work in tandem with a human decision maker who is picking up the slack. If the human decision maker happens to be especially accurate for certain types of examples, the system should be more deferential even if it's otherwise confident; conversely, if there are certain types of examples where the human is particularly bad, the system might simply hazard its best guess even if it's not sure—aiming to optimize not its *own* accuracy, per se, but the accuracy of the human-machine decision-making team as a whole. See Madras, Pitassi, and Zemel, "Predict Responsibly."

In related work, Shun Zhang, Edmund Durfee and Satinder Singh from the University of Michigan have explored the idea of an agent in a gridworld environment that seeks to minimize side effects by asking whether the human user minds certain things being changed or not, and they are able to offer bounds for how to operate safely with the fewest such queries possible. See Zhang, Durfee, and Singh, "Minimax-Regret Querying on Side Effects for Safe Optimality in Factored Markov Decision Processes."

30. Kahn et al., "Uncertainty-Aware Reinforcement Learning for Collision Avoidance."

31. For related work linking uncertainty with unfamiliar environments, see Kenton et al., "Generalizing from a Few Environments in Safety-Critical Reinforcement

Learning." For related work in the context of imitation learning and autonomous cars, see Tigas et al., "Robust Imitative Planning."

32. Holt et al., "An Unconscious Patient with a DNR Tattoo." And see Bever, "A Man Collapsed with 'Do Not Resuscitate' Tattooed on His Chest," and Hersher, "When a Tattoo Means Life or Death," for press accounts.

33. Holt et al., "An Unconscious Patient with a DNR Tattoo."

34. Cooper and Aronowitz, "DNR Tattoos."

35. Holt et al., "An Unconscious Patient with a DNR Tattoo."

36. Bever, "A Man Collapsed with 'Do Not Resuscitate' Tattooed on His Chest."

37. Sunstein, "Irreparability as Irreversibility." See also Sunstein, "Irreversibility."

38. Sunstein, "Beyond the Precautionary Principle." See also Sunstein, *Laws of Fear*.

39. Amodei et al., "Concrete Problems in AI Safety," has an excellent and broad discussion of "avoiding negative side effects" and "impact regularizers," and Taylor et al., "Alignment for Advanced Machine Learning Systems," also discusses various ideas for "impact measures." For a good overview of more recent impact research, see Daniel Filan, "Test Cases for Impact Regularisation Methods," https://www.alignmentforum.org/posts/wzPzPmAsG3BwrBrwy/test-cases-for-impact-regularisation-methods.

Carnegie Mellon PhD student Benjamin Eysenbach looked at a similar idea within the 3D MuJoCo environment. His idea was reversibility, mixed with the hiker and backpacker's ethos of "leave no trace." The idea is to use normal reinforcement-learning methods to develop competence in various tasks, with a crucial proviso. Unlike the ergodic environment of an Atari game, where typically learning involves hundreds of thousands of externally imposed reboots, his agents were responsible for always resetting *themselves* back to their exact starting configuration before making another attempt at whatever it was they were trying to do. Initial results are encouraging, with his stick-figure cheetah, for instance, scooting up to a cliff's edge, then backpedaling—seemingly having internalized that there is no reversing back once it's gone over the edge. See Eysenbach et al., "Leave No Trace." See also the much earlier Weld and Etzioni, "The First Law of Robotics (a Call to Arms)," for a similar idea.

40. For Armstrong's work on low-impact AI agents, see Armstrong and Levinstein, "Low Impact Artificial Intelligences." His papers from 2012 and 2013 were among the earliest to explicitly address this issue: see Armstrong, "The Mathematics of Reduced Impact," and Armstrong, "Reduced Impact AI."

41. Armstrong and Levinstein, "Low Impact Artificial Intelligences."

42. Armstrong and Levinstein.

43. As Eliezer Yudkowsky put it, "If you're going to cure cancer, make sure the patient still dies!" See https://intelligence.org/2016/12/28/ai-alignment-why-its-hard-and-where-to-start/. See also Armstrong and Levinstein, "Low Impact Artificial

Intelligences," which uses the example of an asteroid headed for earth. A system constrained to only take "low-impact" actions might fail to divert it—or, perhaps even worse, a system capable of offsetting might divert the asteroid, saving the planet, and then blow the planet up anyway.

44. Victoria Krakovna, personal interview, December 8, 2017.

45. See Krakovna et al., "Penalizing Side Effects Using Stepwise Relative Reachability." For Krakovna, framing the problem in terms of "side effects," rather than "impact," per se, makes at least some of the paradoxes seem to go away. "If a robot is carrying boxes and bumps into a vase," she says, "breaking the vase is a side effect, because the robot could have easily gone around the vase. On the other hand, a cooking robot that's making an omelette has to break some eggs, so breaking eggs is not a side effect." See also Victoria Krakovna, "Measuring and Avoiding Side Effects Using Relative Reachability," June 5, 2018, https://vkrakovna.wordpress.com/2018/06/05/measuring-and-avoiding-side-effects-using-relative-reachability/.

46. Leike et al., "AI Safety Gridworlds."

47. Victoria Krakovna, personal interview, December 8, 2017.

48. The idea of stepwise baselines was suggested by Alexander Turner in https://www.alignmentforum.org/posts/DvmhXysefEyEvXuXS/overcoming-clinginess-in-impact-measures. The idea of relative reachability is explored in Krakovna et al., "Penalizing Side Effects Using Stepwise Relative Reachability," and Krakovna et al., "Designing Agent Incentives to Avoid Side Effects," *Deep-Mind Safety Research* (blog), https://medium.com/@deepmindsafetyresearch/designing-agent-incentives-to-avoid-side-effects-e1ac80ea6107.

49. Turner, Hadfield-Menell, and Tadepalli, "Conservative Agency via Attainable Utility Preservation." See also Turner's "Reframing Impact" sequence at https://www.alignmentforum.org/s/7CdoznhJaLEKHwvJW and additional discussion in his "Towards a New Impact Measure," https://www.alignmentforum.org/posts/yEa7kwoMpsBgaBCgb/towards-a-new-impact-measure; he writes, "I have a theory that AUP seemingly works for advanced agents not because the content of the attainable set's utilities actually matters, but rather because there exists a common utility achievement currency of *power*." See Turner, "Optimal Far-sighted Agents Tend to Seek Power." For more on the notion of power in an AI safety context, including an information-theoretic account of "empowerment," see Amodei et al., "Concrete Problems in AI Safety," which, in turn, references Salge, Glackin, and Polani, "Empowerment: An Introduction," and Mohamed and Rezende, "Variational Information Maximisation for Intrinsically Motivated Reinforcement Learning."

50. Alexander Turner, personal interview, July 11, 2019.

51. Wiener, "Some Moral and Technical Consequences of Automation."

52. According to Paul Christiano, "corrigibility" as a tenet of AI safety began with the

Machine Intelligence Research Institute's Eliezer Yudkowsky, and the name itself came from Robert Miles. See Christiano's "Corrigibility," https://ai-alignment. com/corrigibility-3039e668638.

53. Dadich, Ito, and Obama, "Barack Obama, Neural Nets, Self-Driving Cars, and the Future of the World."

54. Dylan Hadfield-Menell, personal interview, March 15, 2018.

55. Turing, "Can Digital Computers Think?"

56. Russell, *Human Compatible*. Russell made this point earlier but in different words in "Should We Fear Supersmart Robots?" The better part of a decade before that, Steve Omohundro, in "The Basic AI Drives," noted that "almost all systems [will] protect their utility functions from modification."

57. Soares et al., "Corrigibility." See also the related work by Armstrong: "Motivated Value Selection for Artificial Agents." For other takes on the interesting problems that arise when modifying or interrupting AI agents, see, e.g., Orseau and Armstrong, "Safely Interruptible Agents," and Riedl and Harrison, "Enter the Matrix." For research on robots that *actually ask people not to shut them off*, and whether humans comply, see Horstmann et al., "Do a Robot's Social Skills and Its Objection Discourage Interactants from Switching the Robot Off?"

58. See Nate Soares et al., "Corrigibility," presentation at AAAI-15, January 25, 2015, https://intelligence.org/wp-content/uploads/2015/01/AAAI-15-corrigibility-slides.pdf.

59. Russell, "Should We Fear Supersmart Robots?"

60. Dylan Hadfield-Menell, "The Off-Switch" (lecture), Colloquium Series on Robust and Beneficial AI (CSRBAI), Machine Intelligence Research Institute, Berkeley, CA, June 8, 2016, https://www.youtube.com/watch?v=t06IciZknDg.

61. Milli et al., "Should Robots Be Obedient?". For other work on cases where it may be best for a system to disobey a human command, see, e.g., Coman et al., "Social Attitudes of AI Rebellion," and Aha and Coman, "The AI Rebellion."

62. Smitha Milli, "Approaches to Achieving AI Safety" (interview), Melbourne, Australia, August 2017, https://www.youtube.com/watch?v=l82SQfrbdj4.

63. For more on corrigibility and model misspecification using this paradigm, see also, e.g., Carey, "Incorrigibility in the CIRL Framework."

64. Dylan Hadfield-Menell, personal interview, March 15, 2018.

65. Russell, *Human Compatible*.

66. Hadfield-Menell et al., "Inverse Reward Design."

67. For a related framing and approach to this problem from DeepMind's Tom Everitt, along with a group of collaborators at both DeepMind and the Australian National University, see Everitt et al., "Reinforcement Learning with a Corrupted Reward Channel."

68. Hadfield-Menell et al., "Inverse Reward Design."

69. Prümmer, *Handbook of Moral Theology.*

70. Rousseau, *Emile; or, On Education.*

71. Actually, it appears that almost all of the actual historical debate concerned cases where it was argued whether an act was sinful or not—but not cases where it was also considered possible that *not* taking that action was sinful. For more discussion on this point, see Sepielli, " 'Along an Imperfectly-Lighted Path.' "

72. The original phrasing of this much-quoted adage appeared on the editorial page of the *San Diego Union* on September 20, 1930: "Retail jewelers assert that every man should carry two watches. But a man with one watch knows what time it is, and a man with two watches could never be sure."

73. See Prümmer, *Handbook of Moral Theology,* §§145–56.

74. See Connell, "Probabilism."

75. Prümmer, *Handbook of Moral Theology.*

76. Will MacAskill, personal interview, May 11, 2018.

77. One philosopher responsible for revisiting these questions in recent years is Michigan Technological University's Ted Lockhart. See Lockhart, *Moral Uncertainty and Its Consequences.* As he puts it: "What shall I do when I am uncertain what I morally ought to do? Philosophers have paid little attention to this sort of question."

78. For more on the ideas of effective altruism, see MacAskill, *Doing Good Better,* and Singer, *The Most Good You Can Do.* For more on the history of the term "effective altruism," see MacAskill's "The History of the Term 'Effective Altruism,' " Effective Altruism Forum, http://effective-altruism.com/ea/5w/the_history_of_the_term _effective_altruism/.

79. MacAskill, Bykvist, and Ord, *Moral Uncertainty.* See also the earlier book by Lockhart: *Moral Uncertainty and Its Consequences.*

80. See, e.g., Lockhart, *Moral Uncertainty and Its Consequences,* and Gustafsson and Torpman, "In Defence of My Favourite Theory."

81. There are purely ordinal theories, for instance, and purely deontic theories. Other troubles exist as well; for more discussion, see MacAskill, Bykvist, and Ord, *Moral Uncertainty.*

82. For more on social choice theory, see, e.g., Mueller, *Public Choice III,* and Sen, *Collective Choice and Social Welfare;* and for more on social choice theory from a computational perspective, see, e.g., Brandt et al., *Handbook of Computational Social Choice.*

83. On the idea of a "moral parliament," see Bostrom, "Moral Uncertainty—Towards a Solution?," and on "moral trade," see Ord, "Moral Trade."

84. For one approach, see, e.g., Humphrys, "Action Selection in a Hypothetical House Robot."

85. See "Allocation of Discretionary Funds from Q1 2019," *The GiveWell Blog*, https://blog.givewell.org/2019/06/12/allocation-of-discretionary-funds-from-q1-2019/.

86. For more discussion in this vein, see also Ord, *The Precipice*.

87. Paul Christiano, personal interview, July 1, 2019.

88. On this theme, see also Sepielli, "What to Do When You Don't Know What to Do When You Don't Know What to Do . . ."

89. Shlegeris, "Why I'm Less of a Hedonic Utilitarian Than I Used to Be."

CONCLUSION

1. Bertrand Russell, "The Philosophy of Logical Atomism," in *Logic and Knowledge*.

2. See Knuth, "Structured Programming with *Go to* Statements," and Knuth, "Computer Programming as an Art," both from 1974. The quote has a somewhat convoluted history, with Knuth himself, 15 years later, in 1989, referring to it as "[C.A.R.] Hoare's dictum" in Knuth, "The Errors of TeX." However, there appears not to be any evidence that the quote was Hoare's. When Hoare himself was asked about the quote in 2004, he said he had "no recollection" of where it came from, suggesting perhaps it was the kind of thing that Edsger Dijkstra might have said, and adding, "I think it would be fair for you assume it is common culture or folklore" (Hans Gerwitz, "Premature Optimization Is the Root of All Evil," https://hans.gerwitz.com/2004/08/12/premature-optimization-is-the-root-of-all-evil.html). Knuth, in 2012, conceded, "I did say things like 'Premature optimization is the root of all evil in programming' " (Mark Harrison, "A note from Donald Knuth about TAOCP," http://codehaus.blogspot.com/2012/03/note-from-donald-knuth-about-taocp.html). In all likelihood, the phrase was indeed his own.

3. The US Centers for Disease Control warns about babies getting hypothermia in cold bedrooms, and in 2017 a healthy, middle-aged man in Thailand died in his bedroom after hypothermic shock, simply from leaving fans running on a cold night. See, respectively, Centers for Disease Control and Prevention, "Prevent Hypothermia and Frostbite," https://www.cdc.gov/disasters/winter/staysafe/hypother mia.html, and *Straits Times*, "Thai Man Dies From Hypothermia After Sleeping With 3 Fans Blowing at Him," November 6, 2017, https://www.straitstimes.com/asia/se-asia/thai-man-dies-from-hypothermia-after-sleeping-with-3-fans-blowing-at-him.

4. Wiener, *God and Golem, Inc.* In 2016, MIRI researcher Jessica Taylor explored the related idea of what she calls "quantilizers": agents that, rather than fully optimizing a potentially problematic metric, settle for behavior that is "good enough"; see Taylor, "Quantilizers." This also shares some resemblance to the regularization method known as "early stopping"; see Yao, Rosasco, and Caponnetto, "On

Early Stopping in Gradient Descent Learning." For further discussion, in an AI safety context, about when "a metric which can be used to improve a system is used to such an extent that further optimization is ineffective or harmful," see Manheim and Garrabrant, "Categorizing Variants of Goodhart's Law."

5. "£13.3m Boost for Oxford's Future of Humanity Institute," http://www.ox.ac.uk/ news/2018-10-10-£133m-boost-oxford's-future-humanity-institute.

6. Huxley, *Ends and Means.*

7. For a recent general-audience discussion on gender bias in medicine, see Perez, *Invisible Women.* For academic literature on the subject, see, e.g., Mastroianni, Faden, and Federman, *Women and Health Research,* and Marts and Keitt, "Foreword." There is also concern within the medical field that the elderly—who make up one of the fastest-growing demographic groups at present—are also significantly underrepresented in medical trials; see, e.g., Vitale et al., "Under-Representation of Elderly and Women in Clinical Trials," and Shenoy and Harugeri, "Elderly Patients' Participation in Clinical Trials."

   This is an active area of study in quite a number of fields; for a recent discussion of zoological museum collections, for instance, see Cooper et al., "Sex Biases in Bird and Mammal Natural History Collections."

8. See, e.g., Bara Fintel, Athena T. Samaras, and Edson Carias, "The Thalidomide Tragedy: Lessons for Drug Safety and Regulation," *Helix,* https://helix. northwestern.edu/article/thalidomide-tragedy-lessons-drug-safety-and-regulation; Nick McKenzie and Richard Baker, "The 50-Year Global Cover-up," *Sydney Morning Herald,* July 26, 2012, https://www.smh.com.au/national/the-50-year-global-cover-up-20120725-22r5c.html; and "Thalidomide," Brought to Life, Science Museum, http://broughttolife.sciencemuseum.org.uk/broughttolife/ themes/controversies/thalidomide, along with, e.g., Marts and Keitt, "Foreword."

9. Sometimes this consensus is downright offensive. In 2019, the AI Now Institute's Kate Crawford and artist Trevor Paglen dug into the ImageNet data and found some bizarre and shocking items. See their "Excavating AI": https://www. excavating.ai. Their work led to ImageNet removing six hundred thousand images of people, labeled everything from "kleptomaniac" to "redneck" to "trollop," from the dataset.

10. The original ImageNet data actually contained twenty thousand categories; the ImageNet Large Scale Visual Recognition Challenge (ILSVRC) won in 2012 by AlexNet used a pared-down version of the data containing only one thousand categories. See Deng et al., "ImageNet," and Russakovsky et al., "ImageNet Large Scale Visual Recognition Challenge."

11. Stuart Russell has argued this point, and he suggests using machine learning itself to infer a more nuanced representation for the varying costs of label errors. See, e.g., Russell, *Human Compatible.*

12. See Mikolov et al., "Efficient Estimation of Word Representations in Vector Space."

13. In May 2019, the arXiv paper Nissim, van Noord, and van der Goot, "Fair Is Better Than Sensational," made a bit of a splash; it critiqued, in sharp terms, the "parallelogram" method for analogies. The authors of Bolukbasi et al., "Man Is to Computer Programmer as Woman Is to Homemaker?," replied on Twitter, and an informal discussion can be found at https://twitter.com/adamfungi/status/1133865428663635968. The Bolukbasi et al. paper itself, in its Appendix A and Appendix B, discusses subtle but consequential differences between the 3CosAdd algorithm and the one its authors used.

14. Tversky, "Features of Similarity."

15. See, e.g., Chen, Peterson, and Griffiths, "Evaluating Vector-Space Models of Analogy."

16. On the possibly criminogenic effects of incarceration itself, see, e.g., discussion in Stemen, "The Prison Paradox" (in particular footnote 23), and Roodman, "Aftereffects."

17. See, e.g., Jung et al., "Eliciting and Enforcing Subjective Individual Fairness."

18. See Poursabzi-Sangdeh et al., "Manipulating and Measuring Model Interpretability."

19. Bryson, "Six Kinds of Explanation for AI," for instance, argues that "explanations" in the context of AI ought to include not just the inner workings of the system but the "human actions that led to the system being released and sold as a product and/or operated as a service."

20. See Ghorbani, Abid, and Zou, "Interpretation of Neural Networks Is Fragile."

21. See Mercier and Sperber, "Why Do Humans Reason?" An intriguing research direction in AI alignment involves developing machine-learning systems able to engage in debate with one another; see Irving, Christiano, and Amodei, "AI Safety via Debate."

22. Jan Leike, "General Reinforcement Learning" (lecture), Colloquium Series on Robust and Beneficial AI 2016, Machine Intelligence Research Institute, Berkeley, California, June 9, 2016, https://www.youtube.com/watch?v=hSiuJuvTBoE&t=239s. The idea of reinforcement learning that avoids unrecoverable mistakes is an active area of research; see, e.g., Saunders et al., "Trial Without Error," and Eysenbach et al., "Leave No Trace," for some approaches.

23. See Omohundro, "The Basic AI Drives." See also, for instance, L. M. Montgomery's 1921 Anne of Green Gables novel *Rilla of Ingleside,* in which Rilla muses, "I wouldn't *want* to go back and be the girl I was two years ago, not even if I could. . . . And still. . . . At the end of two more years I might look back and be thankful for the development they had brought me, too; but I don't want it now." Miss Oliver replies to her, "We never do. That is why we are not left to choose our own means and measure of development, I suppose."

24. See Paul, *Transformative Experience.*

25. The subfield of "multi-agent reinforcement learning" works on problems like these. See, e.g., Foerster et al., "Learning to Communicate with Deep Multi-Agent Reinforcement Learning," and Foerster et al., "Learning with Opponent-Learning Awareness."

26. Piaget, *The Construction of Reality in the Child*.

27. Demski and Garrabrant, "Embedded Agency."

28. See, e.g., Evans, Stuhlmüller, and Goodman, "Learning the Preferences of Ignorant, Inconsistent Agents"; Evans and Goodman, "Learning the Preferences of Bounded Agents"; and Bourgin et al., "Cognitive Model Priors for Predicting Human Decisions."

29. See Ziebart et al., "Maximum Entropy Inverse Reinforcement Learning," and Ziebart, Bagnell, and Dey, "Modeling Interaction via the Principle of Maximum Causal Entropy." Much recent work in robotics and autonomous cars uses this same model of human behavior, sometimes referred to as "noisily rational" behavior or "Boltzmann (ir)rationality." See, e.g., Finn, Levine, and Abbeel, "Guided Cost Learning"; Sadigh et al., "Planning for Autonomous Cars That Leverage Effects on Human Actions"; and Kwon et al., "When Humans Aren't Optimal."

30. As Stuart Russell put it in his original 1998 paper, "Can we determine the reward function by observation *during* rather than *after* learning?" Russell, "Learning Agents for Uncertain Environments (Extended Abstract)."

31. This remains very much an open research question. For recent work, see Chan et al., "The Assistive Multi-Armed Bandit."

32. Berkeley's Smitha Milli and Anca Drăgan have explored this question: see Milli and Drăgan, "Literal or Pedagogic Human?"

33. Stefano Ermon, interview by Ariel Conn, Future of Life Institute, January 26, 2017, https://futureoflife.org/2017/01/26/stefano-ermon-interview/.

34. Roman Yampolskiy, interview by Ariel Conn, Future of Life Institute, January 18, 2017, https://futureoflife.org/2017/01/18/roman-yampolskiy-interview/.

35. See, for instance, Arrow, "A Difficulty in the Concept of Social Welfare."

36. See also the efforts in Recht et al., "Do ImageNet Classifiers Generalize to Image-Net?," to reproduce the accuracy of image recognition systems like AlexNet and others on new photos—there was a persistent gap in accuracy, leading the authors to speculate that, no matter how closely they tried to mimic the original CIFAR-10 and ImageNet methodology, the images and human-supplied labels were simply unavoidably a bit different in 2019 than they were in, say, 2012.

37. Latour, *Pandora's Hope*.

38. See Paul Christiano, "What Failure Looks Like," AI Alignment Forum, March 17, 2019, https://www.alignmentforum.org/posts/HBxe6wdjxK239zajf/what-failure-looks-like. "The stereotyped image of AI catastrophe is a powerful, malicious AI system that takes its creators by surprise and quickly achieves a decisive advan-

tage over the rest of humanity. I think this is probably not what failure will look like," he writes. Instead, he worries that "machine learning will increase our ability to 'get what we can measure,' which could cause a slow-rolling catastrophe."

39. National Transportation Safety Board, 2019. *Collison Between Vehicle Controlled by Developmental Automated Driving System and Pedestrian.* Highway Accident Report NTSB/HAR-19/03. Washington, DC.

40. See, e.g., Odell, *How to Do Nothing.*

41. Read, *The Grass Roots of Art.*

42. Turing et al., "Can Automatic Calculating Machines Be Said to Think?"

ACKNOWLEDGMENTS

1. McCulloch, *Finality and Form.*

BIBLIOGRAPHY

Abbeel, Pieter. "Apprenticeship Learning and Reinforcement Learning with Application to Robotic Control." PhD thesis, Stanford University, 2008.

Abbeel, Pieter, Adam Coates, and Andrew Y. Ng. "Autonomous Helicopter Aerobatics Through Apprenticeship Learning." *International Journal of Robotics Research* 29, no. 13 (2010): 1608–39.

Abbeel, Pieter, Adam Coates, Morgan Quigley, and Andrew Y. Ng. "An Application of Reinforcement Learning to Aerobatic Helicopter Flight." In *Advances in Neural Information Processing Systems*, 1–8, 2007.

Abbeel, Pieter, and Andrew Y. Ng. "Apprenticeship Learning via Inverse Reinforcement Learning." In *Proceedings of the 21st International Conference on Machine Learning*. ACM, 2004.

Ackley, David, and Michael Littman. "Interactions Between Learning and Evolution." In *Artificial Life II: SFI Studies in the Sciences of Complexity*, 10:487–509. Addison-Wesley, 1991.

ACLU Foundation. "The War on Marijuana in Black and White," 2013. https://www.aclu.org/report/report-war-marijuana-black-and-white?redirect=criminal-law-reform/war-marijuana-black-and-white.

Adebayo, Julius, Justin Gilmer, Michael Muelly, Ian Goodfellow, Moritz Hardt, and Been Kim. "Sanity Checks for Saliency Maps." In *Advances in Neural Information Processing Systems*, 9505–15, 2018.

Aha, David W., and Alexandra Coman. "The AI Rebellion: Changing the Narrative." In *Thirty-First AAAI Conference on Artificial Intelligence*, 2017.

Akrour, Riad, Marc Schoenauer, and Michèle Sebag. "APRIL: Active Preference-Learning Based Reinforcement Learning." In *Joint European Conference on Machine Learning and Knowledge Discovery in Databases*, 116–31. Springer, 2012.

Akrour, Riad, Marc Schoenauer, Michèle Sebag, and Jean-Christophe Souplet. "Programming by Feedback." In *International Conference on Machine Learning*, 1503–11. JMLR, 2014.

Aksenov, Pavel. "Stanislav Petrov: The Man Who May Have Saved the World." BBC News, September 25, 2013. https://www.bbc.com/news/world-europe-24280831.

Al-Shawaf, Laith, Daniel Conroy-Beam, Kelly Asao, and David M. Buss. "Human Emotions: An Evolutionary Psychological Perspective." *Emotion Review* 8, no. 2 (2016): 173–86.

Alexander, R. McNeill. "The Gaits of Bipedal and Quadrupedal Animals." *International Journal of Robotics Research* 3, no. 2 (1984): 49–59.

———. *The Human Machine: How the Body Works*. Columbia University Press, 1992.

———. *Optima for Animals*. Princeton University Press, 1996.

Alexander, S.A.H. "Sex, Arguments, and Social Engagements in Martial and Premarital Relations." Master's thesis, University of Missouri–Kansas City, 1971.

Amodei, Dario, Chris Olah, Jacob Steinhardt, Paul Christiano, John Schulman, and Dan Mané. "Concrete Problems in AI Safety." arXiv Preprint arXiv:1606.06565, 2016.

Ampère, André-Marie. *Essai sur la philosophie des sciences; ou, Exposition analytique d'une classification naturelle de toutes les connaissances humaines*. Paris: Bachelier, 1834.

Anantharaman, Thomas S. "Evaluation Tuning for Computer Chess: Linear Discriminant Methods." *ICGA Journal* 20, no. 4 (1997): 224–42.

Anderson, James A., and Edward Rosenfeld. *Talking Nets: An Oral History of Neural Networks*. MIT Press, 1998.

Andre, David, and Astro Teller. "Evolving Team Darwin United." In *RoboCup-98*, 346–51. Springer, 1999.

Andrew, A. M. "Machines Which Learn." *New Scientist*, November 27, 1958.

Andrews, D. A. "The Level of Service Inventory (LSI): The First Follow-up." *Ontario Ministry of Correctional Services*. Toronto, 1982.

Andrews, D. A., and J. L. Bonta. "The Level of Service Inventory–Revised." Toronto: Multi-Health Systems, 1995.

Andrews, D. A., James Bonta, and J. Stephen Wormith. "The Recent Past and Near Future of Risk and/or Need Assessment." *Crime & Delinquency* 52, no. 1 (2006): 7–27.

Angelino, Elaine, Nicholas Larus-Stone, Daniel Alabi, Margo Seltzer, and Cynthia Rudin. "Learning Certifiably Optimal Rule Lists for Categorical Data." *Journal of Machine Learning Research* 18 (2018): 1–78.

Angwin, Julia. *Dragnet Nation: A Quest for Privacy, Security, and Freedom in a World of Relentless Surveillance*. Times Books, 2014.

Angwin, Julia, and Jeff Larson. "ProPublica Responds to Company's Critique of Machine Bias Story." ProPublica, July 29, 2016.

Angwin, Julia, Jeff Larson, Surya Mattu, and Lauren Kirchner. "Machine Bias." Pro-Publica, May 23, 2016.

"Another Day the World Almost Ended." RT, May 19, 2010. https://www.rt.com/usa/nuclear-war-stanislav-petrov/.

Anthony, Thomas, Zheng Tian, and David Barber. "Thinking Fast and Slow with Deep Learning and Tree Search." In *Advances in Neural Information Processing Systems*, 5360–70, 2017.

Arendt, Hannah. *The Human Condition*. University of Chicago Press, 1958.

Argall, Brenna D., Sonia Chernova, Manuela Veloso, and Brett Browning. "A Survey of Robot Learning from Demonstration." *Robotics and Autonomous Systems* 57, no. 5 (2009): 469–83.

Armstrong, Stuart. "The Mathematics of Reduced Impact: Help Needed," LessWrong, February 16, 2012. https://www.lesswrong.com/posts/8Nwg7kqAfCM46tuHq/the-mathematics-of-reduced-impact-help-needed.

——. "Motivated Value Selection for Artificial Agents." In *Artificial Intelligence and Ethics: Papers from the 2015 AAAI Workshop*. AAAI Press, 2015.

——. "Reduced Impact AI: No Back Channels," LessWrong, November 11, 2013. https://www.lesswrong.com/posts/gzQT5AAw8oQdzuwBG/reduced-impact-ai-no-back-channels.

Armstrong, Stuart, and Benjamin Levinstein. "Low Impact Artificial Intelligences." arXiv Preprint arXiv:1705.10720, 2017.

Arrow, Kenneth J. "A Difficulty in the Concept of Social Welfare." *Journal of Political Economy* 58, no. 4 (1950): 328–46.

Ashley, Maurice. *Chess for Success*. Broadway Books, 2005.

Athalye, Vivek R., Fernando J. Santos, Jose M. Carmena, and Rui M. Costa. "Evidence for a Neural Law of Effect." *Science* 359, no. 6379 (2018): 1024–29.

Aurelius, Marcus. *The Emperor Marcus Aurelius: His Conversation with Himself*. Translated by Jeremy Collier. London: R. Sare, 1701.

Aytar, Yusuf, Tobias Pfaff, David Budden, Thomas Paine, Ziyu Wang, and Nando de Freitas. "Playing Hard Exploration Games by Watching YouTube." In *Advances in Neural Information Processing Systems*, 2935–45. MIT Press, 2018.

Bailey, Robert E., and J. Arthur Gillaspy Jr. "Operant Psychology Goes to the Fair: Marian and Keller Breland in the Popular Press, 1947–1966." *Behavior Analyst* 28, no. 2 (2005): 143–59.

Bain, Alexander. *The Senses and the Intellect*. London: John W. Parker and Son, 1855.

Baldassarre, Gianluca, and Marco Mirolli, eds. *Intrinsically Motivated Learning in Natural and Artificial Systems*. Springer, 2013.

Balentine, Bruce. *It's Better to Be a Good Machine Than a Bad Person: Speech Recognition and Other Exotic User Interfaces in the Twilight of the Jetsonian Age*. ICMI Press, 2007.

Barabas, Chelsea, Madars Virza, Karthik Dinakar, Joichi Ito, and Jonathan Zittrain. "Interventions over Predictions: Reframing the Ethical Debate for Actuarial Risk Assessment." In *Proceedings of Machine Learning Research*, 81 (2018): 62–76.

Barber, David, and Christopher M. Bishop. "Ensemble Learning in Bayesian Neural Networks." In *Neural Networks and Machine Learning*, 215–37. Springer, 1998.

Barocas, Solon, and Andrew D. Selbst. "Big Data's Disparate Impact." *California Law Review* 104 (2016): 671–732.

Barto, Andrew G. "Intrinsic Motivation and Reinforcement Learning." In *Intrinsically Motivated Learning in Natural and Artificial Systems*, edited by Gianluca Baldassarre and Marco Mirolli, 17–47. Springer, 2013.

Barto, Andrew, Marco Mirolli, and Gianluca Baldassarre. "Novelty or Surprise?" *Frontiers in Psychology* 4 (2013): 907.

Barto, Andrew G., Satinder Singh, and Nuttapong Chentanez. "Intrinsically Motivated Learning of Hierarchical Collections of Skills." In *Proceedings of the 3rd International Conference on Development and Learning*, 112–19, 2004.

Barto, Andrew G., Richard S. Sutton, and Charles W. Anderson. "Neuronlike Adaptive Elements That Can Solve Difficult Learning Control Problems." *IEEE Transactions on Systems, Man, and Cybernetics* 13, no. 5 (1983): 834–46.

Bateson, Melissa, Daniel Nettle, and Gilbert Roberts. "Cues of Being Watched Enhance Cooperation in a Real-World Setting." *Biology Letters* 2, no. 3 (2006): 412–14.

Bellemare, Marc G., Yavar Naddaf, Joel Veness, and Michael Bowling. "The Arcade Learning Environment: An Evaluation Platform for General Agents." *Journal of Artificial Intelligence Research* 47 (2013): 253–79.

Bellemare, Marc G., Sriram Srinivasan, Georg Ostrovski, Tom Schaul, David Saxton, and Rémi Munos. "Unifying Count-Based Exploration and Intrinsic Motivation." In *Advances in Neural Information Processing Systems*, 1471–79, 2016.

Bellman, Richard. *Dynamic Programming.* Princeton, NJ: Princeton University Press, 1957.

Bendale, Abhijit, and Terrance Boult. "Towards Open World Recognition." In *Proceedings of the IEEE Conference on Computer Vision and Pattern Recognition*, 1893–1902. CVPR, 2015.

Bengio, Yoshua. "Neural Net Language Models." *Scholarpedia* 3, no. 1 (2008): 3881. http://www.scholarpedia.org/article/Neural_net_language_models.

Bengio, Yoshua, Réjean Ducharme, Pascal Vincent, and Christian Jauvin. "A Neural Probabilistic Language Model." *Journal of Machine Learning Research* 3 (2003): 1137–55.

Bengio, Yoshua, Jérôme Louradour, Ronan Collobert, and Jason Weston. "Curriculum Learning." In *Proceedings of the 26th Annual International Conference on Machine Learning*, 41–48. ACM, 2009.

Benjamin, Ludy T., Jr., and Darryl Bruce. "From Bottle-Fed Chimp to Bottlenose Dolphin: A Contemporary Appraisal of Winthrop Kellogg." *Psychological Record* 32, no. 4 (1982): 461–82.

Bentham, Jeremy. *An Introduction to the Principles of Morals and Legislation.* London: T. Payne & Son, 1789.

———. "Letter to Jacques Pierre Brissot de Warville." In *The Correspondence of Jeremy Bentham.* Vol. 4, edited by Alexander Taylor Milne. UCL Press, 2017.

———. "Preface," i–vii. *Panopticon: Postscript; Part I.* Mews-Gate, London: T. Payne, 1791.

Berk, Richard. *Criminal Justice Forecasts of Risk: A Machine Learning Approach.* Springer Science & Business Media, 2012.

Berk, Richard, Hoda Heidari, Shahin Jabbari, Michael Kearns, and Aaron Roth. "Fairness in Criminal Justice Risk Assessments: The State of the Art." arXiv Preprint arXiv:1703.09207, 2017.

Berlyne, Daniel E. *Conflict, Arousal, and Curiosity.* McGraw-Hill, 1960.

———. "'Interest' as a Psychological Concept." *British Journal of Psychology, General Section* 39, no. 4 (1949): 184–95.

———. "Uncertainty and Conflict: A Point of Contact Between Information-Theory and Behavior-Theory Concepts." *Psychological Review* 64, no. 6 (1957): 329–39.

Bernoulli, Daniel. "Specimen theoriae novae de mensura sortis." *Comentarii Academiae Scientarum Imperialis Petropolitanae*, 1738, 175–92.

Bernstein, Jeremy. "A.I." *New Yorker,* December 6, 1981.

Berridge, Kent C. "Food Reward: Brain Substrates of Wanting and Liking." *Neuroscience & Biobehavioral Reviews* 20, no. 1 (1996): 1–25.

Berridge, Kent C., Terry E. Robinson, and J. Wayne Aldridge. "Dissecting Components of Reward: 'Liking,' 'Wanting,' and Learning." *Current Opinion in Pharmacology* 9, no. 1 (2009): 65–73.

Bertrand, Marianne, and Sendhil Mullainathan. "Are Emily and Greg More Employable Than Lakisha and Jamal? A Field Experiment on Labor Market Discrimination." *American Economic Review* 94, no. 4 (2004): 991–1013.

Bever, Lindsey. "A Man Collapsed with 'Do Not Resuscitate' Tattooed on His Chest. Doctors Didn't Know What to Do." *Washington Post,* December 1, 2017.

Billard, Aude G., Sylvain Calinon, and Florent Guenter. "Discriminative and Adaptive Imitation in Uni-Manual and Bi-Manual Tasks." *Robotics and Autonomous Systems* 54, no. 5 (2006): 370–84.

Blei, David M., Andrew Y. Ng, and Michael I. Jordan. "Latent Dirichlet Allocation." *Journal of Machine Learning Research* 3, nos. 4–5 (2003): 993–1022.

Blickensderfer, Elizabeth, Janis A. Cannon-Bowers, and Eduardo Salas. "Cross-Training and Team Performance." In *Making Decisions Under Stress: Implications for Individual and Team Training,* 299–311. Washington, DC: American Psychological Association, 1998.

Blomberg, Thomas, William Bales, Karen Mann, Ryan Meldrum, and Joe Nedelec. "Validation of the COMPAS Risk Assessment Classification Instrument." Broward Sheriff's Office, Department of Community Control, 2010.

Blundell, Charles, Julien Cornebise, Koray Kavukcuoglu, and Daan Wierstra. "Weight Uncertainty in Neural Networks." In *Proceedings of the 32nd International Conference on Machine Learning*, 2015.

Bojarski, Mariusz, Davide Del Testa, Daniel Dworakowski, Bernhard Firner, Beat Flepp, Prasoon Goyal, Lawrence D. Jackel, et al. "End to End Learning for Self-Driving Cars." CoRR abs/1604.07316 (2016).

Bolam, J. Paul, and Eleftheria K. Pissadaki. "Living on the Edge with Too Many Mouths to Feed: Why Dopamine Neurons Die." *Movement Disorders* 27, no. 12 (2012): 1478–83.

Bolukbasi, Tolga, Kai-Wei Chang, James Y. Zou, Venkatesh Saligrama, and Adam T. Kalai. "Man Is to Computer Programmer as Woman Is to Homemaker? Debiasing Word Embeddings." In *Advances in Neural Information Processing Systems*, 4349–57, 2016.

Bolukbasi, Tolga, Kai-Wei Chang, James Zou, Venkatesh Saligrama, and Adam Kalai. "Quantifying and Reducing Stereotypes in Word Embeddings." In *2016 ICML Workshop on #Data4good: Machine Learning in Social Good Applications*, 2016.

Bonawitz, Elizabeth Baraff, Tessa J. P. van Schijndel, Daniel Friel, and Laura Schulz. "Children Balance Theories and Evidence in Exploration, Explanation, and Learning." *Cognitive Psychology* 64, no. 4 (2012): 215–34.

Bostrom, Nick. "Astronomical Waste: The Opportunity Cost of Delayed Technological Development." *Utilitas* 15, no. 3 (2003): 308–14.

———. "Moral Uncertainty—Towards a Solution?" *Overcoming Bias*, January 1, 2009. http://www.overcomingbias.com/2009/01/moral-uncertainty-towards-a-solution.html.

———. *Superintelligence: Paths, Dangers, Strategies*. Oxford University Press, 2014.

Bostrom, Nick, and Milan M. Ćirković. *Global Catastrophic Risks*. Oxford: Oxford University Press, 2008.

Bourgin, David D., Joshua C. Peterson, Daniel Reichman, Thomas L. Griffiths, and Stuart J. Russell. "Cognitive Model Priors for Predicting Human Decisions." In *Proceedings of the 36th International Conference on Machine Learning (ICML)*, 2019.

Box, George E. P. "Robustness in the Strategy of Scientific Model Building." In *Robustness in Statistics*, edited by Robert L. Launer and Graham N. Wilkinson, 201–36. Academic Press, 1979.

———. "Science and Statistics." *Journal of the American Statistical Association* 71, no. 356 (December 1976): 791–99.

Brandt, Felix, Vincent Conitzer, Ulle Endriss, Jérôme Lang, and Ariel D Procaccia. *Handbook of Computational Social Choice*. Cambridge University Press, 2016.

Breiman, Leo, Jerome H. Friedman, Richard A. Olshen, and Charles J. Stone. *Classification and Regression Trees*. Chapman & Hall/CRC, 1984.

Breland, Keller, and Marian Breland. "The Misbehavior of Organisms." *American Psychologist* 16, no. 11 (1961): 681–84.

Brennan, T., W. Dieterich, and W. Oliver. "COMPAS: Correctional Offender Management for Alternative Sanctions." Northpointe Institute for Public Management, 2007.

Brennan, Tim, and William Dieterich. "Correctional Offender Management Profiles for Alternative Sanctions (COMPAS)." In *Handbook of Recidivism Risk/Needs Assessment Tools*, 49–75. Wiley Blackwell, 2018.

Brennan, Tim, and Dave Wells. "The Importance of Inmate Classification in Small Jails." *American Jails* 6, no. 2 (1992): 49–52.

Brickman, Philip. "Hedonic Relativism and Planning the Good Society." *Adaptation Level Theory* (1971): 287–301.

Brickman, Philip, Dan Coates, and Ronnie Janoff-Bulman. "Lottery Winners and Accident Victims: Is Happiness Relative?" *Journal of Personality and Social Psychology* 36, no. 8 (1978): 917.

Brinnin, John Malcolm. *The Third Rose: Gertrude Stein and Her World*. Boston: Little, Brown, 1959.

Brooks, Rodney A. *Flesh and Machines: How Robots Will Change Us*. Pantheon, 2003.

———. "Intelligence Without Representation." *Artificial Intelligence* 47 (1991): 139–59.

Brown, Tom B., Nicholas Carlini, Chiyuan Zhang, Catherine Olsson, Paul Christiano, and Ian Goodfellow. "Unrestricted Adversarial Examples." arXiv Preprint arXiv:1809.08352, 2018.

Bryson, Joanna. "Six Kinds of Explanation for AI (One Is Useless)." *Adventures in NI* (blog), September 19, 2019. https://joanna-bryson.blogspot.com/2019/09/six-kinds -of-explanation-for-ai-one-is.html.

Buchsbaum, Daphna, Alison Gopnik, Thomas L. Griffiths, and Patrick Shafto. "Children's Imitation of Causal Action Sequences Is Influenced by Statistical and Pedagogical Evidence." *Cognition* 120, no. 3 (2011): 331–40.

Buolamwini, Joy, and Timnit Gebru. "Gender Shades: Intersectional Accuracy Disparities in Commercial Gender Classification." In *Conference on Fairness, Accountability and Transparency*, 77–91, 2018.

Burda, Yuri, Harri Edwards, Deepak Pathak, Amos Storkey, Trevor Darrell, and Alexei A. Efros. "Large-Scale Study of Curiosity-Driven Learning." *Proceedings of the Seventh International Conference on Learning Representations*, May 2019.

Burda, Yuri, Harrison Edwards, Amos Storkey, and Oleg Klimov. "Exploration by Random Network Distillation." arXiv Preprint arXiv:1810.12894, 2018.

Burgess, Ernest W. "Factors Determining Success or Failure on Parole." In *The Workings of the Indeterminate-Sentence Law and the Parole System in Illinois*, edited by

Andrew A. Bruce, Ernest W. Burgess, Albert J. Harno, and John Landesco. Springfield, IL: Illinois State Board of Parole, 1928.

———. "Prof. Burgess on Parole Reform." *Chicago Tribune,* May 19, 1937.

Burke, Peggy B., ed. *A Handbook for New Parole Board Members.* Association of Paroling Authorities International/National Institute of Corrections, 2003. http://
apaintl.org/Handbook/CEPPParoleHandbook.pdf.

Bykvist, Krister. "Alternative Actions and the Spirit of Consequentialism." *Philosophical Studies* 107, no. 1 (2002): 45–68.

Byrne, Robert. "Chess-Playing Computer Closing in on Champions." *New York Times,*
September 26, 1989.

Çakmak, Maya, and Manuel Lopes. "Algorithmic and Human Teaching of Sequential
Decision Tasks." In *Twenty-Sixth AAAI Conference on Artificial Intelligence,* 2012.

Caldwell, Tommy. *The Push.* Viking, 2017.

Caliskan, Aylin, Joanna J. Bryson, and Arvind Narayanan. "Semantics Derived Automatically from Language Corpora Contain Human-Like Biases." *Science* 356, no.
6334 (2017): 183–86.

Callan, Rachel C., Kristina N. Bauer, and Richard N. Landers. "How to Avoid the
Dark Side of Gamification: Ten Business Scenarios and Their Unintended Consequences." In *Gamification in Education and Business,* 553–68. Springer,
2015.

Campbell, Murray, A. Joseph Hoane Jr., and Feng-hsiung Hsu. "Deep Blue." *Artificial
Intelligence* 134, nos. 1–2 (2002): 57–83.

Cannon-Bowers, Janis A., Eduardo Salas, Elizabeth Blickensderfer, and Clint A. Bow-
ers. "The Impact of Cross-Training and Workload on Team Functioning: A Repli-
cation and Extension of Initial Findings." *Human Factors* 40, no. 1 (1998): 92–101.

Carey, Ryan. "Incorrigibility in the CIRL Framework." In *AIES '18: Proceedings of the
2018 AAAI/ACM Conference on AI, Ethics, and Society,* 30–35. New Orleans:
ACM, 2018.

Carse, James P. *Finite and Infinite Games.* Free Press, 1986

Carter, Shan, Zan Armstrong, Ludwig Schubert, Ian Johnson, and Chris Olah. "Activation Atlas." *Distill,* 2019.

Caruana, Rich, Yin Lou, Johannes Gehrke, Paul Koch, Marc Sturm, and Noémie Elha-
dad. "Intelligible Models for Healthcare: Predicting Pneumonia Risk and Hospital 30-Day Readmission." In *Proceedings of the 21th ACM SIGKDD International
Conference on Knowledge Discovery and Data Mining,* 1721–30. ACM, 2015.

Caruana, Richard. "Explainability in Context—Health." Lecture, Algorithms and
Explanations, NYU School of Law, April 28, 2017.

Caruana, Richard A. "Multitask Learning: A Knowledge-Based Source of Inductive
Bias." In *Machine Learning: Proceedings of the Tenth International Conference,*
41–48, 1993.

Caswell, Estelle. "Color Film Was Built for White People. Here's What It Did to Dark Skin." *Vox*, September 18, 2015. https://www.vox.com/2015/9/18/9348821/photography-race-bias.

Chan, Lawrence, Dylan Hadfield-Menell, Siddhartha Srinivasa, and Anca Drăgan. "The Assistive Multi-Armed Bandit." In *2019 14th ACM/IEEE International Conference on Human-Robot Interaction (HRI)*, 354–63. IEEE, 2019.

"A Chance to Fix Parole in New York." *New York Times*, September 4, 2015.

Chen, Dawn, Joshua C. Peterson, and Thomas L. Griffiths. "Evaluating Vector-Space Models of Analogy." In *Proceedings of the 39th Annual Conference of the Cognitive Science Society*, 2017.

Choi, Jongwook, Yijie Guo, Marcin Moczulski, Junhyuk Oh, Neal Wu, Mohammad Norouzi, and Honglak Lee. "Contingency-Aware Exploration in Reinforcement Learning." In *International Conference on Learning Representations*, 2019.

Chouldechova, Alexandra. "Fair Prediction with Disparate Impact: A Study of Bias in Recidivism Prediction Instruments." *Big Data* 5, no. 2 (2017): 153–63.

———. "Transparency and Simplicity in Criminal Risk Assessment." *Harvard Data Science Review* 2, no. 1 (January 2020).

Chow, C. K. "An Optimum Character Recognition System Using Decision Functions." *IRE Transactions on Electronic Computers* 4 (1957): 247–54.

———. "On Optimum Recognition Error and Reject Tradeoff." *IEEE Transactions on Information Theory* 16, no. 1 (1970): 41–46.

Christian, Brian, and Tom Griffiths. *Algorithms to Live By*. Henry Holt, 2016.

Christiano, Paul F., Jan Leike, Tom Brown, Miljan Martic, Shane Legg, and Dario Amodei. "Deep Reinforcement Learning from Human Preferences." In *Advances in Neural Information Processing Systems*, 4299–4307, 2017.

Christiano, Paul, Buck Shlegeris, and Dario Amodei. "Supervising Strong Learners by Amplifying Weak Experts." arXiv Preprint arXiv:1810.08575, 2018.

Clabaugh, Hinton G. "Foreword." In *The Workings of the Indeterminate-Sentence Law and the Parole System in Illinois*, edited by Andrew A. Bruce, Ernest W. Burgess, Albert J. Harno, and John Landesco. Springfield, IL: Illinois State Board of Parole, 1928.

Clark, Jack, and Dario Amodei. "Faulty Reward Functions in the Wild." *OpenAI Blog*, December 21, 2016. https://blog.openai.com/faulty-reward-functions/.

Coates, Adam, Pieter Abbeel, and Andrew Y. Ng. "Learning for Control from Multiple Demonstrations." In *Proceedings of the 25th International Conference on Machine Learning*, 144–51. ACM, 2008.

Colombetti, Marco, and Marco Dorigo. "Robot Shaping: Developing Situated Agents Through Learning." Berkeley, CA: International Computer Science Institute, 1992.

Coman, Alexandra, Benjamin Johnson, Gordon Briggs, and David W. Aha. "Social Attitudes of AI Rebellion: A Framework." In *Workshops at the Thirty-First AAAI Conference on Artificial Intelligence*, 2017.

Connell, F. J. "Probabilism." In *New Catholic Encyclopedia*, 2nd ed., edited by F. J. Con-
    nell. 11:727. Gale, 2002.

Conway, Flo, and Jim Siegelman. *Dark Hero of the Information Age: In Search of Norbert
    Wiener, the Father of Cybernetics*. Basic Books, 2005.

Cooper, Gregory F., Vijoy Abraham, Constantin F. Aliferis, John M. Aronis, Bruce G.
    Buchanan, Richard Caruana, Michael J. Fine, et al. "Predicting Dire Outcomes of
    Patients with Community Acquired Pneumonia." *Journal of Biomedical Informat-
    ics* 38, no. 5 (2005): 347–66.

Cooper, Gregory F., Constantin F. Aliferis, Richard Ambrosino, John Aronis, Bruce G.
    Buchanan, Richard Caruana, Michael J. Fine, et al. "An Evaluation of Machine-
    Learning Methods for Predicting Pneumonia Mortality." *Artificial Intelligence in
    Medicine* 9, no. 2 (1997): 107–38.

Cooper, Lori, and Paul Aronowitz. "DNR Tattoos: A Cautionary Tale." *Journal of Gen-
    eral Internal Medicine* 27, no. 10 (2012): 1383.

Cooper, Natalie, Alexander L. Bond, Joshua L. Davis, Roberto Portela Miguez, Louise
    Tomsett, and Kristofer M. Helgen. "Sex Biases in Bird and Mammal Natural His-
    tory Collections." *Proceedings of the Royal Society B* 286 (2019): 20192025.

Corbett, Dale, and Roy A. Wise. "Intracranial Self-Stimulation in Relation to the
    Ascending Dopaminergic Systems of the Midbrain: A Moveable Electrode Map-
    ping Study." *Brain Research* 185, no. 1 (1980): 1–15.

Corbett-Davies, Sam. "Algorithmic Decision Making and the Cost of Fairness." Lec-
    ture, Simons Institute for the Theory of Computing, November 17, 2017.

Corbett-Davies, Sam, and Sharad Goel. "The Measure and Mismeasure of Fairness: A
    Critical Review of Fair Machine Learning." arXiv Preprint arXiv:1808.00023, 2018.

Corbett-Davies, Sam, Emma Pierson, Avi Feller, Sharad Goel, and Aziz Huq. "Algo-
    rithmic Decision Making and the Cost of Fairness." In *Proceedings of the 23rd
    ACM SIGKDD International Conference on Knowledge Discovery and Data Min-
    ing*, 797–806. Halifax: ACM, 2017.

Cortes, Corinna, Giulia DeSalvo, and Mehryar Mohri. "Learning with Rejection." In
    International Conference on Algorithmic Learning Theory, 67–82, 2016.

Coulom, Rémi. "Efficient Selectivity and Backup Operators in Monte-Carlo Tree Search."
    In *5th International Conference on Computers and Games*, 72–83. Springer, 2006.

Covey, Mark K., Steve Saladin, and Peter J. Killen. "Self-Monitoring, Surveillance,
    and Incentive Effects on Cheating." *Journal of Social Psychology* 129, no. 5 (1989):
    673–79.

"CPD Welcomes the Opportunity to Comment on Recently Published RAND Review."
    Chicago Police Department news release, August 17, 2016.

Crawford, Kate, and Trevor Paglen. "Excavating AI: The Politics of Images in Machine
    Learning Training Sets," 2019. https://www.excavating.ai.

Creighton, Jolene. "Making AI Safe in an Unpredictable World: An Interview with

Thomas G. Dietterich," September 17, 2018. https://futureoflife.org/2018/09/17/
making-ai-safe-in-an-unpredictable-world-an-interview-with-thomas-g-dietterich/.

Cumming, William W. "A Review of Geraldine Jonçich's *The Sane Positivist: A Biography of Edward L. Thorndike." Journal of the Experimental Analysis of Behavior* 72,
no. 3 (1999): 429–32.

Da, Qing, Yang Yu, and Zhi-Hua Zhou. "Learning with Augmented Class by Exploiting Unlabeled Data." In *Twenty-Eighth AAAI Conference on Artificial Intelligence,*
1760–66, 2014.

Dabkowski, Piotr, and Yarin Gal. "Real Time Image Saliency for Black Box Classifiers."
In *Advances in Neural Information Processing Systems,* 6967–76, 2017.

Dadich, Scott, Joi Ito, and Barack Obama. "Barack Obama, Neural Nets, Self-Driving
Cars, and the Future of the World." *Wired,* October 12, 2016.

Dana, Jason, and Robyn M. Dawes. "The Superiority of Simple Alternatives to Regression for Social Science Predictions." *Journal of Educational and Behavioral Statistics* 29, no. 3 (2004): 317–31.

Dastin, Jeffrey. "Amazon Scraps Secret AI Recruiting Tool That Showed Bias Against
Women." Reuters, October 9, 2018.

Daswani, Mayank, and Jan Leike. "A Definition of Happiness for Reinforcement Learning Agents." In *Artificial General Intelligence,* edited by Jordi Bieger, Ben Goertzel,
and Alexey Potapov, 231–40. Springer, 2015.

Dawes, Robyn M. "A Case Study of Graduate Admissions: Application of Three Principles of Human Decision Making." *American Psychologist* 26, no. 2 (1971): 180–88.

———. "A Look at Analysis." PhD thesis, Harvard University, 1958.

———. "The Robust Beauty of Improper Linear Models in Decision Making." *American
Psychologist* 34, no. 7 (1979): 571–82.

Dawes, Robyn M., and Bernard Corrigan. "Linear Models in Decision Making." *Psychological Bulletin* 81, no. 2 (1974): 95–106.

Dawes, Robyn M., David Faust, and Paul E. Meehl. "Clinical Versus Actuarial Judgment." *Science* 243, no. 4899 (1989): 1668–74.

Dayan, Peter, and Yael Niv. "Reinforcement Learning: The Good, the Bad and the Ugly."
Current Opinion in Neurobiology 18, no. 2 (2008): 185–96.

Dechant, Kathy, and Jack Veiga. "More on the Folly." *Academy of Management Executive* 9, no. 1 (1995): 15–16.

Deci, Edward, and Richard M. Ryan. *Intrinsic Motivation and Self-Determination in
Human Behavior.* Springer Science & Business Media, 1985.

Demski, Abram, and Scott Garrabrant. "Embedded Agency." CoRR abs/1902.09469
(2019).

Deng, Jia, Wei Dong, Richard Socher, Li-Jia Li, Kai Li, and Li Fei-Fei. "ImageNet: A
Large-Scale Hierarchical Image Database." In *2009 IEEE Conference on Computer
Vision and Pattern Recognition,* 248–55. IEEE, 2009.

Denker, John S., and Yann LeCun. "Transforming Neural-Net Output Levels to Probability Distributions." In *Advances in Neural Information Processing Systems 3*,
853–59, 1991.

Denker, John, Daniel Schwartz, Ben Wittner, Sara Solla, Richard Howard, Lawrence
Jackel, and John Hopfield. "Large Automatic Learning, Rule Extraction, and Generalization." *Complex Systems* 1, no. 5 (1987): 877–922.

Desmarais, Sarah L., and Jay P. Singh. "Risk Assessment Instruments Validated and
Implemented in Correctional Settings in the United States," Council of State Governments Jusice Center, 2013.

Deterding, Sebastian, Dan Dixon, Rilla Khaled, and Lennart Nacke. "From Game
Design Elements to Gamefulness: Defining Gamification." In *Proceedings of the
15th International Academic Mindtrek Conference: Envisioning Future Media
Environments*, 9–15. ACM, 2011.

Devlin, Jacob, Ming-Wei Chang, Kenton Lee, and Kristina Toutanova. "BERT: Pre-
Training of Deep Bidirectional Transformers for Language Understanding." In
*Proceedings of the 2019 Conference of the North American Chapter of the Asso-
ciation for Computational Linguistics: Human Language Technologies (NAACL-
HLT)*, 4171–86, 2019.

Dieterich, William, Christina Mendoza, and Tim Brennan. "COMPAS Risk Scales:
Demonstrating Accuracy Equity and Predictive Parity." Northpointe Inc. Research
Department, July 8, 2016.

Dietterich, Thomas G. "Steps Toward Robust Artificial Intelligence." *AI Magazine* 38,
no. 3 (2017): 3–24.

Diuk, Carlos, Andre Cohen, and Michael L. Littman. "An Object-Oriented Representation for Efficient Reinforcement Learning." In *Proceedings of the 25th International Conference on Machine Learning*, 240–47. ACM, 2008.

Doctorow, Cory. "Two Years Later, Google Solves 'Racist Algorithm' Problem by Purging 'Gorilla' Label from Image Classifier." *Boing Boing*, January 11, 2018. https://
boingboing.net/2018/01/11/gorilla-chimp-monkey-unpersone.html.

Dominitz, Jeff, and John Knowles. "Crime Minimisation and Racial Bias: What Can
We Learn From Police Search Data?" *The Economic Journal* 116, no. 515 (2006):
F368–84.

Doshi-Velez, Finale, and Been Kim. "Towards a Rigorous Science of Interpretable
Machine Learning." arXiv Preprint arXiv:1702.08608, 2017.

Douglass, Frederick. "Negro Portraits." *Liberator* 19, no. 16 (April 20, 1849). http://fair-
use.org/the-liberator/1849/04/20/the-liberator-19–16.pdf.

Dowd, Maureen. "Peter Thiel, Trump's Tech Pal, Explains Himself." *New York Times*,
January 11, 2017.

Drăgan, Anca D., Kenton C. T. Lee, and Siddhartha S. Srinivasa. "Legibility and Predictability of Robot Motion." In *8th ACM/IEEE International Conference on Human-Robot Interaction*, 301–08. IEEE, 2013.

Dressel, Julia, and Hany Farid. "The Accuracy, Fairness, and Limits of Predicting Recidivism." *Science Advances* 4 (2018): eaao5580.

Dwork, Cynthia, Moritz Hardt, Toniann Pitassi, Omer Reingold, and Richard Zemel. "Fairness Through Awareness." In *Proceedings of the 3rd Innovations in Theoretical Computer Science Conference*, 214–26. ACM, 2012.

Dwork, Cynthia, Frank McSherry, Kobbi Nissim, and Adam Smith. "Calibrating Noise to Sensitivity in Private Data Analysis." In *Proceedings of the Third Conference on Theory of Cryptography*, edited by Shai Halevi and Tal Rabin, 265–84. Berlin: Springer, 2006.

Easterling, Keller. "Walter Pitts." *Cabinet* 5 (2001–02).

Eaves, Baxter S., Jr., Naomi H. Feldman, Thomas L. Griffiths, and Patrick Shafto. "Infant-Directed Speech Is Consistent with Teaching." *Psychological Review* 123, no. 6 (2016): 758.

Ecoffet, Adrien, Joost Huizinga, Joel Lehman, Kenneth O. Stanley, and Jeff Clune. "Go-Explore: A New Approach for Hard-Exploration Problems." arXiv Preprint arXiv:1901.10995, 2019.

Egan, Greg. *Axiomatic*. Millennium, 1995.

Einhorn, Hillel J. "Expert Measurement and Mechanical Combination." *Organizational Behavior and Human Performance* 7 (1972): 86–106.

Elek, Jennifer, Sara Sapia, and Susan Keilitz. "Use of Court Date Reminder Notices to Improve Court Appearance Rates." Pretrial Justice Center for the Courts, 2017.

Elman, Jeffrey L. "Learning and Development in Neural Networks: The Importance of Starting Small." *Cognition* 48, no. 1 (1993): 71–99.

Engelgau, Michael M., Linda S. Geiss, Jinan B. Saaddine, James P. Boyle, Stephanie M. Benjamin, Edward W. Gregg, Edward F. Tierney, et al. "The Evolving Diabetes Burden in the United States." *Annals of Internal Medicine* 140, no. 11 (2004): 945–50.

Ensign, Danielle, Sorelle A. Friedler, Scott Neville, Carlos Scheidegger, and Suresh Venkatasubramanian. "Runaway Feedback Loops in Predictive Policing." arXiv Preprint arXiv:1706.09847, 2017.

Entwistle, Noah J., and John D. Wilson. *Degrees of Excellence: The Academic Achievement Game*. Hodder & Stoughton, 1977.

Erhan, Dumitru, Yoshua Bengio, Aaron Courville, and Pascal Vincent. "Visualizing Higher-Layer Features of a Deep Network." Université de Montréal, 2009.

Esteva, Andre, Brett Kuprel, Roberto A. Novoa, Justin Ko, Susan M. Swetter, Helen M. Blau, and Sebastian Thrun. "Dermatologist-Level Classification of Skin Cancer with Deep Neural Networks." *Nature* 542, no. 7639 (2017): 115.

Evans, Owain, and Noah D. Goodman. "Learning the Preferences of Bounded Agents." In *NIPS Workshop on Bounded Optimality*, vol. 6, 2015.

Evans, Owain, Andreas Stuhlmüller, and Noah D. Goodman. "Learning the Preferences of Ignorant, Inconsistent Agents." In *Thirtieth AAAI Conference on Artificial Intelligence*, 2016.

Evans, Teresa M., Lindsay Bira, Jazmin Beltran Gastelum, L. Todd Weiss, and Nathan L. Vanderford. "Evidence for a Mental Health Crisis in Graduate Education." *Nature Biotechnology* 36, no. 3 (2018): 282.

Everitt, Tom, Victoria Krakovna, Laurent Orseau, Marcus Hutter, and Shane Legg. "Reinforcement Learning with a Corrupted Reward Channel." In *Proceedings of the Twenty-Sixth International Joint Conference on Artificial Intelligence (IJCAI-17)*, 4705–13, 2017.

Eysenbach, Benjamin, Shixiang Gu, Julian Ibarz, and Sergey Levine. "Leave No Trace: Learning to Reset for Safe and Autonomous Reinforcement Learning." In *International Conference on Learning Representations*, 2018.

Fantz, Robert L. "Visual Experience in Infants: Decreased Attention to Familiar Patterns Relative to Novel Ones." *Science* 146, no. 3644 (1964): 668–70.

Farley, Claire T., and C. Richard Taylor. "A Mechanical Trigger for the Trot-Gallop Transition in Horses." *Science* 253, no. 5017 (1991): 306–08.

Ferrari, Pier F., Elisabetta Visalberghi, Annika Paukner, Leonardo Fogassi, Angela Ruggiero, and Stephen J. Suomi. "Neonatal Imitation in Rhesus Macaques." *PLoS Biology* 4, no. 9 (2006): e302.

Ferster, C. B. "Schedules of Reinforcement with Skinner." In *Festschrift for B. F. Skinner*, edited by P. B. Dews, 37–46. New York: Irvington, 1970.

Ferster, Charles B., and B. F. Skinner. *Schedules of Reinforcement*. East Norwalk, CT: Appleton-Century-Crofts, 1957.

Finn, Chelsea, Sergey Levine, and Pieter Abbeel. "Guided Cost Learning: Deep Inverse Optimal Control via Policy Optimization." In *Proceedings of the 33rd International Conference on Machine Learning*, 49–58. PMLR, 2016.

Firth, John Rupert. *Papers in Linguistics, 1934–1951*. Oxford University Press, 1957.

Fisac, Jaime F., Monica A. Gates, Jessica B. Hamrick, Chang Liu, Dylan Hadfield-Menell, Malayandi Palaniappan, Dhruv Malik, S. Shankar Sastry, Thomas L. Griffiths, and Anca D. Drăgan. "Pragmatic-Pedagogic Value Alignment." In *International Symposium on Robotics Research (ISSR)*. Puerto Varas, Chile, 2017.

Fisac, Jaime F., Chang Liu, Jessica B. Hamrick, Shankar Sastry, J. Karl Hedrick, Thomas L. Griffiths, and Anca D. Drăgan. "Generating Plans That Predict Themselves." In *Workshop on the Algorithmic Foundations of Robotics (WAFR)*, 2016.

Fischer, Bobby, Stuart Margulies, and Don Mosenfelder. *Bobby Fischer Teaches Chess*. Basic Systems, 1966.

Florensa, Carlos, David Held, Markus Wulfmeier, Michael Zhang, and Pieter Abbeel.

"Reverse Curriculum Generation for Reinforcement Learning." In *Proceedings of the 1st Annual Conference on Robot Learning,* edited by Sergey Levine, Vincent Vanhoucke, and Ken Goldberg, 482–95. PMLR, 2017.

Flores, Anthony W., Kristin Bechtel, and Christopher T. Lowenkamp. "False Positives, False Negatives, and False Analyses: A Rejoinder to 'Machine Bias: There's Software Used Across the Country to Predict Future Criminals. and It's Biased Against Blacks.'" *Federal Probation* 80, no. 2 (2016): 38–46.

Foerster, Jakob, Ioannis Alexandros Assael, Nando de Freitas, and Shimon Whiteson. "Learning to Communicate with Deep Multi-Agent Reinforcement Learning." In *Advances in Neural Information Processing Systems,* 2137–45, 2016.

Foerster, Jakob, Richard Y. Chen, Maruan Al-Shedivat, Shimon Whiteson, Pieter Abbeel, and Igor Mordatch. "Learning with Opponent-Learning Awareness." In *Proceedings of the 17th International Conference on Autonomous Agents and Multiagent Systems,* 122–30. International Foundation for Autonomous Agents and Multiagent Systems, 2018.

Fong, Ruth, and Andrea Vedaldi. "Net2Vec: Quantifying and Explaining How Concepts Are Encoded by Filters in Deep Neural Networks." In *Proceedings of the IEEE Conference on Computer Vision and Pattern Recognition,* 8730–38. 2018.

"Four Out of Ten Violate Parole, Says Legislator." *Chicago Tribune,* May 15, 1937.

Frederick, Shane, and George Loewenstein. "Hedonic Adaptation." In *Well-Being: The Foundations of Hedonic Psychology,* edited by Daniel Kahneman, Edward Diener, and Norbert Schwarz, 302–29. New York: Russell Sage Foundation, 1999.

Friedman, Batya, and Helen Nissenbaum. "Bias in Computer Systems." *ACM Transactions on Information Systems (TOIS)* 14, no. 3 (1996): 330–47.

Fu, Justin, John Co-Reyes, and Sergey Levine. "EX^2: Exploration with Exemplar Models for Deep Reinforcement Learning." In *Advances in Neural Information Processing Systems,* 2577–87, 2017.

Furedy, John J., and Christine P. Furedy. "'My First Interest Is Interest': Berlyne as an Exemplar of the Curiosity Drive." In *Advances in Intrinsic Motivation and Aesthetics,* edited by Hy I. Day, 1–18. New York: Plenum Press, n.d.

Fürnkranz, Johannes, and Miroslav Kubat. *Machines That Learn to Play Games.* Huntington, NY: Nova Science, 2001.

Gage, Brian F., Amy D. Waterman, William Shannon, Michael Boechler, Michael W. Rich, and Martha J. Radford. "Validation of Clinical Classification Schemes for Predicting Stroke: Results from the National Registry of Atrial Fibrillation." *JAMA* 285, no. 22 (2001): 2864–70.

Gal, Yarin. "Uncertainty in Deep Learning." PhD thesis, University of Cambridge, 2016.

Gal, Yarin, and Zoubin Ghahramani. "A Theoretically Grounded Application of Dropout in Recurrent Neural Networks." In *Advances in Neural Information Processing Systems 29.* 2016.

——. "Bayesian Convolutional Neural Networks with Bernoulli Approximate Variational Inference." In *4th International Conference on Learning Representations (ICLR) Workshop Track,* 2016.

——. "Dropout as a Bayesian Approximation: Representing Model Uncertainty in Deep Learning." In *Proceedings of the 33rd International Conference on Machine Learning,* 1050–09. PMLR, 2016.

Gal, Yarin, Rowan McAllister, and Carl E. Rasmussen. "Improving PILCO with Bayesian Neural Network Dynamics Models." In *Data-Efficient Machine Learning Workshop, ICML,* 2016.

Gans, Joshua. *Parentonomics: An Economist Dad Looks at Parenting.* MIT Press, 2009.

Gardner, Matthew P. H., Geoffrey Schoenbaum, and Samuel J. Gershman. "Rethinking Dopamine as Generalized Prediction Error." *Proceedings of the Royal Society B* 285, no. 1891 (2018): 20181645.

Garg, Nikhil, Londa Schiebinger, Dan Jurafsky, and James Zou. "Word Embeddings Quantify 100 Years of Gender and Ethnic Stereotypes." *Proceedings of the National Academy of Sciences* 115, no. 16 (2018): E3635–44.

Gebru, Timnit. "Race and Gender." In *The Oxford Handbook of Ethics of AI,* edited by Markus D. Dubber, Frank Pasquale, and Sunit Das. New York: Oxford University Press, 2020.

Gefter, Amanda. "The Man Who Tried to Redeem the World with Logic." *Nautilus* 21 (February 2015).

Gendron-Bellemare, Marc. "Fast, Scalable Algorithms for Reinforcement Learning in High Dimensional Domains." PhD thesis, University of Alberta, 2013.

Gergely, György, Harold Bekkering, and Ildikó Király. "Rational Imitation in Preverbal Infants." *Nature* 415, no. 6873 (2002): 755.

Gershgorn, Dave. "Companies Are on the Hook If Their Hiring Algorithms Are Biased." *Quartz,* October 22, 2018.

Ghahramani, Zoubin. "Probabilistic Machine Learning and Artificial Intelligence." *Nature* 521, no. 7553 (2015): 452–59.

Ghorbani, Amirata, Abubakar Abid, and James Zou. "Interpretation of Neural Networks Is Fragile." In *Proceedings of the AAAI Conference on Artificial Intelligence* 33 (2019): 3681–88.

Gielniak, Michael J., and Andrea L. Thomaz. "Generating Anticipation in Robot Motion." In *2011 RO-MAN,* 449–54. IEEE, 2011.

Giusti, Alessandro, Jérôme Guzzi, Dan C. Cireşan, Fang-Lin He, Juan P. Rodríguez, Flavio Fontana, Matthias Faessler, et al. "A Machine Learning Approach to Visual Perception of Forest Trails for Mobile Robots." *IEEE Robotics and Automation Letters* 1, no. 2 (2015): 661–67.

Glimcher, Paul W. "Understanding Dopamine and Reinforcement Learning: The Dopamine Reward Prediction Error Hypothesis." *Proceedings of the National Academy of Sciences* 108, Supplement 3 (2011): 15647–54.

Goel, Sharad, Justin M. Rao, and Ravi Shroff. "Personalized Risk Assessments in the Criminal Justice System." *American Economic Review* 106, no. 5 (2016): 119–23.

Goldberg, Lewis R. "Man Versus Model of Man: A Rationale, Plus Some Evidence, for a Method of Improving on Clinical Inferences." *Psychological Bulletin* 73, no. 6 (1970): 422–32.

———. "Simple Models or Simple Processes? Some Research on Clinical Judgments." *American Psychologist* 23, no. 7 (1968): 483–96.

Goldin, Claudia, and Cecilia Rouse. "Orchestrating Impartiality: The Impact of 'Blind' Auditions on Female Musicians." *American Economic Review* 90, no. 4 (2000): 715–41.

Goldman, Holly S. "Dated Rightness and Moral Imperfection." *Philosophical Review* 85, no. 4 (1976): 449–487.

———. "Doing the Best One Can." In *Values and Morals: Essays in Honor of William Frankena, Charles Stevenson, and Richard Brandt,* edited by Alvin I. Goldman and Jaegwon Kim, 185–214. D. Reidel, 1978.

Gonen, Hila, and Yoav Goldberg. "Lipstick on a Pig: Debiasing Methods Cover Up Systematic Gender Biases in Word Embeddings but Do Not Remove Them." In *Proceedings of the 2019 Annual Conference of the North American Chapter of the Association for Computational Linguistics,* 2019.

Goodfellow, Ian J., Jonathon Shlens, and Christian Szegedy. "Explaining and Harnessing Adversarial Examples." In *International Conference on Learning Representations,* 2015.

Goodman, Bryce, and Seth Flaxman. "European Union Regulations on Algorithmic Decision-Making and a 'Right to Explanation.'" *AI Magazine* 38, no. 3 (2017).

Gopnik, Alison. "Explanation as Orgasm and the Drive for Causal Knowledge: The Function, Evolution, and Phenomenology of the Theory-Formation System." In *Explanation and Cognition,* edited by Frank C. Keil and Robert A. Wilson, 299–323. MIT Press, 2000.

———. *The Gardener and the Carpenter: What the New Science of Child Development Tells Us About the Relationship Between Parents and Children.* Macmillan, 2016.

Gopnik, Alison, Andrew N. Meltzoff, and Patricia K. Kuhl. *The Scientist in the Crib: Minds, Brains, and How Children Learn.* William Morrow, 1999.

Goswami, Samir. "Unlocking Options for Women: A Survey of Women in Cook County Jail." *University of Maryland Law Journal of Race, Religion, Gender and Class* 2 (2002): 89–114.

Gottfredson, Stephen D., and G. Roger Jarjoura. "Race, Gender, and Guidelines-Based Decision Making." *Journal of Research in Crime and Delinquency* 33, no. 1 (1996): 49–69.

Gouldin, Lauryn P. "Disentangling Flight Risk from Dangerousness." *BYU Law Review,* 2016, 837–98.

Graeber, David. *The Utopia of Rules: On Technology, Stupidity, and the Secret Joys of Bureaucracy.* Melville House, 2015.

Graves, Alex. "Practical Variational Inference for Neural Networks." In *Advances in Neural Information Processing Systems,* 2348–56, 2011.

Graves, Alex, Marc G. Bellemare, Jacob Menick, Remi Munos, and Koray Kavukcuoglu. "Automated Curriculum Learning for Neural Networks." In *Proceedings of the 34th International Conference on Machine Learning,* 1311–20. PMLR, 2017.

Greenwald, Anthony G., Debbie E. McGhee, and Jordan L. K. Schwartz. "Measuring Individual Differences in Implicit Cognition: The Implicit Association Test." *Journal of Personality and Social Psychology* 74, no. 6 (1998): 1464–80.

Greydanus, Sam, Anurag Koul, Jonathan Dodge, and Alan Fern. "Visualizing and Understanding Atari Agents." In *Proceedings of the 35th International Conference on Machine Learning,* edited by Jennifer Dy and Andreas Krause, 1792–1801. PMLR, 2018.

Guo, Chuan, Geoff Pleiss, Yu Sun, and Kilian Q. Weinberger. "On Calibration of Modern Neural Networks." In *Proceedings of the 34th International Conference on Machine Learning,* 1321–30. PMLR, 2017.

Gustafsson, Johan E., and Olle Torpman. "In Defence of My Favourite Theory." *Pacific Philosophical Quarterly* 95, no. 2 (2014): 159–74.

Hadfield-Menell, Dylan, Smitha Milli, Pieter Abbeel, Stuart J. Russell, and Anca D. Drăgan. "Inverse Reward Design." In *Advances in Neural Information Processing Systems,* 6768–77, 2017.

Hadfield-Menell, Dylan, Stuart J. Russell, Pieter Abbeel, and Anca Drăgan. "Cooperative Inverse Reinforcement Learning." In *Advances in Neural Information Processing Systems,* 3909–17, 2016.

Haggbloom, Steven J., Renee Warnick, Jason E. Warnick, Vinessa K. Jones, Gary L. Yarbrough, Tenea M. Russell, Chris M. Borecky, et al. "The 100 Most Eminent Psychologists of the 20th Century." *Review of General Psychology* 6, no. 2 (2002): 139–52.

Hamari, Juho, Jonna Koivisto, and Harri Sarsa. "Does Gamification Work? A Literature Review of Empirical Studies on Gamification." In *47th Hawaii International Conference on System Sciences (HICSS),* 3025–34. IEEE, 2014.

Han, Hu, and Anil K. Jain. "Age, Gender and Race Estimation from Unconstrained Face Images." *MSU Technical Report MSU-CSE-14-5,* 2014.

Hansen, C., M. Tosik, G. Goossen, C. Li, L. Bayeva, F. Berbain, and M. Rotaru. "How to Get the Best Word Vectors for Resume Parsing." In *SNN Adaptive Intelligence/ Symposium: Machine Learning,* 2015.

Harcourt, Bernard E. *Against Prediction: Profiling, Policing, and Punishing in an Actuarial Age.* University of Chicago Press, 2007.

———. "Risk as a Proxy for Race: The Dangers of Risk Assessment." *Federal Sentencing Reporter* 27, no. 4 (2015): 237–43.

égorie{

Hardt, Moritz. "How Big Data Is Unfair: Understanding Unintended Sources of Unfairness in Data Driven Decision Making." *Medium,* September 26, 2014. https://medium.com/@mrtz/how-big-data-is-unfair-9aa544d739de.

Hardt, Moritz, Eric Price, and Nathan Srebro. "Equality of Opportunity in Supervised Learning." In *Advances in Neural Information Processing Systems,* 3315–23, 2016.

Harlow, Harry F. "Learning and Satiation of Response in Intrinsically Motivated Complex Puzzle Performance by Monkeys." *Journal of Comparative and Physiological Psychology* 43, no. 4 (1950): 289.

Harlow, Harry F., Margaret Kuenne Harlow, and Donald R. Meyer. "Learning Motivated by a Manipulation Drive." *Journal of Experimental Psychology* 40, no. 2 (1950): 228.

Hass, Robert. *Time and Materials: Poems 1997–2005.* Ecco, 2007.

Hastie, Trevor, and Robert Tibshirani. "Generalized Additive Models." *Statistical Science* 1, no. 3 (1986): 297–318.

Hebb, Donald Olding. *The Organization of Behavior: A Neuropsychological Theory.* New York: John Wiley & Sons, 1949.

Heims, Steve Joshua. *The Cybernetics Group.* MIT Press, 1991.

———. *John von Neumann and Norbert Wiener.* MIT Press, 1980.

Heine, Steven J., Timothy Takemoto, Sophia Moskalenko, Jannine Lasaleta, and Joseph Henrich. "Mirrors in the Head: Cultural Variation in Objective Self-Awareness." *Personality and Social Psychology Bulletin* 34, no. 7 (2008): 879–87.

Hendrycks, Dan, Mantas Mazeika, and Thomas Dietterich. "Deep Anomaly Detection with Outlier Exposure." In *International Conference on Learning Representations,* 2019.

Hernández-Lobato, José Miguel, and Ryan Adams. "Probabilistic Backpropagation for Scalable Learning of Bayesian Neural Networks." In *International Conference on Machine Learning,* 1861–69. 2015.

Hersher, Rebecca. "When a Tattoo Means Life or Death. Literally." *Weekend Edition Sunday,* NPR, January 21, 2018.

Hester, Todd, Matej Večerík, Olivier Pietquin, Marc Lanctot, Tom Schaul, Bilal Piot, Dan Horgan, et al. "Deep Q-Learning from Demonstrations." In *Thirty-Second AAAI Conference on Artificial Intelligence,* 2018.

"Heuristics." *New Yorker,* August 29, 1959, 22–23.

Hilton, Lisette. "The Artificial Brain as Doctor." *Dermatology Times,* January 15, 2018.

Hinton, Geoffrey E. "Connectionist Learning Procedures." In *Artificial Intelligence* 40 (1989): 185–234.

———. "Learning Distributed Representations of Concepts." In *Proceedings of the Eighth Annual Conference of the Cognitive Science Society,* 1–12. 1986.

Hinton, Geoffrey, and Drew Van Camp. "Keeping Neural Networks Simple by Minimizing the Description Length of the Weights." In *Proceedings of the 6th Annual ACM Conference on Computational Learning Theory*, 5–13. 1993.

Ho, Mark K., Fiery A. Cushman, Michael L. Littman, and Joseph L. Austerweil. 2019. "Communication in Action: Belief-directed Planning and Pragmatic Action Interpretation in Communicative Demonstrations." PsyArXiv. February 19. doi:10.31234/osf.io/a8sxk.

Ho, Mark K., Michael Littman, James MacGlashan, Fiery Cushman, and Joseph L. Austerweil. "Showing Versus Doing: Teaching by Demonstration." In *Advances in Neural Information Processing Systems*, 3027–35, 2016.

Hobbes, Thomas. *Leviathan*. Andrew Crooke, 1651.

Hoeting, Jennifer A., David Madigan, Adrian E. Raftery, and Chris T. Volinsky. "Bayesian Model Averaging: A Tutorial." *Statistical Science* 14, no. 4 (1999): 382–401.

Hoffman, David. "'I Had a Funny Feeling in My Gut.'" *Washington Post*, February 10, 1999. https://www.washingtonpost.com/wp-srv/inatl/longterm/coldwar/soviet10.htm.

Hofstadter, Douglas R. *Gödel, Escher, Bach: An Eternal Golden Braid*. Basic Books, 1979.

Hogan, Neville. "An Organizing Principle for a Class of Voluntary Movements." *Journal of Neuroscience* 4, no. 11 (1984): 2745–54.

Holmes, Oliver Wendell, and Frederick Pollock. *Holmes-Pollock Letters: The Correspondence of Mr. Justice Holmes and Sir Frederick Pollock, 1874–1932*. Vol. 1. Harvard University Press, 1941.

Holt, Gregory E., Bianca Sarmento, Daniel Kett, and Kenneth W. Goodman. "An Unconscious Patient with a DNR Tattoo." *New England Journal of Medicine* 377, no. 22 (2017): 2192–93.

Holte, Robert C. "Very Simple Classification Rules Perform Well on Most Commonly Used Datasets." *Machine Learning* 11, no. 1 (1993): 63–90.

Horner, Victoria, and Andrew Whiten. "Causal Knowledge and Imitation/Emulation Switching in Chimpanzees (*Pan troglodytes*) and Children (*Homo sapiens*)." *Animal Cognition* 8, no. 3 (2005): 164–81.

Hornik, Kurt, Maxwell Stinchcombe, and Halbert White. "Multilayer Feedforward Networks Are Universal Approximators." *Neural Networks* 2, no. 5 (1989): 359–66.

Horstmann, Aike C., Nikolai Bock, Eva Linhuber, Jessica M. Szczuka, Carolin Straßmann, and Nicole C. Krämer. "Do a Robot's Social Skills and Its Objection Discourage Interactants from Switching the Robot Off?" *PLoS One* 13, no. 7 (2018): e0201581.

Hosu, Ionel-Alexandru, and Traian Rebedea. "Playing Atari Games with Deep Reinforcement Learning and Human Checkpoint Replay." arXiv Preprint arXiv:1607.05077, 2016.

Houthooft, Rein, Xi Chen, Yan Duan, John Schulman, Filip De Turck, and Pieter Abbeel. "VIME: Variational Information Maximizing Exploration." In *Advances in Neural Information Processing Systems 29,* edited by D. D. Lee, M. Sugiyama, U. V. Luxburg, I. Guyon, and R. Garnett, 1109–17, 2016.

Howard, Andrew G. "Some Improvements on Deep Convolutional Neural Network Based Image Classification." arXiv Preprint arXiv:1312.5402, 2013.

Howard, John W., and Robyn M. Dawes. "Linear Prediction of Marital Happiness." *Personality and Social Psychology Bulletin* 2, no. 4 (1976): 478–80.

Hoyt, Donald F., and C. Richard Taylor. "Gait and the Energetics of Locomotion in Horses." *Nature* 292, no. 5820 (1981): 239–40.

Hsu, Feng-hsiung. "IBM's Deep Blue Chess Grandmaster Chips." *IEEE Micro* 19, no. 2 (1999): 70–81.

Hsu, Feng-hsiung, Thomas Anantharaman, Murray Campbell, and Andreas Nowatzyk. "A Grandmaster Chess Machine." *Scientific American* 263, no. 4 (1990): 44–51.

Huang, Gary B., Marwan Mattar, Tamara Berg, and Eric Learned-Miller. "Labeled Faces in the Wild: A Database for Studying Face Recognition in Unconstrained Environments." In *Workshop on Faces in "Real-Life" Images: Detection, Alignment, and Recognition,* 2008.

Huang, Xin, Stephen G. McGill, Brian C. Williams, Luke Fletcher, and Guy Rosman. "Uncertainty-Aware Driver Trajectory Prediction at Urban Intersections." In *2019 International Conference on Robotics and Automation,* 9718–24. 2019.

Hubinger, Evan, Chris van Merwijk, Vladimir Mikulik, Joar Skalse, and Scott Garrabrant. "Risks from Learned Optimization in Advanced Machine Learning Systems." arXiv Preprint arXiv:1906.01820, 2019.

Humphrys, Mark. "Action Selection in a Hypothetical House Robot: Using Those RL Numbers." In *Proceedings of the First International ICSC Symposia on Intelligent Industrial Automation (IIA-96) and Soft Computing (SOCO-96),* 216–22. ICSC Academic Press, 1996.

Hursthouse, Rosalind. "Normative Virtue Ethics." In *How Should One Live?: Essays on the Virtues,* edited by Roger Crisp, 19–36. Oxford University Press, 1996.

Huxley, Aldous. *Ends and Means: An Enquiry into the Nature of Ideals and into the Methods Employed for Their Realization.* Chatto & Windus, 1937.

Ilyas, Andrew, Shibani Santurkar, Dimitris Tsipras, Logan Engstrom, Brandon Tran, and Aleksander Mądry. "Adversarial Examples Are Not Bugs, They Are Features." In *Advances in Neural Information Processing Systems,* 125–36. 2019.

Irving, Geoffrey, Paul Christiano, and Dario Amodei. "AI Safety via Debate." arXiv Preprint arXiv:1805.00899, 2018.

Jackson, Frank. "Procrastinate Revisited." *Pacific Philosophical Quarterly* 95, no. 4 (2014): 634–47.

Jackson, Frank, and Robert Pargetter. "Oughts, Options, and Actualism." *Philosophical Review* 95, no. 2 (1986): 233–55.

James, William. *Psychology: Briefer Course*. New York: Macmillan, 1892.

Jaynes, Edwin T. "Information Theory and Statistical Mechanics." *Physical Review* 106, no. 4 (1957): 620–30.

Jefferson, Thomas. *Notes on the State of Virginia*. Paris, 1785.

Jelinek, Fred, and Robert L. Mercer. "Interpolated Estimation of Markov Source Parameters from Sparse Data." In *Proceedings, Workshop on Pattern Recognition in Practice*, edited by Edzard S. Gelsema and Laveen N. Kanal, 381–97. 1980.

Jeon, Hong Jun, Smitha Milli, and Anca D. Drăgan. "Reward-Rational (Implicit) Choice: A Unifying Formalism for Reward Learning." arXiv Preprint arXiv:2002.04833, 2020.

Joffe-Walt, Chana. "Allowance Economics: Candy, Taxes and Potty Training." *Planet Money*, September 2010.

Johnson, Deborah G., and Helen Nissenbaum. "Computers, Ethics & Social Values." Upper Saddle River, NJ: Prentice Hall, 1995.

Johnson, Robert N. "Virtue and Right." *Ethics* 113, no. 4 (2003): 810–34.

Jonçich, Geraldine M. *The Sane Positivist: A Biography of Edward L. Thorndike*. Middletown, CT: Wesleyan University Press, 1968.

Jung, Christopher, Michael Kearns, Seth Neel, Aaron Roth, Logan Stapleton, and Zhiwei Steven Wu. "Eliciting and Enforcing Subjective Individual Fairness." arXiv Preprint arXiv:1905.10660, 2019.

Kaelbling, Leslie Pack. *Learning in Embedded Systems*. MIT Press, 1993.

Kaelbling, Leslie Pack, Michael L. Littman, and Andrew W. Moore. "Reinforcement Learning: A Survey." *Journal of Artificial Intelligence Research* 4 (1996): 237–85.

Kahn, Gregory, Adam Villaflor, Vitchyr Pong, Pieter Abbeel, and Sergey Levine. "Uncertainty-Aware Reinforcement Learning for Collision Avoidance." arXiv Preprint arXiv:1702.01182, 2017.

Kahneman, Daniel. *Thinking, Fast and Slow*. Farrar, Straus and Giroux, 2011.

Kakade, Sham, and Peter Dayan. "Dopamine: Generalization and Bonuses." *Neural Networks* 15, nos. 4–6 (2002): 549–59.

Kálmán, Rudolf Emil. "When Is a Linear Control System Optimal?" *Journal of Basic Engineering* 86, no. 1 (1964): 51–60.

Kamin, Leon J. "Predictability, Surprise, Attention, and Conditioning." In *Punishment and Aversive Behavior*. New York: Appleton-Century-Crofts, 1969.

Kang, Daniel, Yi Sun, Dan Hendrycks, Tom Brown, and Jacob Steinhardt. "Testing Robustness Against Unforeseen Adversaries." arXiv Preprint arXiv:1908.08016, 2019.

Kashdan, Todd B., and Paul J. Silvia. "Curiosity and Interest: The Benefits of Thriving on Novelty and Challenge." *Oxford Handbook of Positive Psychology* 2 (2009): 367–74.

Kasparov, Garry. *How Life Imitates Chess: Making the Right Moves, from the Board to the Boardroom.* Bloomsbury USA, 2007.

Katz, Slava. "Estimation of Probabilities from Sparse Data for the Language Model Component of a Speech Recognizer." *IEEE Transactions on Acoustics, Speech, and Signal Processing* 35, no. 3 (1987): 400–01.

Kellogg, Winthrop Niles, and Luella Agger Kellogg. *The Ape and the Child: A Comparative Study of the Environmental Influence upon Early Behavior.* Whittlesey House, 1933.

Kenton, Zachary, Angelos Filos, Owain Evans, and Yarin Gal. "Generalizing from a Few Environments in Safety-Critical Reinforcement Learning." In *ICLR Workshop on Safe Machine Learning,* 2019.

Kerr, Steven. "On the Folly of Rewarding A, While Hoping for B." *Academy of Management Journal* 18, no. 4 (1975): 769–83.

Kim, Been, Martin Wattenberg, Justin Gilmer, Carrie Cai, James Wexler, Fernanda Viegas, and Rory Sayres. "Interpretability Beyond Feature Attribution: Quantitative Testing with Concept Activation Vectors (TCAV)." In *Proceedings of the 35th International Conference on Machine Learning,* edited by Jennifer Dy and Andreas Krause, 2668–77. PMLR, 2018.

Kimball, Spencer, and Paayal Zaveri. "Tim Cook on Facebook's Data-Leak Scandal: 'I Wouldn't Be in This Situation.'" CNBC, March 28, 2018. https://www.cnbc.com /2018/03/28/tim-cook-on-facebooks-scandal-i-wouldnt-be-in-this-situation.html.

Kindermans, Pieter-Jan, Sara Hooker, Julius Adebayo, Maximilian Alber, Kristof T. Schütt, Sven Dähne, Dumitru Erhan, and Been Kim. "The (Un)reliability of Saliency Methods." arXiv Preprint arXiv:1711.00867, 2017.

Kinsley, Philip. "What Convict Will Do If Paroled Is Study Illinois Man Presents to Science Parley." *Chicago Tribune,* January 4, 1936.

Kiros, Ryan, Yukun Zhu, Russ R. Salakhutdinov, Richard Zemel, Raquel Urtasun, Antonio Torralba, and Sanja Fidler. "Skip-Thought Vectors." In *Advances in Neural Information Processing Systems 28,* edited by C. Cortes, N. D. Lawrence, D. D. Lee, M. Sugiyama, and R. Garnett, 3294–3302, 2015.

Klare, Brendan F., Ben Klein, Emma Taborsky, Austin Blanton, Jordan Cheney, Kristen Allen, Patrick Grother, Alan Mah, and Anil K. Jain. "Pushing the Frontiers of Unconstrained Face Detection and Recognition: IARPA Janus Benchmark A." In *Proceedings of the Twenty-Eighth IEEE Conference on Computer Vision and Pattern Recognition,* 1931–39. CVPR, 2015.

Kleinberg, Jon, Himabindu Lakkaraju, Jure Leskovec, Jens Ludwig, and Sendhil Mullainathan. "Human Decisions and Machine Predictions." *Quarterly Journal of Economics* 133, no. 1 (2018): 237–93.

Kleinberg, Jon, Jens Ludwig, Sendhil Mullainathan, and Ashesh Rambachan. "Algorithmic Fairness." In *AEA Papers and Proceedings* 108 (2018): 22–27.

Kleinberg, Jon, Sendhil Mullainathan, and Manish Raghavan. "Inherent Trade-offs in the Fair Determination of Risk Scores." In *The 8th Innovations in Theoretical Computer Science Conference*. Berkeley, 2017.

Kleinberg, Jon, and Manish Raghavan. "Selection Problems in the Presence of Implicit Bias." In *Proceedings of the 9th Conference on Innovations in Theoretical Computer Science (ITCS)*. 2018.

Klopf, A. Harry. "Brain Function and Adaptive Systems: A Heterostatic Theory." Bedford, MA: Air Force Cambridge Research Laboratories, 1972.

Knox, W. Bradley, Peter Stone, and Cynthia Breazeal. "Training a Robot via Human Feedback: A Case Study." In *International Conference on Social Robotics*, 460–70. Springer, 2013.

Knuth, Donald E. "Computer Programming as an Art." *Communications of the ACM* 17, no. 12 (1974): 667–73.

———. "The Errors of TeX." *Software: Practice and Experience* 19, no. 7 (1989): 607–85.

———. "Structured Programming with *Go to* Statements." *Computing Surveys* 6, no. 4 (1974): 261–301.

Kobayashi, Hiromi, and Shiro Kohshima. "Unique Morphology of the Human Eye and Its Adaptive Meaning: Comparative Studies on External Morphology of the Primate Eye." *Journal of Human Evolution* 40, no. 5 (2001): 419–35.

Konečni, Vladimir J. "Daniel E. Berlyne: 1924–1976." *American Journal of Psychology* 91, no. 1 (March 1978): 133–37.

Krakovna, Victoria, Laurent Orseau, Ramana Kumar, Miljan Martic, and Shane Legg. "Penalizing Side Effects Using Stepwise Relative Reachability." arXiv Preprint arXiv:1806.01186, 2019.

Kroll, Joshua A., Solon Barocas, Edward W. Felten, Joel R. Reidenberg, David G. Robinson, and Harlan Yu. "Accountable Algorithms." *University of Pennsylvania Law Review* 165 (2016): 633–705.

Kurita, Keita, Nidhi Vyas, Ayush Pareek, Alan W. Black, and Yulia Tsvetkov. "Measuring Bias in Contextualized Word Representations." arXiv Preprint arXiv:1906.07337, 2019.

Kwon, Minae, Erdem Biyik, Aditi Talati, Karan Bhasin, Dylan P. Losey, and Dorsa Sadigh. "When Humans Aren't Optimal: Robots That Collaborate with Risk-Aware Humans." In *ACM/IEEE International Conference on Human-Robot Interaction (HRI)*. 2020.

Lage, Isaac, Andrew Ross, Samuel J. Gershman, Been Kim, and Finale Doshi-Velez. "Human-in-the-Loop Interpretability Prior." In *Advances in Neural Information Processing Systems*, 10159–68, 2018.

Lakkaraju, Himabindu, Stephen H. Bach, and Jure Leskovec. "Interpretable Decision Sets: A Joint Framework for Description and Prediction." In *Proceedings of the 22nd ACM SIGKDD International Conference on Knowledge Discovery and Data Mining*, 1675–84. ACM, 2016.

Lakshminarayanan, Balaji, Alexander Pritzel, and Charles Blundell. "Simple and Scalable Predictive Uncertainty Estimation Using Deep Ensembles." In *Advances in Neural Information Processing Systems*, 6402–13, 2017.

Landauer, Thomas K., Peter W. Foltz, and Darrell Laham. "An Introduction to Latent Semantic Analysis." *Discourse Processes* 25 (1998): 259–84.

Landecker, Will. "Interpretable Machine Learning and Sparse Coding for Computer Vision." PhD thesis, Portland State University, 2014.

Landecker, Will, Michael D. Thomure, Luís M. A. Bettencourt, Melanie Mitchell, Garrett T. Kenyon, and Steven P. Brumby. "Interpreting Individual Classifications of Hierarchical Networks." In *2013 IEEE Symposium on Computational Intelligence and Data Mining (CIDM)*, 32–38. IEEE, 2013.

Lansing, Sharon. "New York State COMPAS-Probation Risk and Need Assessment Study: Examining the Recidivism Scale's Effectiveness and Predictive Accuracy." Division of Criminal Justice Services, Office of Justice Research and Performance, September 2012. http://www.criminaljustice.ny.gov/crimnet/ojsa/opca/compas_probation_report_2012.pdf.

Larson, Jeff, and Julia Angwin. "Technical Response to Northpointe." ProPublica, July 29, 2016.

Larson, Jeff, Surya Mattu, Lauren Kirchner, and Julia Angwin. "How We Analyzed the COMPAS Recidivism Algorithm." ProPublica, May 23, 2016.

Latour, Bruno. *Pandora's Hope: Essays on the Reality of Science Studies.* Harvard University Press, 1999.

Le, Quoc, and Tomáš Mikolov. "Distributed Representations of Sentences and Documents." In *International Conference on Machine Learning*, 1188–96. 2014.

LeCun, Yann, Bernhard Boser, John S. Denker, Donnie Henderson, Richard E. Howard, Wayne Hubbard, and Lawrence D. Jackel. "Backpropagation Applied to Handwritten Zip Code Recognition." *Neural Computation* 1, no. 4 (1989): 541–51.

Lee, Chen-Yu, Saining Xie, Patrick Gallagher, Zhengyou Zhang, and Zhuowen Tu. "Deeply-Supervised Nets." In *Proceedings of the Eighteenth International Conference on Artificial Intelligence and Statistics*, 562–70. 2015.

Legg, Shane, and Marcus Hutter. "A Collection of Definitions of Intelligence." *Frontiers in Artificial Intelligence and Applications* 157 (2007): 17–24.

———. "Universal Intelligence: A Definition of Machine Intelligence." *Minds and Machines* 17, no. 4 (December 2007): 391–444.

Legg, Shane, and Joel Veness. "An Approximation of the Universal Intelligence Measure." In *Papers from the Ray Solomonoff 85th Memorial Conference.* Melbourne, Australia, 2011.

Leibig, Christian, Vaneeda Allken, Murat Seçkin Ayhan, Philipp Berens, and Siegfried Wahl. "Leveraging Uncertainty Information from Deep Neural Networks for Disease Detection." *Scientific Reports* 7, no. 1 (2017): 17816.

Leighty, Robert D. "DARPA ALV (Autonomous Land Vehicle) Summary." Army Engineer Topographic Laboratories, Fort Belvoir, VA, 1986.

Leike, Jan, and Marcus Hutter. "Bad Universal Priors and Notions of Optimality." In *Proceedings of the 28th Conference on Learning Theory*, 1244–59. PMLR: 2015.

Leike, Jan, David Krueger, Tom Everitt, Miljan Martic, Vishal Maini, and Shane Legg. "Scalable Agent Alignment via Reward Modeling: A Research Direction." arXiv Preprint arXiv:1811.07871, 2018.

Leike, Jan, Miljan Martic, Victoria Krakovna, Pedro A. Ortega, Tom Everitt, Andrew Lefrancq, Laurent Orseau, and Shane Legg. "AI Safety Gridworlds." arXiv Preprint arXiv:1711.09883, 2017.

Lenson, David. *On Drugs*. University of Minnesota Press, 1995.

Letham, Benjamin, Cynthia Rudin, Tyler H. McCormick, David Madigan. "Interpretable Classifiers Using Rules and Bayesian Analysis: Building a Better Stroke Prediction Model." *Annals of Applied Statistics* 9, no. 3 (2015): 1350–71.

Lewis, Sarah, ed. "Vision & Justice." *Aperture* 223 (2016).

Li, Lihong, Michael L. Littman, Thomas J. Walsh, and Alexander L. Strehl. "Knows What It Knows: A Framework for Self-Aware Learning." *Machine Learning* 82, no. 3 (2011): 399–443.

Li, Xin, and Fuxin Li. "Adversarial Examples Detection in Deep Networks with Convolutional Filter Statistics." In *Proceedings of the IEEE International Conference on Computer Vision*, 5764–72. 2017.

Lieder, Falk. "Gamify Your Goals: How Turning Your Life into a Game Can Help You Make Better Decisions and Be More Productive." LessWrong, February 3, 2016. https://www.lesswrong.com/posts/g2355pBbaYSfPMk2m/gamify-your-goals-how -turning-your-life-into-a-game-can-help.

Lieder, Falk, Owen X. Chen, Paul M. Krueger, and Thomas L. Griffiths. "Cognitive Prostheses for Goal Achievement." *Nature Human Behaviour* 3, no. 10 (2019): 1096–1106.

Lieder, Falk, Noah D. Goodman, and Quentin J. M. Huys. "Learned Helplessness and Generalization." In *Proceedings of the 35th Annual Meeting of the Cognitive Science Society*, 900–05. Austin, TX, 2013.

Lin, Min, Qiang Chen, and Shuicheng Yan. "Network in Network." arXiv Preprint arXiv:1312.4400, 2013.

Lip, Gregory Y. H., Robby Nieuwlaat, Ron Pisters, Deirdre A. Lane, and Harry J.G.M. Crijns. "Refining Clinical Risk Stratification for Predicting Stroke and Thromboembolism in Atrial Fibrillation Using a Novel Risk Factor–Based Approach: The Euro Heart Survey on Atrial Fibrillation." *Chest* 137, no. 2 (2010): 263–72.

Lipsey, Richard G., and Kelvin Lancaster. "The General Theory of Second Best." *Review of Economic Studies* 24, no. 1 (1956): 11–32.

Liu, Anqi, and Brian Ziebart. "Robust Classification Under Sample Selection Bias." In *Advances in Neural Information Processing Systems*, 37–45, 2014.

Liu, Lydia T., Sarah Dean, Esther Rolf, Max Simchowitz, and Moritz Hardt. "Delayed Impact of Fair Machine Learning." In *Proceedings of the 35th International Conference on Machine Learning*, 2018.

Liu, Si, Risheek Garrepalli, Thomas G. Dietterich, Alan Fern, and Dan Hendrycks. "Open Category Detection with PAC Guarantees." In *Proceedings of the 35th International Conference on Machine Learning*, 2018.

Liu, Si, Risheek Garrepalli, Alan Fern, and Thomas G. Dietterich. "Can We Achieve Open Category Detection with Guarantees?" In *Workshops at the Thirty-Second AAAI Conference on Artificial Intelligence*, 2018.

Lockhart, Ted. *Moral Uncertainty and Its Consequences*. Oxford University Press, 2000.

Lou, Yin, Rich Caruana, Johannes Gehrke, and Giles Hooker. "Accurate Intelligible Models with Pairwise Interactions." In *Proceedings of the 19th ACM SIGKDD International Conference on Knowledge Discovery and Data Mining*, 623–31. ACM, 2013.

Lowell, Josh, and Peter Mortimer, directors. *The Dawn Wall*. Red Bull Films, 2018.

Lum, Kristian, and William Isaac. "To Predict and Serve?" *Significance* 13, no. 5 (2016): 14–19.

Lyons, Derek E., Andrew G. Young, and Frank C. Keil. "The Hidden Structure of Over-imitation." *Proceedings of the National Academy of Sciences* 104, no. 50 (2007): 19751–56.

MacAskill, William. *Doing Good Better: Effective Altruism and a Radical New Way to Make a Difference*. Guardian Faber, 2015.

MacAskill, William, Krister Bykvist, and Toby Ord. *Moral Uncertainty*. Oxford University Press, 2020.

Machado, Marlos C., Marc G. Bellemare, Erik Talvitie, Joel Veness, Matthew Hausknecht, and Michael Bowling. "Revisiting the Arcade Learning Environment: Evaluation Protocols and Open Problems for General Agents." *Journal of Artificial Intelligence Research* 61 (2018): 523–62.

MacKay, David J. C. "A Practical Bayesian Framework for Backpropagation Networks." *Neural Computation* 4, no. 3 (1992): 448–72.

Madras, David, Toniann Pitassi, and Richard Zemel. "Predict Responsibly: Increasing Fairness by Learning to Defer." In *Neural Information Processing Systems*, 2018.

Mądry, Aleksander, Aleksandar Makelov, Ludwig Schmidt, Dimitris Tsipras, and Adrian Vladu. "Towards Deep Learning Models Resistant to Adversarial Attacks." *International Conference on Learning Representations* (ICLR), 2018.

Maier, Steven F., and Martin E. Seligman. "Learned Helplessness: Theory and Evidence." *Journal of Experimental Psychology: General* 105, no. 1 (1976): 3–46.

Malik, Dhruv, Malayandi Palaniappan, Jaime Fisac, Dylan Hadfield-Menell, Stuart Russell, and Anca Drăgan. "An Efficient, Generalized Bellman Update for Cooperative Inverse Reinforcement Learning." In *Proceedings of the 35th International Conference on Machine Learning*, edited by Jennifer Dy and Andreas Krause, 3394–3402. PMLR, 2018.

Malone, Thomas W. "Toward a Theory of Intrinsically Motivating Instruction." *Cognitive Science* 4 (1981): 333–70.

———. "What Makes Computer Games Fun?" *Byte* 6 (1981): 258–77.

Malone, Thomas W., and Mark R. Lepper. "Making Learning Fun: A Taxonomy of Intrinsic Motivations for Learning." In *Aptitude, Learning and Instruction III: Conative and Affective Process Analyses*, edited by Richard E. Snow and Marshall J. Farr, 223–53. Hillsdale, NJ: Lawrence Erlbaum Associates, 1987.

Manheim, David, and Scott Garrabrant. "Categorizing Variants of Goodhart's Law." arXiv Preprint arXiv:1803.04585, 2019.

Manning, Christopher. "Lecture 2: Word Vector Representations: Word2vec," April 3, 2017. https://www.youtube.com/watch?v=ERibwqs9p38.

Manning, Christopher D., and Hinrich Schütze. *Foundations of Statistical Natural Language Processing*. MIT Press, 1999.

Marewski, Julian N., and Gerd Gigerenzer. "Heuristic Decision Making in Medicine." *Dialogues in Clinical Neuroscience* 14, no. 1 (2012): 77–89.

Marks, Michelle A., Mark J. Sabella, C. Shawn Burke, and Stephen J. Zaccaro. "The Impact of Cross-Training on Team Effectiveness." *Journal of Applied Psychology* 87, no. 1 (2002): 3–13.

Marts, Sherry A., and Sarah Keitt. "Foreword: A Historical Overview of Advocacy for Research in Sex-Based Biology." In *Advances in Molecular and Cell Biology*, 34 (2004): v–xiii.

Mastroianni, Anna C., Ruth Faden, and Daniel Federman, eds. *Women and Health Research: Ethical and Legal Issues of Including Women in Clinical Studies*. National Academies Press, 1994.

Matarić, Maja J. "Reward Functions for Accelerated Learning." In *Machine Learning: Proceedings of the Eleventh International Conference*, 181–89. 1994.

Maxwell, James Clerk. "On Governors." *Proceedings of the Royal Society of London* 16 (1868): 270–83.

Mayson, Sandra G. "Bias in, Bias Out." *Yale Law Journal* 128 (2019): 2218–2300.

———. "Dangerous Defendants." *Yale Law Journal* 127 (2017): 490–568.

McCarthy, John. "Computer-Controlled Cars," 1969. http://www-formal.stanford.edu/jmc/progress/cars/cars.html.

———. "What Is Artificial Intelligence?" 1998. http://www-formal.stanford.edu/jmc/whatisai.pdf.

McCarthy, John, and Edward A. Feigenbaum. "In Memoriam: Arthur Samuel: Pioneer in Machine Learning." *AI Magazine* 11, no. 3 (1990): 10.

McCulloch, Warren S. *The Collected Works of Warren S. McCulloch.* Edited by Rook McCulloch. Salinas, CA: Intersystems Publications, 1989.

———. *Finality and Form: In Nervous Activity.* C. C. Thomas, 1952.

———. "Recollections of the Many Sources of Cybernetics." In *ASC Forum* 6, no. 2 (1974): 5–16.

McCulloch, Warren S., and Walter Pitts. "A Logical Calculus of Ideas Immanent in Nervous Activity." *Bulletin of Mathematical Biophysics* 4 (1943).

McFadden, Syreeta. "Teaching the Camera to See My Skin." BuzzFeed News, April 2, 2014. https://www.buzzfeednews.com/article/syreetamcfadden/teaching-the-camera -to-see-my-skin.

McGonigal, Jane. *Reality Is Broken: Why Games Make Us Better and How They Can Change the World.* Penguin, 2011.

———. *SuperBetter: A Revolutionary Approach to Getting Stronger, Happier, Braver and More Resilient.* Penguin, 2015.

McGuigan, Nicola, and Murray Graham. "Cultural Transmission of Irrelevant Tool Actions in Diffusion Chains of 3- and 5-Year-Old Children." *European Journal of Developmental Psychology* 7, no. 5 (2010): 561–77.

Meehl, Paul E. "Causes and Effects of My Disturbing Little Book." *Journal of Personality Assessment* 50, no. 3 (1986): 370–75.

———. *Clinical Versus Statistical Prediction.* Minneapolis: University of Minnesota Press, 1954.

Meltzoff, Andrew N. "Born to Learn: What Infants Learn from Watching Us." In *The Role of Early Experience in Infant Development,* 145–64. Skillman, NJ: Pediatric Institute Publications, 1999.

———. "Foundations for Developing a Concept of Self: The Role of Imitation in Relating Self to Other and the Value of Social Mirroring, Social Modeling, and Self Practice in Infancy." In *The Self in Transition: Infancy to Childhood,* 139–64. Chicago: University of Chicago Press, 1990.

———. "'Like Me': A Foundation for Social Cognition." *Developmental Science* 10, no. 1 (2007): 126–34.

———. "Understanding the Intentions of Others: Re-enactment of Intended Acts by 18-Month-Old Children." *Developmental Psychology* 31, no. 5 (1995): 838.

Meltzoff, Andrew N., and M. Keith Moore. "Imitation of Facial and Manual Gestures by Human Neonates." *Science* 198, no. 4312 (1977): 75–78.

———. "Newborn Infants Imitate Adult Facial Gestures." *Child Development* 54, no. 3 (1983): 702–09.

Meltzoff, Andrew N., Lynne Murray, Elizabeth Simpson, Mikael Heimann, Emese

Nagy, Jacqueline Nadel, Eric J Pedersen, et al. "Re-examination of Oostenbroek et al. (2016): Evidence for Neonatal Imitation of Tongue Protrusion." *Developmental Science* 21, no. 4 (2018): e12609.

Meltzoff, Andrew N., Anna Waismeyer, and Alison Gopnik. "Learning About Causes from People: Observational Causal Learning in 24-Month-Old Infants." *Developmental Psychology* 48, no. 5 (2012): 1215.

Mercier, Hugo, and Dan Sperber. "Why Do Humans Reason? Arguments for an Argumentative Theory." *Behavioral and Brain Sciences* 34 (2011): 57–111.

Merler, Michele, Nalini Ratha, Rogerio S. Feris, and John R Smith. "Diversity in Faces." arXiv Preprint arXiv:1901.10436, 2019.

Metz, Cade. "Is Ethical A.I. Even Possible?" *New York Times*, March 1, 2019.

———. "We Teach A.I. Systems Everything, Including Our Biases." *New York Times*, November 11, 2019.

Mikolov, Tomáš. "Learning Representations of Text Using Neural Networks." In *NIPS Deep Learning Workshop 2013*, 2013. https://drive.google.com/file/d/0B7Xk CwpI5KDYRWRnd1RzWXQ2TWc/edit.

Mikolov, Tomáš, Kai Chen, Greg Corrado, and Jeffrey Dean. "Efficient Estimation of Word Representations in Vector Space." arXiv Preprint arXiv:1301.3781, 2013.

Mikolov, Tomáš, Quoc V. Le, and Ilya Sutskever. "Exploiting Similarities Among Languages for Machine Translation." arXiv Preprint arXiv:1309.4168, 2013.

Mikolov, Tomáš, Ilya Sutskever, and Quoc Le. "Learning the Meaning Behind Words." *Google Open Source Blog*, August 14, 2013. https://opensource.googleblog. com/2013/08/learning-meaning-behind-words.html.

Mikolov, Tomáš, Wen-tau Yih, and Geoffrey Zweig. "Linguistic Regularities in Continuous Space Word Representations." In *Proceedings of the 2013 Conference of the North American Chapter of the Association for Computational Linguistics: Human Language Technologies*, 746–51. 2013.

Miller, Geoffrey. "Reconciling Evolutionary Psychology and Ecological Psychology: How to Perceive Fitness Affordances." *Acta Psychologica Sinica* 39, no. 3 (2007): 546–55.

Milli, Smitha, and Anca D. Drăgan. "Literal or Pedagogic Human? Analyzing Human Model Misspecification in Objective Learning." In *The Conference on Uncertainty in Artificial Intelligence (UAI)*, 2019.

Milli, Smitha, Dylan Hadfield-Menell, Anca Drăgan, and Stuart Russell. "Should Robots Be Obedient?" In *26th International Joint Conference on Artificial Intelligence*. 2017.

Minsky, Marvin. "Steps Toward Artificial Intelligence." *Proceedings of the Institute of Radio Engineers* 49, no. 1 (1961): 8–30.

———. "Theory of Neural-Analog Reinforcement Systems and Its Application to the Brain Model Problem." PhD thesis, Princeton University, 1954.

Minsky, Marvin L., and Seymour A. Papert. *Perceptrons: An Introduction to Computational Geometry*. MIT Press, 1969.

Mirenowicz, Jacques, and Wolfram Schultz. "Importance of Unpredictability for Reward Responses in Primate Dopamine Neurons." *Journal of Neurophysiology* 72, no. 2 (1994): 1024–27.

Mnih, Volodymyr, Koray Kavukcuoglu, David Silver, Alex Graves, Ioannis Antonoglou, Daan Wierstra, and Martin Riedmiller. "Playing Atari with Deep Reinforcement Learning." In *Deep Learning Workshop, NIPS 2013*. 2013.

Mnih, Volodymyr, Koray Kavukcuoglu, David Silver, Andrei A. Rusu, Joel Veness, Marc G. Bellemare, Alex Graves, et al. "Human-Level Control Through Deep Reinforcement Learning." *Nature* 518, no. 7540 (2015): 529–33.

Mohamed, Shakir, and Danilo Jimenez Rezende. "Variational Information Maximisation for Intrinsically Motivated Reinforcement Learning." In *Advances in Neural Information Processing Systems 28*, 2125–33. 2015.

Mombaur, Katja, Anh Truong, and Jean-Paul Laumond. "From Human to Humanoid Locomotion—an Inverse Optimal Control Approach." *Autonomous Robots* 28, no. 3 (2010): 369–83.

Monahan, John, and Jennifer L. Skeem. "Risk Assessment in Criminal Sentencing." *Annual Review of Clinical Psychology* 12 (2016): 489–513.

Montague, P. Read, Peter Dayan, Christophe Person, and Terrence J. Sejnowski. "Bee Foraging in Uncertain Environments Using Predictive Hebbian Learning." *Nature* 377, no. 6551 (1995): 725.

Montague, P. Read, Peter Dayan, and Terrence J. Sejnowski. "A Framework for Mesencephalic Dopamine Systems Based on Predictive Hebbian Learning." *Journal of Neuroscience* 16, no. 5 (1996): 1936–47.

Moravec, Hans. "Obstacle Avoidance and Navigation in the Real World by a Seeing Robot Rover." PhD thesis, Stanford University, 1980.

Mordvintsev, Alexander, Christopher Olah, and Mike Tyka. "DeepDream—a Code Example for Visualizing Neural Networks." *Google AI Blog*, July 1, 2015. https://ai.googleblog.com/2015/07/deepdream-code-example-for-visualizing.html.

———. "Inceptionism: Going Deeper into Neural Networks." *Google AI Blog*, June 17, 2015. https://ai.googleblog.com/2015/06/inceptionism-going-deeper-into-neural.html.

Morgan, Conway Lloyd. *An Introduction to Comparative Psychology*. 1894. 2nd ed., Charles Scribner's Sons, 1904.

Moss-Racusin, Corinne A., John F. Dovidio, Victoria L. Brescoll, Mark J. Graham, and Jo Handelsman. "Science Faculty's Subtle Gender Biases Favor Male Students." *Proceedings of the National Academy of Sciences* 109, no. 41 (2012): 16474–79.

Mueller, Benjamin, Robert Gebeloff, and Sahil Chinoy. "Surest Way to Face Marijuana Charges in New York: Be Black or Hispanic." *New York Times*, May 13, 2018.

Mueller, Dennis C. *Public Choice III*. Cambridge University Press, 2003.

Munro, Robert. "Diversity in AI Is Not Your Problem, It's Hers." *Medium*, November 11, 2019. https://medium.com/@robert.munro/bias-in-ai-3ea569f79d6a.

Murdoch, Iris. *The Bell*. Chatto & Windus, 1958.

Naddaf, Yavar. "Game-Independent AI Agents for Playing Atari 2600 Console Games." Master's thesis, University of Alberta, 2010.

Nair, Ashvin, Bob McGrew, Marcin Andrychowicz, Wojciech Zaremba, and Pieter Abbeel. "Overcoming Exploration in Reinforcement Learning with Demonstrations." In *2018 IEEE International Conference on Robotics and Automation (ICRA)*, 6292–99. IEEE, 2018.

Nalisnick, Eric, Bhaskar Mitra, Nick Craswell, and Rich Caruana. "Improving Document Ranking with Dual Word Embeddings." In *Proceedings of the 25th International World Wide Web Conference*, 83–84. Montreal: International World Wide Web Conferences Steering Committee, 2016.

Narla, Akhila, Brett Kuprel, Kavita Sarin, Roberto Novoa, and Justin Ko. "Automated Classification of Skin Lesions: From Pixels to Practice." *Journal of Investigative Dermatology* 138, no. 10 (2018): 2108–10.

Neal, Radford M. "Bayesian Learning for Neural Networks." PhD thesis, University of Toronto, 1995.

Nematzadeh, Aida, Stephan C. Meylan, and Thomas L. Griffiths. "Evaluating Vector-Space Models of Word Representation; or, The Unreasonable Effectiveness of Counting Words Near Other Words." In *Proceedings of the 39th Annual Conference of the Cognitive Science Society*, 2017.

"New Navy Device Learns by Doing." *New York Times*, July 8, 1958.

"New York's Broken Parole System." *New York Times*, February 16, 2014.

Ng, Andrew Y. "Shaping and Policy Search in Reinforcement Learning." PhD thesis, University of California, Berkeley, 2003.

Ng, Andrew Y., Adam Coates, Mark Diel, Varun Ganapathi, Jamie Schulte, Ben Tse, Eric Berger, and Eric Liang. "Autonomous Inverted Helicopter Flight via Reinforcement Learning." In *Experimental Robotics IX*, 363–72. Springer, 2006.

Ng, Andrew Y., Daishi Harada, and Stuart Russell. "Policy Invariance Under Reward Transformations: Theory and Application to Reward Shaping." In *ICML* 99 (1999): 278–87.

Ng, Andrew Y., H. Jin Kim, Michael I. Jordan, and Shankar Sastry. "Autonomous Helicopter Flight via Reinforcement Learning." In *Advances in Neural Information Processing Systems*, 799–806, 2004.

Ng, Andrew Y., and Stuart J. Russell. "Algorithms for Inverse Reinforcement Learning." In *Proceedings of the 17th International Conference on Machine Learning*. Stanford, CA: Morgan Kaufmann, 2000.

Nguyen, Anh, Alexey Dosovitskiy, Jason Yosinski, Thomas Brox, and Jeff Clune. "Synthesizing the Preferred Inputs for Neurons in Neural Networks via Deep Generator Networks." In *Advances in Neural Information Processing Systems 29*, edited by D. D. Lee, M. Sugiyama, U. V. Luxburg, I. Guyon, and R. Garnett, 3387–95, 2016.

Nguyen, Anh, Jason Yosinski, and Jeff Clune. "Deep Neural Networks Are Easily Fooled: High Confidence Predictions for Unrecognizable Images." In *Proceedings of the IEEE Conference on Computer Vision and Pattern Recognition (CVPR2015)*, 427–36. CVPR, 2015.

Nikolaidis, Stefanos, Przemyslaw Lasota, Ramya Ramakrishnan, and Julie Shah. "Improved Human-Robot Team Performance Through Cross-Training: An Approach Inspired by Human Team Training Practices." *International Journal of Robotics Research* 34, no. 14 (2015): 1711–30.

Nissim, Malvina, Rik van Noord, and Rob van der Goot. "Fair Is Better Than Sensational: Man Is to Doctor as Woman Is to Doctor." arXiv Preprint arXiv:1905.09866, 2019.

Niv, Yael. "Reinforcement Learning in the Brain." *Journal of Mathematical Psychology* 53, no. 3 (2009): 139–54.

Niv, Yael, Daphna Joel, Isaac Meilijson, and Eytan Ruppin. "Evolution of Reinforcement Learning in Foraging Bees: A Simple Explanation for Risk Averse Behavior." *Neurocomputing* 44 (2001): 951–56.

"Northpointe's Response to ProPublica: Demonstrating Accuracy Equity and Predictive Parity," [2016]. http://www.equivant.com/blog/response-to-propublica-demonstrating -accuracy-equity-and-predictive-parity; original version available at https://web .archive.org/web/20160802190300/http:/www.northpointeinc.com/northpointe -analysis.

Odell, Jenny. *How to Do Nothing: Resisting the Attention Economy.* Melville House, 2019.

O'Doherty, John P. "Beyond Simple Reinforcement Learning: The Computational Neurobiology of Reward-Learning and Valuation." *European Journal of Neuroscience* 35, no. 7 (2012): 987–90.

Ohlin, Lloyd E. *Selection for Parole: A Manual of Parole Prediction.* Russell Sage Foundation, 1951.

Olah, Chris, Alexander Mordvintsev, and Ludwig Schubert. "Feature Visualization." *Distill*, 2017.

Olah, Chris, Arvind Satyanarayan, Ian Johnson, Shan Carter, Ludwig Schubert, Katherine Ye, and Alexander Mordvintsev. "The Building Blocks of Interpretability." *Distill*, 2018.

Olds, James. "Pleasure Centers in the Brain." *Scientific American* 195, no. 4 (1956): 105–17.

———. "Pleasure Centers in the Brain." *Engineering and Science* 33, no. 7 (1970): 22–31.

Olds, James, and Peter Milner. "Positive Reinforcement Produced by Electrical Stimulation of Septal Area and Other Regions of Rat Brain." *Journal of Comparative and
Physiological Psychology* 47, no. 6 (1954): 419–27.

Omohundro, Stephen M. "The Basic AI Drives." In *Artificial General Intelligence 2008:
Proceedings of the First AGI Conference,* edited by Pei Wang, Ben Goertzel, and
Stan Franklin, 483–92. Amsterdam: IOS Press, 2008.

O'Neil, Cathy. *Weapons of Math Destruction: How Big Data Increases Inequality and
Threatens Democracy.* Crown, 2016.

Oostenbroek, Janine, Thomas Suddendorf, Mark Nielsen, Jonathan Redshaw, Siobhan
Kennedy-Costantini, Jacqueline Davis, Sally Clark, and Virginia Slaughter. "Com-
prehensive Longitudinal Study Challenges the Existence of Neonatal Imitation in
Humans." *Current Biology* 26, no. 10 (2016): 1334–48.

Ord, Toby. "Moral Trade." *Ethics* 126, no. 1 (2015): 118–38.

———. *The Precipice: Existential Risk and the Future of Humanity.* Hachette Books,
2020.

Orseau, Laurent. "Universal Knowledge-Seeking Agents." In *Algorithmic Learning Theory: 22nd International Conference,* 353–367. Springer, 2011.

Orseau, Laurent, and Stuart Armstrong. "Safely Interruptible Agents." In *Proceedings
of the Thirty-Second Uncertainty in Artificial Intelligence Conference,* 557–66.
AUAI Press, 2016.

Orseau, Laurent, Tor Lattimore, and Marcus Hutter. "Universal Knowledge-Seeking
Agents for Stochastic Environments." In *International Conference on Algorithmic
Learning Theory,* 158–72. 2013.

Ostrovski, Georg, Marc G. Bellemare, Aäron van den Oord, and Rémi Munos. "Count-
Based Exploration with Neural Density Models." arXiv Preprint arXiv:1703.01310,
June 14, 2017.

O'Toole, Garson. "There's a Way to Do It Better—Find It." Quote Investigator, July 16,
2013. https://quoteinvestigator.com/2013/07/16/do-it-better/.

Oudeyer, Pierre-Yves, and Frederic Kaplan. "What Is Intrinsic Motivation? A Typology
of Computational Approaches." *Frontiers in Neurorobotics* 1, no. 6 (2007): 1–13.

Pathak, Deepak, Pulkit Agrawal, Alexei A. Efros, and Trevor Darrell. "Curiosity-
Driven Exploration by Self-Supervised Prediction." In *International Conference on
Machine Learning (ICML),* 2017.

Paul, L. A. *Transformative Experience.* Oxford University Press, 2014.

Paulus, Markus, Sabine Hunnius, Marlies Vissers, and Harold Bekkering. "Imitation
in Infancy: Rational or Motor Resonance?" *Child Development* 82, no. 4 (2011):
1047–57.

Pavlov, I. P. *Conditioned Reflexes: An Investigation of the Physiological Activity of the
Cerebral Cortex.* Translated by G. V. Anrep. London: Oxford University Press,
1927.

Pearl, Judea. "The Seven Tools of Causal Inference, with Reflections on Machine Learning." *Communications of the ACM* 62, no. 3 (2019): 54–60.

Pedreshi, Dino, Salvatore Ruggieri, and Franco Turini. "Discrimination-Aware Data Mining." In *Proceedings of the 14th ACM SIGKDD International Conference on Knowledge Discovery and Data Mining*, 560–68. ACM, 2008.

Peng, Xue Bin, Pieter Abbeel, Sergey Levine, and Michiel van de Panne. "DeepMimic: Example-Guided Deep Reinforcement Learning of Physics-Based Character Skills." *ACM Transactions on Graphics (TOG)* 37, no. 4 (2018): 143.

Perez, Caroline Criado. *Invisible Women: Data Bias in a World Designed for Men.* Abrams, 2019.

Persico, Nicola. "Racial Profiling, Fairness, and Effectiveness of Policing." *American Economic Review* 92, no. 5 (2002): 1472–97.

Peterson, Gail B. "A Day of Great Illumination: B. F. Skinner's Discovery of Shaping." *Journal of the Experimental Analysis of Behavior* 82, no. 3 (2004): 317–28.

Piaget, Jean. *The Construction of Reality in the Child.* Translated by Margaret Cook. Basic Books, 1954.

Piccinini, Gualtiero. "The First Computational Theory of Mind and Brain: A Close Look at McCulloch and Pitts's 'Logical Calculus of Ideas Immanent in Nervous Activity.'" *Synthese* 141, no. 2 (2004): 175–215.

Plato. *Protagoras and Meno.* Translated by W.K.C. Guthrie. Penguin Books, 1956.

Podkopacz, Marcy R. "Building and Validating the 2007 Hennepin County Adult Pretrial Scale." Minneapolis, MN: Fourth Judicial District Research Division, 2010. http://www.mncourts.gov/mncourtsgov/media/assets/documents/4/reports/Validation_of_the_New_2007_Hennepin_County_Pretrial_Scale.pdf.

Podkopacz, Marcy R., Deborah Eckberg, and Gina Kubits. "Fourth Judicial District Pretrial Evaluation: Scale Validation Study." Minneapolis, MN: Fourth Judicial District Research Division, 2006. http://www.mncourts.gov/Documents/4/Public/Research/PreTrial_Scale_Validation_(2006).pdf.

Pomerleau, Dean A. "ALVINN: An Autonomous Land Vehicle in a Neural Network." In *Advances in Neural Information Processing Systems*, 305–13. 1989.

———. "Knowledge-Based Training of Artificial Neural Networks for Autonomous Robot Driving." In *Robot Learning*, 19–43. Kluwer Academic Publishers, 1993.

Poplin, Ryan, Avinash V. Varadarajan, Katy Blumer, Yun Liu, Michael V. McConnell, Greg S. Corrado, Lily Peng, and Dale R. Webster. "Prediction of Cardiovascular Risk Factors from Retinal Fundus Photographs via Deep Learning." *Nature Biomedical Engineering* 2, no. 3 (2018): 158–64.

Poursabzi-Sangdeh, Forough, Daniel G. Goldstein, Jake M. Hofman, Jenn Wortman Vaughan, and Hanna Wallach. "Manipulating and Measuring Model Interpretability." arXiv Preprint arXiv:1802.07810, 2018.

Prost, Flavien, Nithum Thain, and Tolga Bolukbasi. "Debiasing Embeddings for Reduced Gender Bias in Text Classification." arXiv Preprint arXiv:1908.02810, 2019.

Prümmer, Dominic M., OP. *Handbook of Moral Theology* Translated by Reverend Gerald W. Shelton, STL. 5th ed. Cork: Mercier Press, 1956.

Quinlan, J. Ross. *C4.5: Programs for Machine Learning.* San Mateo, CA: Morgan Kaufmann Publishers, 1993.

Radford, Alec, Jeffrey Wu, Rewon Child, David Luan, Dario Amodei, and Ilya Sutskever. "Language Models Are Unsupervised Multitask Learners." *OpenAI Blog,* 2019. https://openai.com/blog/better-language-models/.

Ramakrishnan, Ramya, Chongjie Zhang, and Julie Shah. "Perturbation Training for Human-Robot Teams." *Journal of Artificial Intelligence Research* 59 (2017): 495–541.

Ramírez, Naja Ferjan, Sarah Roseberry Lytle, and Patricia K. Kuhl. "Parent Coaching Increases Conversational Turns and Advances Infant Language Development." *Proceedings of the National Academy of Sciences* 117, no. 7 (2020): 3484–91.

Randløv, Jette, and Preben Alstrøm. "Learning to Drive a Bicycle Using Reinforcement Learning and Shaping." In *Proceedings of the 15th International Conference on Machine Learning,* 463–71. Morgan Kaufmann Publishers, 1998.

Read, Herbert. *The Grass Roots of Art: Four Lectures on Social Aspects of Art in an Industrial Age.* World Publishing, 1961.

Recht, Benjamin, Rebecca Roelofs, Ludwig Schmidt, and Vaishaal Shankar. "Do ImageNet Classifiers Generalize to ImageNet?" In *Proceedings of the 36th International Conference on Machine Learning,* edited by Kamalika Chaudhuri and Ruslan Salakhutdinov, 5389–5400. PMLR, 2019.

"Report on Algorithmic Risk Assessment Tools in the U.S. Criminal Justice System." Partnership on AI, 2019. https://www.partnershiponai.org/wp-content/uploads/2019/04/Report-on-Algorithmic-Risk-Assessment-Tools.pdf.

Rescorla, Robert A., and Allan R. Wagner. "A Theory of Pavlovian Conditioning: Variations in the Effectiveness of Reinforcement and Nonreinforcement." In *Classical Conditioning II: Current Research and Theory,* 64–99. Appleton-Century-Crofts, 1972.

Rezaei, Ashkan, Rizal Fathony, Omid Memarrast, and Brian Ziebart. "Fairness for Robust Log Loss Classification." In *Proceedings of the Thirty-Fourth AAAI Conference on Artificial Intelligence,* 2020.

Ribeiro, Marco Tulio, Sameer Singh, and Carlos Guestrin. "Why Should I Trust You? Explaining the Predictions of Any Classifier." In *Proceedings of the 22nd ACM SIGKDD International Conference on Knowledge Discovery and Data Mining,* 1135–44. ACM, 2016.

Riedl, Mark O., and Brent Harrison. "Enter the Matrix: A Virtual World Approach to Safely Interruptable Autonomous Systems." CoRR abs/1703.10284 (2017). http://arxiv.org/abs/1703.10284.

Ring, Mark B., and Laurent Orseau. "Delusion, Survival, and Intelligent Agents." In *Artificial General Intelligence: 4th International Conference, AGI 2011*, edited by Jürgen Schmidhuber, Kristinn R. Thórisson, and Moshe Looks, 11–20. Springer, 2011.

"Rival." *New Yorker*, December 6, 1958.

Rivest, Ronald L. "Learning Decision Lists." *Machine Learning* 2, no. 3 (1987): 229–46.

Roder, Beverly J., Emily W. Bushnell, and Anne Marie Sasseville. "Infants' Preferences for Familiarity and Novelty During the Course of Visual Processing." *Infancy* 1, no. 4 (2000): 491–507.

Rohde, Douglas L. T., and David C. Plaut. "Language Acquisition in the Absence of Explicit Negative Evidence: How Important Is Starting Small?" *Cognition* 72, no. 1 (1999): 67–109.

Roland, Alex, and Philip Shiman. *Strategic Computing: DARPA and the Quest for Machine Intelligence, 1983–1993*. MIT Press, 2002.

Romanes, George John. *Animal Intelligence*. New York: D. Appleton, 1882.

Romo, Ranulfo, and Wolfram Schultz. "Dopamine Neurons of the Monkey Midbrain: Contingencies of Responses to Active Touch During Self-Initiated Arm Movements." *Journal of Neurophysiology* 63, no. 3 (1990): 592–606.

Rong, Xin. "Word2vec Parameter Learning Explained." arXiv Preprint arXiv:1411.2738, 2014.

Roodman, David. "Aftereffects: In the U.S., Evidence Says Doing More Time Typically Leads to More Crime After." *Open Philanthropy Blog*, 2017. https://www.openphilanthropy.org/blog/aftereffects-us-evidence-says-doing-more-time-typically-leads-more-crime-after.

Rosenberg, Charles R., and Terrence J. Sejnowski. "NETtalk: A Parallel Network That Learns to Read Aloud." Johns Hopkins University, 1986.

Rosenblatt, Frank. "The Perceptron: A Probabilistic Model for Information Storage and Organization in the Brain." *Psychological Review* 65, no. 6 (1958): 386.

———. "Principles of Neurodynamics: Perceptrons and the Theory of Brain Mechanisms." Buffalo, NY: Cornell Aeronautical Laboratory, March 15, 1961.

Rosenblueth, Arturo, Norbert Wiener, and Julian Bigelow. "Behavior, Purpose and Teleology." *Philosophy of Science* 10, no. 1 (1943): 18–24.

Ross, Stéphane, and J. Andrew Bagnell. "Efficient Reductions for Imitation Learning." In *Proceedings of the 13th International Conference on Artificial Intelligence and Statistics (AISTATS)*, 661–68. 2010.

Ross, Stéphane, Geoffrey J. Gordon, and J. Andrew Bagnell. "A Reduction of Imitation Learning and Structured Prediction to No-Regret Online Learning." In *Proceedings of the Fourteenth International Conference on Artificial Intelligence and Statistics*, 627–35. 2011.

Roth, Lorna. "Looking at Shirley, the Ultimate Norm: Colour Balance, Image Technologies, and Cognitive Equity." *Canadian Journal of Communication* 34, no. 1 (2009).

Rousseau, Jean-Jacques. *Emile; or, On Education.* Translated by Barbara Foxley. E. P. Dutton, 1921.

Rudd, Ethan M., Lalit P. Jain, Walter J. Scheirer, and Terrance E. Boult. "The Extreme Value Machine." *IEEE Transactions on Pattern Analysis and Machine Intelligence* 40, no. 3 (2017): 762–68.

Ruder, Sebastian. "An Overview of Multi-Task Learning in Deep Neural Networks." arXiv Preprint arXiv:1706.05098, 2017.

Rudin, Cynthia, and Joanna Radin. "Why Are We Using Black Box Models in AI When We Don't Need To? A Lesson From An Explainable AI Competition." *Harvard Data Science Review*, 2019.

Rudin, Cynthia, and Berk Ustun. "Optimized Scoring Systems: Toward Trust in Machine Learning for Healthcare and Criminal Justice." *Interfaces* 48, no. 5 (2018): 449–66.

Rudin, Cynthia, Caroline Wang, and Beau Coker. "The Age of Secrecy and Unfairness in Recidivism Prediction." *Harvard Data Science Review* 2, no. 1 (2020).

Rumelhart, D. E., G. E. Hinton, and R. J. Williams. "Learning Internal Representations by Error Propagation." In *Parallel Distributed Processing: Explorations in the Microstructure of Cognition*, 1:318–62. MIT Press, 1986.

Rumelhart, David E., and James L. McClelland. *Parallel Distributed Processing: Explorations in the Microstructure of Cognition.* MIT Press, 1986.

Rummery, Gavin A., and Mahesan Niranjan. "On-Line Q-Learning Using Connectionist Systems." Cambridge University Engineering Department, 1994.

Russakovsky, Olga, Jia Deng, Hao Su, Jonathan Krause, Sanjeev Satheesh, Sean Ma, Zhiheng Huang, et al. "ImageNet Large Scale Visual Recognition Challenge." *International Journal of Computer Vision* 115, no. 3 (2015): 211–52.

Russell, Bertrand. "Ideas That Have Harmed Mankind." In *Unpopular Essays.* MPG Books, 1950.

———. *Logic and Knowledge: Essays 1901–1950.* George Allen & Unwin, 1956.

Russell, Stuart. *Human Compatible.* Viking, 2019.

———. "Learning Agents for Uncertain Environments (Extended Abstract)." In *Proceedings of the Eleventh Annual ACM Workshop on Computational Learning Theory (COLT-98)*, 101–03. Madison, WI: ACM Press, 1998.

———. "Should We Fear Supersmart Robots?" *Scientific American* 314, no. 6 (2016): 58–59.

Russell, Stuart J., and Peter Norvig. *Artificial Intelligence: A Modern Approach.* 3rd ed. Upper Saddle River, NJ: Pearson, 2010.

Rust, John. "Do People Behave According to Bellman's Principle of Optimality?" Working Papers in Economics. Hoover Institution, Stanford University, 1992.

———. "Structural Estimation of Markov Decision Processes." *Handbook of Econometrics* 4 (1994): 3081–3143.

Rutledge, Robb B., Nikolina Skandali, Peter Dayan, and Raymond J. Dolan. "A Computational and Neural Model of Momentary Subjective Well-Being." *Proceedings of the National Academy of Sciences* 111, no. 33 (August 19, 2014): 12252–57.

Saayman, Graham, Elinor Wardwell Ames, and Adrienne Moffett. "Response to Novelty as an Indicator of Visual Discrimination in the Human Infant." *Journal of Experimental Child Psychology* 1, no. 2 (1964): 189–98.

Sadigh, Dorsa, S. Shankar Sastry, Sanjit A. Seshia, and Anca D. Drăgan. "Planning for Autonomous Cars That Leverage Effects on Human Actions." In *Proceedings of Robotics: Science and Systems (RSS)*, 2016. https://doi.org/10.15607/RSS.2016.XII.029.

Sahami, Mehran, Marti Hearst, and Eric Saund. "Applying the Multiple Cause Mixture Model to Text Categorization." In *ICML-96: Proceedings of the Thirteenth International Conference on Machine Learning*, 435–43. 1996.

Saksida, Lisa M., Scott M. Raymond, and David S. Touretzky. "Shaping Robot Behavior Using Principles from Instrumental Conditioning." *Robotics and Autonomous Systems* 22, nos. 3–4 (1997): 231–49.

Salge, Christoph, Cornelius Glackin, and Daniel Polani. "Empowerment: An Introduction." In *Guided Self-Organization: Inception*, 67–114. Springer, 2014.

Salimans, Tim, and Richard Chen. "Learning Montezuma's Revenge from a Single Demonstration." arXiv Preprint arXiv:1812.03381, 2018.

Samuel, Arthur L. "Some Studies in Machine Learning Using the Game of Checkers." *IBM Journal of Research and Development* 3, no. 3 (1959): 210–29.

Samuelson, Paul A. "A Note on Measurement of Utility." *The Review of Economic Studies* 4, no. 2 (1937): 155–61.

Sarbin, Theodore R. "A Contribution to the Study of Actuarial and Individual Methods of Prediction." *American Journal of Sociology* 48, no. 5 (1943): 593–602.

Sargent, Thomas J. "Estimation of Dynamic Labor Demand Schedules Under Rational Expectations." *Journal of Political Economy* 86, no. 6 (1978): 1009–44.

Saund, Eric. "A Multiple Cause Mixture Model for Unsupervised Learning." *Neural Computation* 7, no. 1 (1995): 51–71.

Saunders, Jessica. "Pitfalls of Predictive Policing." *U.S. News & World Report*, October 7, 2016.

Saunders, Jessica, Priscillia Hunt, and John S. Hollywood. "Predictions Put into Practice: A Quasi-Experimental Evaluation of Chicago's Predictive Policing Pilot." *Journal of Experimental Criminology* 12, no. 3 (2016): 347–71.

Saunders, William, Girish Sastry, Andreas Stuhlmueller, and Owain Evans. "Trial Without Error: Towards Safe Reinforcement Learning via Human Intervention." In *Proceedings of the 17th International Conference on Autonomous Agents and Multiagent Systems*, 2067–99. International Foundation for Autonomous Agents and Multiagent Systems, 2018.

Savage, Tony. "Shaping: The Link Between Rats and Robots." *Connection Science* 10, nos. 3–4 (1998): 321–40.

Sayre-McCord, Geoff. "Moral Realism." *The Stanford Encyclopedia of Philosophy*, edited by Edward N. Zalta. Entry revised February 3, 2015: https://plato.stanford.edu/entries/moral-realism/.

Schaeffer, Jonathan, Joseph Culberson, Norman Treloar, Brent Knight, Paul Lu, and Duane Szafron. "A World Championship Caliber Checkers Program." *Artificial Intelligence* 53, nos. 2–3 (1992): 273–89.

Schauer, Frederick. "Giving Reasons." *Stanford Law Review* 47 (1995): 633–59.

Scheirer, Walter J., Anderson Rocha, Archana Sapkota, and Terrance E. Boult. "Toward Open Set Recognition." *IEEE Transactions on Pattern Analysis and Machine Intelligence* 35, no. 7 (2013): 1757–72.

Schmidhuber, Jürgen. "Curious Model-Building Control Systems." In *Proceedings of the International Joint Conference on Neural Networks*, 2:1458–63. Singapore: IEEE, 1991.

———. "Formal Theory of Creativity, Fun, and Intrinsic Motivation (1990–2010)." *IEEE Transactions on Autonomous Mental Development* 2, no. 3 (September 2010): 230–47.

Schmidt, Ben. "Rejecting the Gender Binary: A Vector-Space Operation." *Ben's Bookworm Blog*, October 30, 2015. http://bookworm.benschmidt.org/posts/2015–10–30-rejecting-the-gender-binary.html.

Schrage, D. P., Y. K. Yillikci, S. Liu, J.V.R. Prasad, and S. V. Hanagud. "Instrumentation of the Yamaha R-50/RMAX Helicopter Testbeds for Airloads Identification and Follow-on Research." In *Twenty-Fifth European Rotorcraft Forum*, P4-1–P4-13. 1999.

Schrittwieser, Julian, Ioannis Antonoglou, Thomas Hubert, Karen Simonyan, Laurent Sifre, Simon Schmitt, Arthur Guez, et al. "Mastering Atari, Go, Chess and Shogi by Planning with a Learned Model." arXiv Preprint arXiv:1911.08265, 2019.

Schultz, Wolfram. "Multiple Dopamine Functions at Different Time Courses." *Annual Review of Neuroscience* 30 (2007): 259–88.

Schultz, Wolfram, Paul Apicella, and Tomas Ljungberg. "Responses of Monkey Dopamine Neurons to Reward and Conditioned Stimuli During Successive Steps of Learning a Delayed Response Task." *Journal of Neuroscience* 13, no. 3 (March 1993): 900–13.

Schultz, Wolfram, Peter Dayan, and P. Read Montague. "A Neural Substrate of Prediction and Reward." *Science* 275, no. 5306 (1997): 1593–99.

Schulz, Laura. "Infants Explore the Unexpected." *Science* 348, no. 6230 (2015): 42–43.

Schulz, Laura E., and Elizabeth Baraff Bonawitz. "Serious Fun: Preschoolers Engage in More Exploratory Play When Evidence Is Confounded." *Developmental Psychology* 43, no. 4 (2007): 1045–50.

Schwitzgebel, Eric, and Mara Garza. "A Defense of the Rights of Artificial Intelligences." *Midwest Studies in Philosophy* 39, no. 1 (2015): 98–119.

Selbst, Andrew D., and Julia Powles. "Meaningful Information and the Right to Explanation." *International Data Privacy Law* 7, no. 4 (2017): 233–42.

Selfridge, Oliver G., Richard S. Sutton, and Andrew G. Barto. "Training and Tracking in Robotics." In *Proceedings of the 9th International Joint Conference on Artificial Intelligence*, 670–72. San Francisco: Morgan Kaufmann, 1985.

Selvaraju, Ramprasaath R., Abhishek Das, Ramakrishna Vedantam, Michael Cogswell, Devi Parikh, and Dhruv Batra. "Grad-Cam: Why Did You Say That?" arXiv Preprint arXiv:1611.07450, 2016.

Sen, Amartya. *Collective Choice and Social Welfare*. Harvard University Press, 2018.

Sepielli, Andrew. "What to Do When You Don't Know What to Do When You Don't Know What to Do . . ." *Noûs* 48, no. 3 (2014): 521–44.

Sepielli, Andrew Christopher. "'Along an Imperfectly-Lighted Path': Practical Rationality and Normative Uncertainty." PhD thesis, Rutgers University, 2010.

Shafto, Patrick, Noah D. Goodman, and Thomas L. Griffiths. "A Rational Account of Pedagogical Reasoning: Teaching by, and Learning from, Examples." *Cognitive Psychology* 71 (2014): 55–89.

Shannon, Claude E. "A Mathematical Theory of Communication." *Bell System Technical Journal* 27, no. 3 (July 1948): 379–423.

Shaw, J. Cliff, Allen Newell, Herbert A. Simon, and T. O. Ellis. "A Command Structure for Complex Information Processing." In *Proceedings of the May 6–8, 1958, Western Joint Computer Conference: Contrasts in Computers*, 119–28. Los Angeles, CA, 1958.

Shead, Sam. "DeepMind's Human-Bashing AlphaGo AI Is Now Even Stronger." *Business Insider*, October 18, 2017. https://www.businessinsider.com/deepminds-alphago-ai-gets-alphago-zero-upgrade-2017-10.

Shenoy, Premnath, and Anand Harugeri. "Elderly Patients' Participation in Clinical Trials." *Perspectives in Clinical Research* 6, no. 4 (2015): 184–89.

Shlegeris, Buck. "Why I'm Less of a Hedonic Utilitarian Than I Used to Be," February 19, 2019. http://shlegeris.com/2019/02/21/hedonic.

Silver, David, Aja Huang, Chris J. Maddison, Arthur Guez, Laurent Sifre, George van den Driessche, Julian Schrittwieser, et al. "Mastering the Game of Go with Deep Neural Networks and Tree Search." *Nature* 529 (January 28, 2016): 484–89.

Silver, David, Thomas Hubert, Julian Schrittwieser, Ioannis Antonoglou, Matthew Lai, Arthur Guez, Marc Lanctot, et al. "A General Reinforcement Learning Algorithm That Masters Chess, Shogi, and Go Through Self-Play." *Science* 362, no. 6419 (2018): 1140–44.

Silver, David, Julian Schrittwieser, Karen Simonyan, Ioannis Antonoglou, Aja Huang, Arthur Guez, Thomas Hubert, et al. "Mastering the Game of Go Without Human Knowledge." *Nature* 550, no. 7676 (2017): 354.

Silvia, Paul J. *Exploring the Psychology of Interest.* Oxford University Press, 2006.

Simoiu, Camelia, Sam Corbett-Davies, and Sharad Goel. "The Problem of Infra-Marginality in Outcome Tests for Discrimination." *Annals of Applied Statistics* 11, no. 3 (2017): 1193–1216.

Simon, Herbert A. "The Cat That Curiosity Couldn't Kill." *Carnegie Mellon Magazine* 10, no. 1 (1991): 35–36.

Simonite, Tom. "When It Comes to Gorillas, Google Photos Remains Blind." *Wired,* January 11, 2018. https://www.wired.com/story/when-it-comes-to-gorillas-google-photos-remains-blind/.

Simonyan, Karen, Andrea Vedaldi, and Andrew Zisserman. "Deep Inside Convolutional Networks: Visualising Image Classification Models and Saliency Maps." arXiv Preprint arXiv:1312.6034, 2013.

Simonyan, Karen, and Andrew Zisserman. "Very Deep Convolutional Networks for Large-Scale Image Recognition." arXiv Preprint arXiv:1409.1556, 2014.

Singer, Peter. "The Drowning Child and the Expanding Circle." *New Internationalist* 289 (1997).

———. "Famine, Affluence, and Morality." *Philosophy and Public Affairs* 1, no. 1 (1972): 229–43.

———. *The Most Good You Can Do: How Effective Altruism Is Changing Ideas About Living Ethically.* Yale University Press, 2015.

Singh, Satinder Pal. "Transfer of Learning by Composing Solutions of Elemental Sequential Tasks." *Machine Learning* 8, nos. 3–4 (1992): 323–39.

Singh, Satinder, Nuttapong Chentanez, and Andrew G. Barto. "Intrinsically Motivated Reinforcement Learning." In *Advances in Neural Information Processing Systems,* 1281–88, 2005.

Singh, Satinder, Richard L. Lewis, and Andrew G. Barto. "Where Do Rewards Come From?" In *Proceedings of the Annual Conference of the Cognitive Science Society,* 2601–06. 2009.

Singh, Satinder, Richard L. Lewis, Jonathan Sorg, Andrew G. Barto, and Akram Helou. "On Separating Agent Designer Goals from Agent Goals: Breaking the Preferences–Parameters Confound." CiteSeerX, 2010.

Skeem, Jennifer L., and Christopher T. Lowenkamp. "Risk, Race, and Recidivism: Predictive Bias and Disparate Impact." *Criminology* 54, no. 4 (2016): 680–712.

Skinner, B. F. "Has Gertrude Stein a Secret?" *Atlantic Monthly* 153 (January 1934): 50–57.

———. "How to Teach Animals." *Scientific American* 185, no. 6 (1951): 26–29.

———. *A Matter of Consequences: Part Three of an Autobiography.* Knopf, 1983.

———. "Pigeons in a Pelican." *American Psychologist* 15, no. 1 (1960): 28.

———. "The Rate of Establishment of a Discrimination." *Journal of General Psychology* 9, no. 2 (1933): 302–50.

———. "Reinforcement Today." *American Psychologist* 13, no. 3 (1958): 94–99.

———. *The Shaping of a Behaviorist: Part Two of an Autobiography.* Knopf, 1979.

———. "Some Relations Between Behavior Modification and Basic Research." In *Behavior Modification: Issues and Extensions,* edited by Sidney W. Bijou and Emilio Ribes-Inesta, 1–6. Academic Press, 1972.

———. *Walden Two.* Hackett Publishing, 1948.

Skinner, Burrhus Frederic. "A Case History in Scientific Method." *American Psychologist* 11, no. 5 (1956): 221–33.

Smalheiser, Neil R. "Walter Pitts." *Perspectives in Biology and Medicine* 43, no. 2 (2000): 217–26.

Smilkov, Daniel, Nikhil Thorat, Been Kim, Fernanda Viégas, and Martin Wattenberg. "Smoothgrad: Removing Noise by Adding Noise." arXiv Preprint arXiv:1706.03825, 2017.

Smith, Holly M. "Possibilism." In *Encyclopedia of Ethics,* 2nd ed., vol. 3, edited by Lawrence C. Becker and Charlotte B. Becker. Routledge, 2001.

Smith, Lewis, and Yarin Gal. "Understanding Measures of Uncertainty for Adversarial Example Detection." In *Uncertainty in Artificial Intelligence: Proceedings of the Thirty-Fourth Conference* (2018).

Smith, Mitch. "In Wisconsin, a Backlash Against Using Data to Foretell Defendants' Futures." *New York Times,* June 22, 2016.

Snyder, Mark. *Public Appearances, Private Realities: The Psychology of Self-Monitoring.* W. H. Freeman, 1987.

Soares, Nate, Benja Fallenstein, Eliezer Yudkowsky, and Stuart Armstrong. "Corrigibility." In *Artificial Intelligence and Ethics: Papers from the 2015 AAAI Workshop.* AAAI Press, 2015.

Sobel, Jordan Howard. "Utilitarianism and Past and Future Mistakes." *Noûs* 10, no. 2 (1976): 195–219.

Solomons, Leon M., and Gertrude Stein. "Normal Motor Automatism." *Harvard Psychological Review* 3, no. 5 (1896): 492–512.

Sommers, Christina Hoff. "Blind Spots in the 'Blind Audition' Study." *Wall Street Journal,* October 20, 2019.

Sorg, Jonathan, Richard L. Lewis, and Satinder P. Singh. "Reward Design via Online Gradient Ascent." In *Advances in Neural Information Processing Systems,* 2190–98, 2010.

Sorg, Jonathan, Satinder P. Singh, and Richard L. Lewis. "Internal Rewards Mitigate Agent Boundedness." In *Proceedings of the 27th International Conference on Machine Learning (ICML-10),* 1007–14. Omnipress, 2010.

Sorg, Jonathan Daniel. "The Optimal Reward Problem: Designing Effective Reward for Bounded Agents." PhD thesis, University of Michigan, 2011.

Spielvogel, Carl. "Advertising: Promoting a Negative Quality." *New York Times,* September 4, 1957.

Spignesi, Stephen J. *The Woody Allen Companion.* Andrews McMeel, 1992.

Srivastava, Nitish, Geoffrey Hinton, Alex Krizhevsky, Ilya Sutskever, and Ruslan Sal-akhutdinov. "Dropout: A Simple Way to Prevent Neural Networks from Overfit-ting." *Journal of Machine Learning Research* 15, no. 1 (2014): 1929–58.

Stadie, Bradly C., Sergey Levine, and Pieter Abbeel. "Incentivizing Exploration in Rein-forcement Learning with Deep Predictive Models." In *NIPS 2015 Workshop on Deep Reinforcement Learning,* 2015.

Stahl, Aimee E., and Lisa Feigenson. "Observing the Unexpected Enhances Infants' Learning and Exploration." *Science* 348, no. 6230 (2015): 91–94.

Stanley-Jones, D., and K. Stanley-Jones. *Kybernetics of Natural Systems: A Study in Pat-terns of Control.* Pergamon, 1960.

Stauffer, John, Zoe Trodd, and Celeste-Marie Bernier. *Picturing Frederick Douglass: An Illustrated Biography of the Nineteenth Century's Most Photographed Ameri-can.* W. W. Norton, 2015.

Steel, Emily, and Julia Angwin. "On the Web's Cutting Edge, Anonymity in Name Only." *Wall Street Journal,* August 3, 2010.

Steel, Piers. "The Nature of Procrastination: A Meta-Analytic and Theoretical Review of Quintessential Self-Regulatory Failure." *Psychological Bulletin* 133, no. 1 (2007): 65.

Stefik, Mark. "Strategic Computing at DARPA: Overview and Assessment." *Communi-cations of the ACM* 28, no. 7 (1985): 690–704.

Stein, Gertrude. *The Autobiography of Alice B. Toklas.* Harcourt, Brace, 1933.

Steinhardt, Jacob, and Percy S. Liang. "Unsupervised Risk Estimation Using Only Con-ditional Independence Structure." In *Advances in Neural Information Processing Systems,* 3657–65, 2016.

Stemen, Don. "The Prison Paradox: More Incarceration Will Not Make Us Safer." *For the Record Evidence Brief Series,* 2017.

Stock, Pierre, and Moustapha Cisse. "ConvNets and ImageNet Beyond Accuracy: Understanding Mistakes and Uncovering Biases." In *The European Conference on Computer Vision (ECCV),* 2018.

Strabala, Kyle, Min Kyung Lee, Anca Drăgan, Jodi Forlizzi, Siddhartha S. Srinivasa, Maya Çakmak, and Vincenzo Micelli. "Toward Seamless Human-Robot Hando-vers." *Journal of Human-Robot Interaction* 2, no. 1 (2013): 112–32.

"Strategic Computing: New-Generation Computing Technology: A Strategic Plan for Its Development and Application to Critical Problems in Defense." Defense Advanced Research Projects Agency, 1983.

Strehl, Alexander L., and Michael L. Littman. "An Analysis of Model-Based Interval Estimation for Markov Decision Processes." *Journal of Computer and System Sci-ences* 74, no. 8 (2008): 1309–31.

Struck, Aaron F., Berk Ustun, Andres Rodriguez Ruiz, Jong Woo Lee, Suzette M. LaRoche, Lawrence J. Hirsch, Emily J. Gilmore, Jan Vlachy, Hiba Arif Haider, Cynthia Rudin, and M. Brandon Westover. "Association of an Electroencephalography-Based Risk Score With Seizure Probability in Hospitalized Patients." *JAMA Neurology* 74, no. 12 (2017): 1419–24.

Subramanian, Kaushik, Charles L. Isbell Jr., and Andrea L. Thomaz. "Exploration from Demonstration for Interactive Reinforcement Learning." In *Proceedings of the 2016 International Conference on Autonomous Agents & Multiagent Systems*, 447–56. International Foundation for Autonomous Agents and Multiagent Systems, 2016.

Sundararajan, Mukund, Ankur Taly, and Qiqi Yan. "Axiomatic Attribution for Deep Networks." In *Proceedings of the 34th International Conference on Machine Learning*, 3319–28. JMLR, 2017.

Sunstein, Cass R. "Beyond the Precautionary Principle." *University of Pennsylvania Law Review* 151 (2003): 1003–58.

———. "Irreparability as Irreversibility." *Supreme Court Review*, 2019.

———. "Irreversibility." *Law, Probability & Risk* 9 (2010): 227–45.

———. *Laws of Fear: Beyond the Precautionary Principle*. Cambridge University Press, 2005.

Sutton, Richard S. "Integrated Architectures for Learning, Planning, and Reacting Based on Approximating Dynamic Programming." In *Machine Learning: Proceedings of the Seventh International Conference*, 216–24. San Mateo, CA: Morgan Kaufmann, 1990.

———. "Learning to Predict by the Methods of Temporal Differences." *Machine Learning* 3, no. 1 (1988): 9–44.

———. "Reinforcement Learning Architectures for Animats." In *From Animals to Animats: Proceedings of the First International Conference on Simulation of Adaptive Behavior*, 288–96. 1991.

———. "Temporal Credit Assignment in Reinforcement Learning." PhD thesis, University of Massachusetts, Amherst, 1984.

———. "A Unified Theory of Expectation in Classical and Instrumental Conditioning." Bachelor's thesis, Stanford University, 1978.

Sutton, Richard S., and Andrew G. Barto. *Reinforcement Learning: An Introduction*. 2nd ed. MIT Press, 2018.

Sweeney, Latanya. "Discrimination in Online Ad Delivery." *Communications of the ACM* 56, no. 5 (2013): 44–54.

———. "Simple Demographics Often Identify People Uniquely." *Carnegie Mellon University, Data Privacy Working Paper 3* (Pittsburgh), 2000.

Szegedy, Christian, Wojciech Zaremba, Ilya Sutskever, Joan Bruna, Dumitru Erhan, Ian Goodfellow, and Rob Fergus. "Intriguing Properties of Neural Networks." In *International Conference on Learning Representations*, 2014.

Takayama, Leila, Doug Dooley, and Wendy Ju. "Expressing Thought: Improving Robot

Readability with Animation Principles." In *2011 6th ACM/IEEE International Conference on Human-Robot Interaction (HRI)*, 69–76. IEEE, 2011.

Tang, Haoran, Rein Houthooft, Davis Foote, Adam Stooke, Xi Chen, Yan Duan, John Schulman, Filip De Turck, and Pieter Abbeel. "# Exploration: A Study of Count-Based Exploration for Deep Reinforcement Learning." In *Advances in Neural Information Processing Systems*, 2753–62. 2017.

Tarleton, Nick. "Coherent Extrapolated Volition: A Meta-Level Approach to Machine Ethics." Singularity Institute, 2010.

Taylor, Jessica. "Quantilizers: A Safer Alternative to Maximizers for Limited Optimization." In *Workshops at the Thirtieth AAAI Conference on Artificial Intelligence*, 2016.

Taylor, Jessica, Eliezer Yudkowsky, Patrick LaVictoire, and Andrew Critch. "Alignment for Advanced Machine Learning Systems." Machine Intelligence Research Institute, July 27, 2016.

Tesauro, Gerald. "Practical Issues in Temporal Difference Learning." In *Advances in Neural Information Processing Systems*, 259–66, 1992.

———. "TD-Gammon, a Self-Teaching Backgammon Program, Achieves Master-Level Play." *Neural Computation* 6, no. 2 (1994): 215–19. https://doi.org/10.1162/neco.1994.6.2.215.

———. "Temporal Difference Learning and TD-Gammon." *Communications of the ACM* 38, no. 3 (March 1995): 58–68. https://doi.org/10.1145/203330.203343.

Thorndike, Edward L. "Animal Intelligence: An Experimental Study of the Associative Processes in Animals." *Psychological Review: Monograph Supplements* 2, no. 4 (1898): i.

———. *The Elements of Psychology*. New York: A. G. Seiler, 1905.

———. "Fundamental Theorems in Judging Men." *Journal of Applied Psychology* 2 (1918): 67–76.

———. *The Psychology of Learning*. Vol. 2. Teachers College, Columbia University, 1913.

———. "A Theory of the Action of the After-Effects of a Connection upon It." *Psychological Review* 40, no. 5 (1933): 434–39.

Tibshirani, Robert. "Regression Shrinkage and Selection via the Lasso." *Journal of the Royal Statistical Society, Series B (Methodological)* 58, no. 1 (1996): 267–88.

Tigas, Panagiotis, Angelos Filos, Rowan McAllister, Nicholas Rhinehart, Sergey Levine, and Yarin Gal. "Robust Imitative Planning: Planning from Demonstrations Under Uncertainty." In *NeurIPS 2019 Workshop on Machine Learning for Autonomous Driving*. 2019.

Timmerman, Travis. "Effective Altruism's Underspecification Problem." In *Effective Altruism: Philosophical Issues*, edited by Hilary Greaves and Theron Pummer. Oxford University Press, 2019.

Timmerman, Travis, and Yishai Cohen. "Actualism and Possibilism in Ethics." In *The*

Stanford Encyclopedia of Philosophy, edited by Edward N. Zalta, Summer 2019. https://plato.stanford.edu/archives/sum2019/entries/actualism-possibilism-ethics/.

Todorov, Emanuel, Tom Erez, and Yuval Tassa. "MuJoCo: A Physics Engine for Model-Based Control." In *2012 IEEE/RSJ International Conference on Intelligent Robots and Systems*, 5026–33. IEEE, 2012.

Tolman, Edward Chace. "The Determiners of Behavior at a Choice Point." *Psychological Review* 45, no. 1 (1938): 1.

Tomasello, Michael. "Do Apes Ape?" In *Social Learning in Animals: The Roots of Culture*, edited by Cecilia M. Heyes and Bennett G. Galef Jr., 319–46. San Diego: Academic Press, 1996.

Tomasello, Michael, Malinda Carpenter, Josep Call, Tanya Behne, and Henrike Moll. "Understanding and Sharing Intentions: The Origins of Cultural Cognition." *Behavioral and Brain Sciences* 28, no. 5 (2005): 675–91.

Tomasello, Michael, Brian Hare, Hagen Lehmann, and Josep Call. "Reliance on Head Versus Eyes in the Gaze Following of Great Apes and Human Infants: The Cooperative Eye Hypothesis." *Journal of Human Evolution* 52, no. 3 (2007): 314–20.

Tomasik, Brian. "Do Artificial Reinforcement-Learning Agents Matter Morally?" arXiv Preprint arXiv:1410.8233, 2014.

Turing, A. M. "Computing Machinery and Intelligence." *Mind* 59, no. 236 (1950): 433–60.

———. "Intelligent Machinery." In *The Essential Turing*, edited by B. Jack Copeland, 410–32. 1948. Reprint, Oxford University Press, 2004.

Turing, Alan. "Can Digital Computers Think?" *BBC Third Programme*, May 15, 1951.

Turing, Alan, Richard Braithwaite, Geoffrey Jefferson, and Max Newman. "Can Automatic Calculating Machines Be Said to Think?" *BBC Third Programme*, January 14, 1952.

Turner, Alexander Matt. "Optimal Farsighted Agents Tend to Seek Power." arXiv preprint, arXiv:1912.01683, 2019.

Turner, Alexander Matt, Dylan Hadfield-Menell, and Prasad Tadepalli. "Conservative Agency via Attainable Utility Preservation." In *AAAI/ACM Conference on AI, Ethics, and Society*. 2020.

Tversky, Amos. "Features of Similarity." *Psychological Review* 84, no. 4 (1977): 327–52.

Uno, Yoji, Mitsuo Kawato, and Rika Suzuki. "Formation and Control of Optimal Trajectory in Human Multijoint Arm Movement: Minimum Torque-Change Model." *Biological Cybernetics* 61 (1989): 89–101.

Ustun, Berk, and Cynthia Rudin. "Optimized Risk Scores." In *Proceedings of the 23rd ACM SIGKDD International Conference on Knowledge Discovery and Data Mining*, 1125–34. ACM, 2017.

———. "Supersparse Linear Integer Models for Optimized Medical Scoring Systems." *Machine Learning* 102, no. 3 (2016): 349–91.

Ustun, Berk, Stefano Tracà, and Cynthia Rudin. "Supersparse Linear Integer Models for Predictive Scoring Systems." In *Proceedings of AAAI Late Breaking Track*, 2013.

Ustun, Berk, M. Brandon Westover, Cynthia Rudin, and Matt T. Bianchi. "Clinical Prediction Models for Sleep Apnea: The Importance of Medical History over Symptoms." *Journal of Clinical Sleep Medicine* 12, no. 2 (2016): 161–68.

Varon, Elizabeth R. "Most Photographed Man of His Era: Frederick Douglass." *Washington Post*, January 29, 2016. https://www.washingtonpost.com/opinions/most-photographed-man-of-his-era-frederick-douglass/2016/01/29/4879a766-af1c-11e5-b820-eea4d64be2a1_story.html.

Veasey, Sigrid C., and Ilene M. Rosen. "Obstructive Sleep Apnea in Adults." *New England Journal of Medicine* 380, no. 15 (2019): 1442–49.

Večerík, Matej, Todd Hester, Jonathan Scholz, Fumin Wang, Olivier Pietquin, Bilal Piot, Nicolas Heess, Thomas Rothörl, Thomas Lampe, and Martin Riedmiller. "Leveraging Demonstrations for Deep Reinforcement Learning on Robotics Problems with Sparse Rewards." arXiv Preprint arXiv:1707.08817, 2017.

Vincent, James. "Google 'Fixed' Its Racist Algorithm by Removing Gorillas from Its Image-Labeling Tech." *Verge*, January 11, 2018. https://www.theverge.com/2018/1/12/16882408/google-racist-gorillas-photo-recognition-algorithm-ai.

Visalberghi, Elisabetta, and Dorothy Fragaszy. "'Do Monkeys Ape?' Ten Years After." In *Imitation in Animals and Artifacts*, 471–99. MIT Press, 2002.

Visalberghi, Elisabetta, and Dorothy Munkenbeck Fragaszy. "Do Monkeys Ape?" In *"Language" and Intelligence in Monkeys and Apes: Comparative Developmental Perspectives*, edited by Sue Taylor Parker and Kathleen Rita Gibson, 247–73. Cambridge University Press, 1990.

Visser, Margaret. *Much Depends on Dinner: The Extraordinary History and Mythology, Allure and Obsessions, Perils and Taboos of an Ordinary Meal*. Grove Press, 1986.

Vitale, Cristiana, Massimo Fini, Ilaria Spoletini, Mitja Lainscak, Petar Seferovic, and Giuseppe M. C. Rosano. "Under-Representation of Elderly and Women in Clinical Trials." *International Journal of Cardiology* 232 (2017): 216–21.

von Neumann, John. First Draft of a Report on the EDVAC. Moore School of Electrical Engineering, University of Pennsylvania, June 30, 1945.

von Neumann, John, and Oskar Morgenstern. *Theory of Games and Economic Behavior*. 3rd ed. Princeton University Press, 1953.

Wachter, Sandra, Brent Mittelstadt, and Luciano Floridi. "Why a Right to Explanation of Automated Decision-Making Does Not Exist in the General Data Protection Regulation." *International Data Privacy Law* 7, no. 2 (2017): 76–99.

Wainer, Howard. "Estimating Coefficients in Linear Models: It Don't Make No Nevermind." *Psychological Bulletin* 83, no. 2 (1976): 213–17.

Warneken, Felix, and Michael Tomasello. "Altruistic Helping in Human Infants and Young Chimpanzees." *Science* 311, no. 5765 (2006): 1301–03.

———. "Helping and Cooperation at 14 Months of Age." *Infancy* 11, no. 3 (2007): 271–94.

Watkins, Christopher J.C.H., and Peter Dayan. "Q-Learning." *Machine Learning* 8, nos. 3–4 (1992): 279–92.

Watkins, Christopher John Cornish Hellaby. "Learning from Delayed Rewards." PhD thesis, King's College, Cambridge, 1989.

Weber, Bruce. "What Deep Blue Learned in Chess School." *New York Times*, May 18, 1997.

Weld, Daniel S., and Oren Etzioni. "The First Law of Robotics (a Call to Arms)." In *Proceedings of the Twelfth National Conference on Artificial Intelligence (AAAI-1994)*, 1042–47. Seattle, 1994.

Whitehead, Alfred North, and Bertrand Russell. *Principia Mathematica*. 2nd ed. 3 vols. Cambridge University Press, 1927.

Whiten, Andrew, Nicola McGuigan, Sarah Marshall-Pescini, and Lydia M. Hopper. "Emulation, Imitation, Over-Imitation and the Scope of Culture for Child and Chimpanzee." *Philosophical Transactions of the Royal Society B: Biological Sciences* 364, no. 1528 (2009): 2417–28.

Whiteson, Shimon, Brian Tanner, and Adam White. "The Reinforcement Learning Competitions." *AI Magazine* 31, no. 2 (2010): 81–94.

Widrow, Bernard, Narendra K. Gupta, and Sidhartha Maitra. "Punish/Reward: Learning with a Critic in Adaptive Threshold Systems." *IEEE Transactions on Systems, Man, and Cybernetics* 5 (1973): 455–65.

Wiener, Norbert. *Cybernetics; or, Control and Communication in the Animal and the Machine*. 2nd ed. MIT Press, 1961.

———. *God and Golem, Inc: A Comment on Certain Points Where Cybernetics Impinges on Religion*. MIT Press, 1964.

———. "Some Moral and Technical Consequences of Automation." *Science* 131, no. 3410 (1960): 1355–58.

Wiewiora, Eric. "Potential-Based Shaping and Q-Value Initialization Are Equivalent." *Journal of Artificial Intelligence Research* 19 (2003): 205–08.

Wilson, Aaron, Alan Fern, and Prasad Tadepalli. "A Bayesian Approach for Policy Learning from Trajectory Preference Queries." In *Advances in Neural Information Processing Systems 25*, 1133–41. 2012.

Wirth, Christian, Riad Akrour, Gerhard Neumann, and Johannes Fürnkranz. "A Survey of Preference-Based Reinforcement Learning Methods." *Journal of Machine Learning Research* 18, no. 1 (2017): 4945–90.

Wise, Roy. "Reinforcement." *Scholarpedia* 4, no. 8 (2009): 2450. http://www.scholarpedia.org/article/Reinforcement.

Wise, Roy A. "Dopamine and Reward: The Anhedonia Hypothesis 30 Years On." *Neurotoxicity Research* 14, nos. 2–3 (2008): 169–83.

———. "Neuroleptics and Operant Behavior: The Anhedonia Hypothesis." *Behavioral and Brain Sciences* 5, no. 1 (1982): 39–53.

Wise, Roy A., Joan Spindler, Harriet DeWit, and Gary J. Gerberg. "Neuroleptic-Induced 'Anhedonia' in Rats: Pimozide Blocks Reward Quality of Food." *Science* 201, no. 4352 (1978): 262–64.

Wittmann, Bianca C., Nathaniel D. Daw, Ben Seymour, and Raymond J. Dolan. "Striatal Activity Underlies Novelty-Based Choice in Humans." *Neuron* 58, no. 6 (2008): 967–73.

Wolf, Susan. "Moral Saints." *The Journal of Philosophy* 79, no. 8 (1982): 419–39.

Wood, Tom. "Google Images 'Racist Algorithm' Has a Fix but It's Not a Great One." LADbible, January 12, 2018. http://www.ladbible.com/technology/news-google-images-racist-algorithm-has-a-fix-but-its-not-a-great-one-20180112.

Wright, Thomas A., John Hollwitz, Richard W. Stackman, Arthur S. De Groat, Sally A. Baack, and Jeffrey P. Shay. "40 Years (and Counting): Steve Kerr Reflections on the 'Folly.'" *Journal of Management Inquiry* 27, no. 3 (2017): 309–15.

Wulfmeier, Markus, Peter Ondrúška, and Ingmar Posner. "Maximum Entropy Deep Inverse Reinforcement Learning." arXiv Preprint arXiv:1507.04888, 2015.

Wulfmeier, Markus, Dominic Zeng Wang, and Ingmar Posner. "Watch This: Scalable Cost-Function Learning for Path Planning in Urban Environments." In *2016 IEEE/RSJ International Conference on Intelligent Robots and Systems (IROS)*, 2089–95. IEEE, 2016.

Xie, Chiang, Yuxin Wu, Laurens van der Maaten, Alan L. Yuille, and Kaiming He. "Feature Denoising for Improving Adversarial Robustness." In *Proceedings of the IEEE Conference on Computer Vision and Pattern Recognition*, 501–09. 2019.

Yao, Yuan, Lorenzo Rosasco, and Andrea Caponnetto. "On Early Stopping in Gradient Descent Learning." *Constructive Approximation* 26, no. 2 (2007): 289–315.

Yu, Yang, Wei-Yang Qu, Nan Li, and Zimin Guo. "Open-Category Classification by Adversarial Sample Generation." In *International Joint Conference on Artificial Intelligence*. 2017.

Yudkowsky, Eliezer. "Coherent Extrapolated Volition." Singularity Institute, 2004.

Zaki, W. Mimi Diyana W., M. Asyraf Zulkifley, Aini Hussain, W. Haslina W. A. Halim, N. Badariah A. Mustafa, and Lim Sin Ting. "Diabetic Retinopathy Assessment: Towards an Automated System." *Biomedical Signal Processing and Control* 24 (2016): 72–82.

Zeiler, Matthew D., and Rob Fergus. "Visualizing and Understanding Convolutional Networks." In *European Conference on Computer Vision*, 818–33. Springer, 2014.

Zeiler, Matthew D., Dilip Krishnan, Graham W. Taylor, and Rob Fergus. "Deconvolutional Networks." In *2010 IEEE Computer Society Conference on Computer Vision and Pattern Recognition (CVPR)*, 2528–35. IEEE, 2010.

Zeiler, Matthew D., Graham W. Taylor, and Rob Fergus. "Adaptive Deconvolutional Networks for Mid and High Level Feature Learning." In *2011 International Conference on Computer Vision*, 2018–25. IEEE, 2011.

Zeng, Jiaming, Berk Ustun, and Cynthia Rudin. "Interpretable Classification Models for Recidivism Prediction." *Journal of the Royal Statistical Society: Series A (Statistics in Society)* 180, no. 3 (2017): 689–722.

Zhang, Shun, Edmund H. Durfee, and Satinder P. Singh. "Minimax-Regret Querying on Side Effects for Safe Optimality in Factored Markov Decision Processes." In *Proceedings of the Twenty-Seventh International Joint Conference on Artificial Intelligence,* 4867–73, IJCAI, 2018.

Zheng, Zeyu, Junhyuk Oh, and Satinder Singh. "On Learning Intrinsic Rewards for Policy Gradient Methods." In *Advances in Neural Information Processing Systems,* 4644–54. 2018.

Zhong, Chen-Bo, Vanessa K. Bohns, and Francesca Gino. "Good Lamps Are the Best Police: Darkness Increases Dishonesty and Self-Interested Behavior." *Psychological Science* 21, no. 3 (2010): 311–14.

Zhu, Lingxue, and Nikolay Laptev. "Engineering Uncertainty Estimation in Neural Networks for Time Series Prediction at Uber." *Uber Engineering* (blog), 2017. https://eng.uber.com/neural-networks-uncertainty-estimation/.

Ziebart, Brian D., J. Andrew Bagnell, and Anind K. Dey. "Modeling Interaction via the Principle of Maximum Causal Entropy." In *Proceedings of the 27th International Conference on Machine Learning,* 1255–62. 2010.

Ziebart, Brian D., Andrew L. Maas, J. Andrew Bagnell, and Anind K. Dey. "Maximum Entropy Inverse Reinforcement Learning." In *Proceedings of the Twenty-Third AAAI Conference on Artificial Intelligence,* 1433–38. AAAI Press, 2008.

Ziegler, Daniel M., Nisan Stiennon, Jeffrey Wu, Tom B. Brown, Alec Radford, Dario Amodei, Paul Christiano, and Geoffrey Irving. "Fine-Tuning Language Models from Human Preferences." arXiv Preprint arXiv:1909.08593, 2019.

INDEX

ABOUT THE AUTHOR

Brian Christian is the author of *The Most Human Human*, a *Wall Street Journal* best seller, *New York Times* Editors' Choice, and *New Yorker* favorite book of the year. He is the author, with Tom Griffiths, of *Algorithms to Live By*, a No. 1 Audible nonfiction best seller, Amazon best science book of the year, and *MIT Technology Review* book of the year.

Christian's writing has been translated into nineteen languages and has appeared in the *New Yorker*, the *Atlantic*, *Wired*, the *Wall Street Journal*, the *Guardian*, the *Paris Review*, and in scientific journals such as *Cognitive Science*. Christian has been featured on *The Daily Show with Jon Stewart*, *Radiolab*, and *Charlie Rose* and has lectured at Google, Facebook, Microsoft, the Santa Fe Institute, and the London School of Economics. He is a recent laureate of the San Francisco Public Library, and his work has won him several awards, including fellowships at Yaddo and the MacDowell Colony, publication in *The Best American Science and Nature Writing*, and an award from the Academy of American Poets.

Born in Wilmington, Delaware, Christian holds degrees in computer science, philosophy, and poetry from Brown University and the University of Washington and is a visiting scholar at the University of California, Berkeley. He lives in San Francisco.